18 0278910 X
PC UW
LLYFRGELL
LIBRARY
ABERYSTWYTH
DILEWYD O STOC
WITHDRAWN FROM STOCK

AF616161

Selected Titles in This Series

41 **David Aldous and James Propp, Editors,** Microsurveys in Discrete Probability

40 **Panos M. Pardalos and Dingzhu Du, Editors,** Network Design: Connectivity and Facilities Location

39 **Paul W. Beame and Samuel R Buss, Editors,** Proof Complexity and Feasible Arithmetics

38 **Rebecca N. Wright and Peter G. Neumann, Editors,** Network Threats

37 **Boris Mirkin, F. R. McMorris, Fred S. Roberts, and Andrey Rzhetsky, Editors,** Mathematical Hierarchies and Biology

36 **Joseph G. Rosenstein, Deborah S. Franzblau, and Fred S. Roberts, Editors,** Discrete Mathematics in the Schools

35 **Dingzhu Du, Jun Gu, and Panos M. Pardalos, Editors,** Satisfiability Problem: Theory and Applications

34 **Nathaniel Dean, Editor,** African Americans in Mathematics

33 **Ravi B. Boppana and James F. Lynch, Editors,** Logic and Random Structures

32 **Jean-Charles Grégoire, Gerard J. Holzmann, and Doron A. Peled, Editors,** The SPIN Verification System

31 **Neil Immerman and Phokion G. Kolaitis, Editors,** Descriptive Complexity and Finite Models

30 **Sandeep N. Bhatt, Editor,** Parallel Algorithms: Third DIMACS Implementation Challenge

29 **Doron A. Peled, Vaughan R. Pratt, and Gerard J. Holzmann, Editors,** Partial Order Methods in Verification

28 **Larry Finkelstein and William M. Kantor, Editors,** Groups and Computation II

27 **Richard J. Lipton and Eric B. Baum, Editors,** DNA Based Computers

26 **David S. Johnson and Michael A. Trick, Editors,** Cliques, Coloring, and Satisfiability: Second DIMACS Implementation Challenge

25 **Gilbert Baumslag, David Epstein, Robert Gilman, Hamish Short, and Charles Sims, Editors,** Geometric and Computational Perspectives on Infinite Groups

24 **Louis J. Billera, Curtis Greene, Rodica Simion, and Richard P. Stanley, Editors,** Formal Power Series and Algebraic Combinatorics/Séries Formelles et Combinatoire Algébrique, 1994

23 **Panos M. Pardalos, David I. Shalloway, and Guoliang Xue, Editors,** Global Minimization of Nonconvex Energy Functions: Molecular Conformation and Protein Folding

22 **Panos M. Pardalos, Mauricio G. C. Resende, and K. G. Ramakrishnan, Editors,** Parallel Processing of Discrete Optimization Problems

21 **D. Frank Hsu, Arnold L. Rosenberg, and Dominique Sotteau, Editors,** Interconnection Networks and Mapping and Scheduling Parallel Computations

20 **William Cook, László Lovász, and Paul Seymour, Editors,** Combinatorial Optimization

19 **Ingemar J. Cox, Pierre Hansen, and Bela Julesz, Editors,** Partitioning Data Sets

18 **Guy E. Blelloch, K. Mani Chandy, and Suresh Jagannathan, Editors,** Specification of Parallel Algorithms

17 **Eric Sven Ristad, Editor,** Language Computations

16 **Panos M. Pardalos and Henry Wolkowicz, Editors,** Quadratic Assignment and Related Problems

15 **Nathaniel Dean and Gregory E. Shannon, Editors,** Computational Support for Discrete Mathematics

14 **Robert Calderbank, G. David Forney, Jr., and Nader Moayeri, Editors,** Coding and Quantization: DIMACS/IEEE Workshop

13 **Jin-Yi Cai, Editor,** Advances in Computational Complexity Theory

(*Continued in the back of this publication*)

DIMACS

Series in Discrete Mathematics
and Theoretical Computer Science

Volume 41

Microsurveys in Discrete Probability

DIMACS Workshop
June 2–6, 1997

David Aldous
James Propp
Editors

NSF Science and Technology Center
in Discrete Mathematics and Theoretical Computer Science
A consortium of Rutgers University, Princeton University,
AT&T Labs, Bell Labs, and Bellcore

American Mathematical Society

This DIMACS volume contains the lectures of speakers at the workshop "Microsurveys in Discrete Probability" held at the Institute for Advanced Study, Princeton, NJ, on June 2–6, 1997. The workshop was part of the DIMACS Special Year on Discrete Probability, and the lectures were selected to give a cross-section of current activity in this field.

1991 *Mathematics Subject Classification.* Primary 60C05; Secondary 60J10, 05C05.

Library of Congress Cataloging-in-Publication Data

Microsurveys in discrete probability : DIMACS workshop, June 2–6, 1997 / David Aldous, James Propp, editors.

p. cm. — (DIMACS series in discrete mathematics and theoretical computer science, ISSN 1052-1798 ; v. 41)

Proceedings of the workshop held at the Institute for Advanced Study, Princeton, N.J., June 2–6, 1997.

Includes bibliographical references.

ISBN 0-8218-0827-3 (hardcover : alk. paper)

1. Probabilities—Congresses. I. Aldous, D. J. (David J.) II. Propp, James, 1960– . III. Series.

QA273.A1M53 1998

519.2–dc21

98-4520

CIP

∞ The paper used in this book is acid-free and falls within the guidelines
established to ensure permanence and durability.
Visit the AMS home page at URL: http://www.ams.org/

10 9 8 7 6 5 4 3 2 1 03 02 01 00 99 98

Contents

Foreword

The workshop on Microsurveys in Discrete Probability was held in June 1997 at the Institute for Advanced Study in Princeton, NJ. We would like to express our appreciation to David Aldous and Jim Propp for their efforts to organize and plan this successful workshop.

The workshop was part of the broader Special Year on Discrete Probability. We extend our thanks to Enrico Bonbieri, Jeffrey Kahn and Peter Winkler for their work over many months as special year organizers.

The workshop focused on microsurveys of recently active areas within discrete probability.

DIMACS gratefully acknowledges the generous support that makes these programs possible. The National Science Foundation, through its Science and Technology Center program, the New Jersey Commission on Science and Technology, DIMACS partners at Rutgers, Princeton, AT&T Labs, Bell Labs and Bellcore generously supported the special year.

Fred S. Roberts
Director

Bernard Chazelle
Co-Director for Princeton

Preface

These are the Proceedings of a workshop held at the Institute for Advanced Study in Princeton, New Jersey, June 2–6, 1997, as part of the 1996–97 DIMACS year "Focus on Discrete Probability". The workshop, entitled "Microsurveys in Discrete Probability", was centered around ten focused survey talks, selected to give a cross-section of current activity. We asked each speaker to treat a small-but-growing topic on which two to ten research papers had been written but for which no survey paper yet existed; it was our belief that it is at just such a stage in its development that a line of research can most benefit from the kind of exposure our workshop offered.

All speakers were invited to write up their lectures, and you hold in your hands the results of the speakers' diligence and a small bit of editorial prodding. We hope that the bound Proceedings will make these emerging topics in discrete probability theory accessible to a wider audience. Other information about the workshop can be found (at least during the late twentieth and early twenty-first centuries) at the workshop Web site: `http://dimacs.rutgers.edu/Workshops/Microsurveys/`.

We would like to thank Mary Jane Hayes and the IAS staff for their efficient and cheerful local organization, Steve Mahaney and the DIMACS Rutgers crew for administering the Focus on Discrete Probability year 1996–97, and Christine Thivierge for overseeing publication of these Proceedings.

David Aldous and James Propp
January 1998

DIMACS Series in Discrete Mathematics
and Theoretical Computer Science
Volume 41, 1998

Tree-valued Markov chains and Poisson-Galton-Watson distributions

David Aldous

ABSTRACT. The Poisson-Galton-Watson distribution on finite trees, and the related $\mathrm{PGW}^{\infty}(1)$ distribution on infinite trees with one end, arise in several contexts, in particular as $n \to \infty$ weak limits within various size-n combinatorial models. We review this topic, introducing slick notation for describing such distributions. We then describe a family of continuous-time Markov chains whose marginal distributions are of Poisson-Galton-Watson type. Different chains in this family have connections with different parts of the extensive literature on tree and forest-valued chains (the random graph process, stochastic coalescence, spanning tree chains) and also illustrate the classical state-space-compactification theory of continuous-time countable-state chains.

1. Introduction

Tree- and forest-valued Markov chains have been studied in several areas.

- Evolution of tree-based data structures (Knuth [**13**], Mahmoud [**16**]).
- The *Markov chain tree theorem* (Pemantle [**18**], Lyons-Peres [**15**]), which to a finite-state Markov chain associates a chain on the set of spanning trees.
- The classical random graph process (Bollobás [**7**], Alon-Spencer [**5**]). By ignoring edges created within components, the components grow as a forest-valued process.
- Reversible models of polymerization (Whittle [**20**], Pittel et al. [**17**]).
- Irreversible models of stochastic coalescence (Aldous [**3**]).
- Spin systems on infinite trees. If the $\{0,1\}$-valued r.v.'s are assigned to edges instead of vertices, we may regard the system as a forest-valued process.

These areas have different specific motivations and have largely developed independently. Setting aside applications, consider the question

> amongst the unlimited number of tree-valued chains one could define, which seem mathematically fundamental?

The purpose of this paper is to describe (sections 3 and 4) one family of six or seven related chains which do seem fundamental, and which have connections with several of the areas above. The chains in our family have the common features
(a) their marginal (i.e. fixed time) distributions are related to Poisson-Galton-Watson distributions.

1991 *Mathematics Subject Classification.* Primary 60J27, Secondary 60C05, 60J80.

Research supported in part by N.S.F. Grant DMS 9622859.

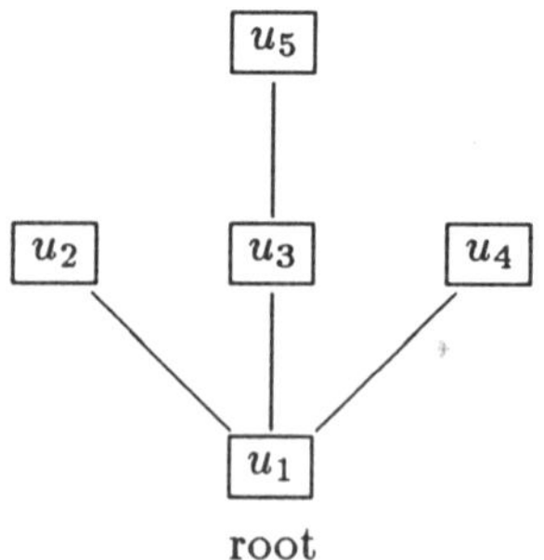

FIGURE 1. A u-labeled tree $\mathbf{t}$

(b) They arise within $n \to \infty$ weak limits of various Markov chains whose state spaces are sets of size-n combinatorial objects.
In other words, our chains provide "time-evolving" extensions of known "static" results asserting that Poisson-Galton-Watson distributions arise as weak limits within combinatorial models, so we start by reviewing that topic in section 2.

Our presentation is somewhat informal, for reasons explained in section 3.2. The chains of sections 3.4 and 3.5 are studied rigorously in Aldous-Pitman [**4**]. The chains in section 4 are novel and may be worthy of rigorous study in future.

2. Poisson-Galton-Watson distributions as weak limits

For $0 \leq \lambda \leq 1$ the *Poisson-Galton-Watson* distribution PGW(λ) on finite trees is the distribution of the family tree of a Galton-Watson branching process with one progenitor and Poisson(λ) offspring distribution. From some viewpoints these are the "canonical" distributions on finite trees, and the related distribution $\mathrm{PGW}^{\infty}(1)$ on a set of infinite trees ("trees with one end") is the canonical distribution on that set. In this section we describe how these distributions arise as weak limits of distributions on size-n structures. The material is essentially known and straightforward: the main novelty is the slick[1] notation we now introduce.

2.1. Notation for distributions on trees. In this paper, trees (and tree-components of forests) are rooted and unordered (i.e. in the usual parent-child interpretation of edges, we do not distinguish birth order of children). There are two standard conventions to specify what is meant by the set of all n-vertex trees. One can regard vertices as labeled by the integers $\{1, 2, \ldots, n\}$, with Cayley's formula asserting there exist n^{n-1} distinct such trees. Or one can regard trees as unlabeled, giving a smaller number of "shapes" of trees, where two trees have the same shape iff there is a root-preserving bijection between vertices which preserves edges. For discussing tree-valued chains, the second convention is unsatisfactory because, in the joint distribution of the chain at times s_1 and s_2, one cannot determine whether a particular vertex at time s_1 is the "same vertex" as some vertex at time s_2. The first convention cannot be used directly because our trees have varying size. Instead we introduce the idea of labeling vertices by distinct *real numbers* $u \in (0,1)$. Define $\mathbf{T}$ to be the set of finite rooted trees in which vertices i (including the root) are labeled with distinct $u_i \in (0,1)$. Call an element $\mathbf{t} \in \mathbf{T}$ a *u-labeled tree*. Write $|\mathbf{t}|$ for the number of vertices of $\mathbf{t}$. Figure 1 illustrates a typical element $\mathbf{t}$ of $\mathbf{T}$.

[1](for non-native English speakers) *slick* has two meanings. (i) clever, concise (ii) slippery, untrustworthy

Given a random rooted tree $\mathcal{T}$, we may regard it as a random element of $\mathbf{T}$ by assigning independent $U(0,1)$ (i.e. uniform on $(0,1)$) random labels to the vertices. The ultimate purpose is that, when we formalize tree-valued Markov chains as $\mathbf{T}$-valued chains $(\mathcal{T}_s, 0 \leq s)$, we can look at joint distributions $(\mathcal{T}_{s_1}, \mathcal{T}_{s_2})$ and tell whether a specific vertex of $\mathcal{T}_{s_2}$ is also a vertex of $\mathcal{T}_{s_1}$. But it turns out that formalizing random trees as u-labeled trees leads to a very useful way of describing distributions on trees, at least in settings related to Poisson-Galton-Watson distributions. Let $f : \mathbf{T} \to [0, \infty)$ be a function such that $f(\mathbf{t})$ depends only on the shape of $\mathbf{t}$, not on the values (u_i). Say a random tree $\mathcal{T}$ has *density* $f(\mathbf{t})$ if the chance that $\mathcal{T}$ is a tree with the same shape as $\mathbf{t}$ and with labels in $[u_i, u_i + du_i]$, $1 \leq i \leq |\mathbf{t}|$, equals $f(\mathbf{t})\ du_1 \ldots du_{|\mathbf{t}|}$. This definition is difficult to assimilate at first; we hope that repeated use throughout the paper will make it seem less mysterious. Here is a "discrete" reformulation. Given a random finite rooted tree $\mathcal{T}$, introduce an artificial parameter N, label vertices of $\mathcal{T}$ by sampling without replacement from $\{1, 2, \ldots, N\}$ (with some unimportant convention for the case $|\mathcal{T}| > N$) and call the resulting random tree $\mathcal{T}_{(N)}$. For given $\mathbf{t} \in \mathbf{T}$ write $\mathbf{t}_{\text{int}}$ for the tree obtained from $\mathbf{t}$ by an arbitrary relabeling of vertices as $\{1, 2, \ldots, |\mathbf{t}|\}$. Then the property

$$\forall \mathbf{t} \in \mathbf{T} : \quad P(\mathcal{T}_{(N)} = \mathbf{t}_{\text{int}}) \sim N^{-|\mathbf{t}|} f(\mathbf{t}) \text{ as } N \to \infty \tag{2.1}$$

is equivalent to saying $\mathcal{T}$ has density $f(\cdot)$. Let us illustrate by calculating the density for a PGW(λ) distributed tree $\mathcal{T}$. Take $\mathbf{t}$ as in Figure 1 and let $\mathbf{t}_{\text{int}}$ be the relabeling of $\{u_1, \ldots, u_5\}$ as $\{1, 2, \ldots, 5\}$. Then

$$P(\mathcal{T}_{(N)} = \mathbf{t}_{\text{int}}) = \frac{1}{N} \times \frac{\lambda^3 e^{-\lambda}}{3!}$$

$$\times \frac{3!}{(N-1)(N-2)(N-3)} \times (e^{-\lambda} \times \lambda e^{-\lambda} \times e^{-\lambda}) \times \frac{1}{N-4} \times e^{-\lambda}$$

where successive terms represent: chance root gets label 1, chance of 3 children, chance children labeled as $2, 3, 4$ in some order, chance children have $(0, 1, 0)$ grandchildren, chance grandchild gets label 5, chance it has 0 children. Thus $P(\mathcal{T}_{(N)} = \mathbf{t}_{\text{int}}) = \frac{\lambda^4 e^{-5\lambda}}{N(N-1)\ldots(N-4)}$ and so this particular $\mathbf{t}$ has $f(\mathbf{t}) = \lambda^4 e^{-5\lambda}$. Applying the same argument to a general tree $\mathbf{t}$ we see that the PGW(λ) density is

$$f_\lambda(\mathbf{t}) = \lambda^{(|\mathbf{t}|-1)} e^{-\lambda|\mathbf{t}|} \tag{2.2}$$

and in particular for $\lambda = 1$

$$f_1(\mathbf{t}) = e^{-|\mathbf{t}|}. \tag{2.3}$$

The focus of this paper is that this density notation (loosely reminiscent of statistical physics) turns out to be very useful for specifying or verifying transition rates for certain tree-valued chains (look ahead to section 3.1). It is not intended to facilitate explicit calculations, though as an exercise the reader might deduce from (2.2) the well-known fact that, for $\mathcal{T}$ with PGW(λ) distribution, the size $|\mathcal{T}|$ has *Borel*(λ) distribution

$$P(|\mathcal{T}| = i) = \frac{(\lambda i)^{i-1}}{i!}\, e^{-\lambda i},\ i \geq 1. \tag{2.4}$$

As abuse of notation, when a random tree $\mathcal{T}$ has density $f(\cdot)$ we sometimes write $P(\mathcal{T} = \mathbf{t})$ in place of $f(\mathbf{t})$.

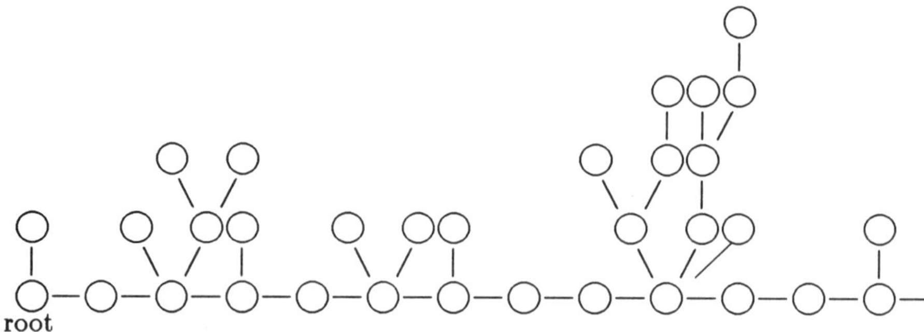

FIGURE 2. A sin-tree $\mathbf{t}$

2.2. PGW$^\infty$(1). Consider a locally finite rooted u-labeled tree $\mathbf{t}$ with an infinite number of vertices, which has a unique path (root $= u_0, u_1, u_2, \ldots$) from the root to infinity. Call such a $\mathbf{t}$ a *sin*-tree (for single infinite path; alternatively, a "tree with one end"). The set $\mathbf{T}^\infty$ of sin-trees may be identified as $\mathbf{T} \times \mathbf{T} \times \ldots$, where $\mathbf{T}$ is the set of finite rooted trees, via the bijection $\mathbf{t} \leftrightarrow (\mathbf{t}_0, \mathbf{t}_1, \mathbf{t}_2, \ldots)$ where the $(\mathbf{t}_i)$ are the trees rooted at u_i obtained by deleting edges of the path $(u_0, u_1, u_2, \ldots)$ in $\mathbf{t}$ (Figure 2). We can use this bijection to define a probability distribution PGW$^\infty$(1) on $\mathbf{T}^\infty$ as the product distribution PGW(1)$\times$PGW(1)$\times \ldots$.

The PGW$^\infty$(1) distribution is in many ways the "canonical" probability distribution on $\mathbf{T}^\infty$. It arises in several $n \to \infty$ settings (see Lemmas 2.3 and 2.2 and section 2.7). We first give a technically useful equivalent description. The *height* of a vertex in a tree is its distance from the root; the height of a tree is the maximum vertex height. Let $\mathbf{T}^h$ be the set of finite rooted trees with height at most h. There is a natural *restriction map* $r_h : \mathbf{T} \to \mathbf{T}^h$, taking $\mathbf{t}$ to the subtree $\mathbf{t}^h$ consisting of vertices at height $\leq h$, and there is a corresponding map $r_h : \mathbf{T}^\infty \to \mathbf{T}^h$, where in $r_h(\mathbf{t})$ the original "path towards infinity" is no longer distinguished.

LEMMA 2.1. *Let $\lambda \leq 1$. Write f_λ^h and f_*^h for the densities of the probability distributions on $\mathbf{T}^h$ induced by r_h from $PGW(\lambda)$ and from $PGW^\infty(1)$. Let $n(h, \mathbf{t})$ denote the number of vertices of $\mathbf{t}$ at height h. Then for $\mathbf{t} \in \mathbf{T}^h$*

$$f_\lambda^h(\mathbf{t}) = e^{\lambda n(h,\mathbf{t})} f_\lambda(\mathbf{t}) = \lambda^{|\mathbf{t}|-1} e^{\lambda(n(h,\mathbf{t})-|\mathbf{t}|)} \tag{2.5}$$

$$f_1^h(\mathbf{t}) = e^{n(h,\mathbf{t})-|\mathbf{t}|} \tag{2.6}$$

$$f_*^h(\mathbf{t}) = n(h, \mathbf{t}) f_1^h(\mathbf{t}). \tag{2.7}$$

PROOF. Let $(u_1, \ldots, u_m)$ be a subset of leaves of a tree $\mathbf{t}$, and write $A(\mathbf{t}; u_1, \ldots, u_m)$ for the set of trees which coincide with $\mathbf{t}$ except that vertices u_i may have adjoined subtrees. Then

$$\frac{f_\lambda(\mathbf{t})}{\sum_{\mathbf{t}' \in A(\mathbf{t}; u_1, \ldots, u_m)} f_\lambda(\mathbf{t}')} = e^{-\lambda m}$$

because at each u_i the PGW(λ) process has chance $e^{-\lambda}$ to be a leaf (and the random trees are finite). By considering $\mathbf{t} \in \mathbf{T}^h$ and the set of all its vertices at height h,

$$\frac{f_\lambda(\mathbf{t})}{f_\lambda^h(\mathbf{t})} = e^{-\lambda n(h,\mathbf{t})}$$

which is (2.5). And (2.6) is just the $\lambda = 1$ case of (2.5). To verify (2.7), fix a height-h vertex u_h of $\mathbf{t} \in \mathbf{T}^h$, and write $\mathbf{t}^{(u_h)}$ for the tree with u_h as distinguished vertex. Consider the path (root $= u_0, u_1, u_2, \ldots, u_h$), and write $(\mathbf{t}_0, \mathbf{t}_1, \ldots, \mathbf{t}_h)$ for the trees obtained by breaking edges of that path. From the definition of PGW$^\infty$(1) and (2.6), the chance that the restriction of PGW$^\infty$(1), with the height-h vertex on the original path to infinity distinguished, is $\mathbf{t}^{(u_h)}$, equals

$$\prod_{i=0}^{h} e^{n(h-i,\mathbf{t}_i)-|\mathbf{t}_i|} = e^{n(h,\mathbf{t})-|\mathbf{t}|}.$$

So (2.7) follows by undistinguishing u_h. □

In words, (2.6,2.7) say that the restriction of PGW$^\infty$(1) to the first h generations can be obtained from the restriction of PGW(1) by size-biasing according to the number of generation-h individuals. Lemma 2.1 and extensions for general Galton-Watson trees go back to Kesten [**12**]. More general such size-biasing relations have recently been emphasized by Lyons et al [**14**] as systematic ways of deriving classical branching process asymptotics. We shall soon see how, in our setting, (2.5 - 2.7) are similarly effective in yielding simple proofs of asymptotics.

2.3. Weak convergence. On the countable set $\mathcal{T}_{\text{unlabeled}}$ of all rooted finite unlabeled trees, convergence in distribution $\mathcal{T}_n \xrightarrow{d} \mathcal{T}$ is just pointwise convergence

$$P(\mathcal{T}_n = \mathbf{t}_{\text{unlabeled}}) \to P(\mathcal{T} = \mathbf{t}_{\text{unlabeled}}) \ \forall \mathbf{t}_{\text{unlabeled}} \in \mathcal{T}_{\text{unlabeled}}. \tag{2.8}$$

Our convention of creating u-labeled trees by assigning i.i.d. $U(0,1)$ labels to vertices makes no difference to weak convergence: we can interpret $\mathcal{T}_n \xrightarrow{d} \mathcal{T}$ for u-labeled trees as (2.8) for unlabeled trees or equivalently in terms of densities as

$$P(\mathcal{T}_n = \mathbf{t}) \to P(\mathcal{T} = \mathbf{t}) \ \forall \mathbf{t} \in \mathbf{T}.$$

To handle sin-trees, recall the restriction map $r_h : \mathbf{T} \cup \mathbf{T}^\infty \to \mathbf{T}^h$, and define $\mathcal{T}_n \xrightarrow{d} \mathcal{T}$ for finite or sin-trees via

$$r_h(\mathcal{T}_n) \xrightarrow{d} r_h(\mathcal{T}) \text{ for all } h < \infty. \tag{2.9}$$

Informally, this is "local" convergence.

The next three sections describe three fundamental examples. We write out proofs to illustrate our "density" notation.

2.4. PGW(λ) as a limit within $G(n, \lambda/n)$. In the random graph $G(n, \lambda/n)$ there are n vertices, and each of the $\binom{n}{2}$ possible edges is present independently with probability λ/n. The next result is implicit in much of classical random graph theory [**7**] and explicit (in different language) in [**1**].

LEMMA 2.2. *Let $\mathcal{G}_\lambda^n$ be the component of $G(n, \lambda/n)$ containing vertex* 1, *considered as a random rooted graph. Then $\mathcal{G}_\lambda^n \xrightarrow{d}$ PGW(λ) as $n \to \infty$.*

PROOF. We claim that $\mathcal{G}_\lambda^n$ has density (on $\mathbf{T}$)

$$\frac{n(n-1)\dots(n-|\mathbf{t}|+1)}{n^{|\mathbf{t}|}}\lambda^{|\mathbf{t}|-1}\left(1-\frac{\lambda}{n}\right)^{Q(n,|\mathbf{t}|)}$$

where

$$Q(n,\mathbf{t}) = |\mathbf{t}|(n-|\mathbf{t}|) + \tfrac{1}{2}|\mathbf{t}|(|\mathbf{t}|-1) - (|\mathbf{t}|-1).$$

This implies that as $n \to \infty$ for fixed $\mathbf{t}$,

$$P(\mathcal{G}_\lambda^n = \mathbf{t}) \to \lambda^{|\mathbf{t}|-1}e^{-\lambda|\mathbf{t}|} = f_\lambda(\mathbf{t})$$

establishing the lemma. To derive the formula, consider the chance that the u-labeling of $\mathcal{G}_\lambda^n$ gives a graph which contains the tree $\mathbf{t}$ of Figure 1, with labels in $[u_i, u_i + du_i]$. The chance is

$$du_1 \times (n-1)(n-2)(n-3)(\lambda/n)^3 du_2 du_3 du_4 \times (n-4)(\lambda/n)du_5$$

where the first term is the chance the root gets label in $[u_1, u_1 + du_1]$, the second is the chance of 3 edges to vertices having labels in $[u_i, u_i + du_i], i = 2, 3, 4$, and the third is the chance that u_3 has an edge to some vertex given label in $[u_5, u_5 + du_5]$. Applying this argument for general $\mathbf{t}$, the chance $\mathcal{G}_\lambda^n$ contains $\mathbf{t}$ is

$$\frac{n(n-1)\dots(n-|\mathbf{t}|+1)}{n^{|\mathbf{t}|}}\lambda^{|\mathbf{t}|-1}\, du_1 du_2 \dots du_{|\mathbf{t}|}.$$

In order for $\mathcal{G}_\lambda^n$ be equal $\mathbf{t}$, there must be no other edges, and $Q(n,\mathbf{t})$ counts the number of forbidden edges. □

2.5. PGW$^\infty$(1) as a limit within the Wright-Fisher model. In the Wright-Fisher model (Ewens [9]) each generation of a population has n female individuals, and each individual in generation g (for $-\infty < g < \infty$) is the daughter of a uniform random mother in generation $g-1$. Thus if we pick an individual in generation 0 as the root, then this individual and her descendants (indicated by • in the Figure 3) form a random tree $\mathcal{T}_n$; the root individual also has an infinite line of descent through generations $-1, -2, \dots$, and the family tree linking all individuals (in generations $-\infty$ through n, say) defines a sin-tree $\mathcal{T}_n^*$. Near-relatives of the root are indicated by ⊗ in Figure 3.

Because each individual has Binomial$(n, 1/n)$ children, the following lemma is intuitively clear.

LEMMA 2.3. *$\mathcal{T}_n \xrightarrow{d} PGW(1)$ and $\mathcal{T}_n^* \xrightarrow{d} PGW^\infty(1)$.*

Arguing as in Lemma 2.2, it is straightforward to state the density of $\mathcal{T}_n$ after u-labeling as

$$P(\mathcal{T}_n = \mathbf{t}) = \prod_{i \ge 1} b(n, n(i,\mathbf{t}))\left(1 - \tfrac{n(i-1,\mathbf{t})}{n}\right)^{n-n(i,\mathbf{t})}$$

where

$$b(n,m) = \tfrac{n(n-1)\dots(n-m+1)}{n^m}, \qquad b(n,0) = 1.$$

Here the term $b(n, n(i,\mathbf{t}))$ reflects existence of edges in $\mathbf{t}$ from generation i to generation $i-1$, and the term $(1 - \frac{n(i-1,\mathbf{t})}{n})^{n-n(i,\mathbf{t})}$ reflects non-existence of other edges. Then as $n \to \infty$

$$P(\mathcal{T}_n = \mathbf{t}) \to \exp(-\sum_{i\ge 1} n(i-1,\mathbf{t})) = e^{-|\mathbf{t}|} = f_1(\mathbf{t}).$$

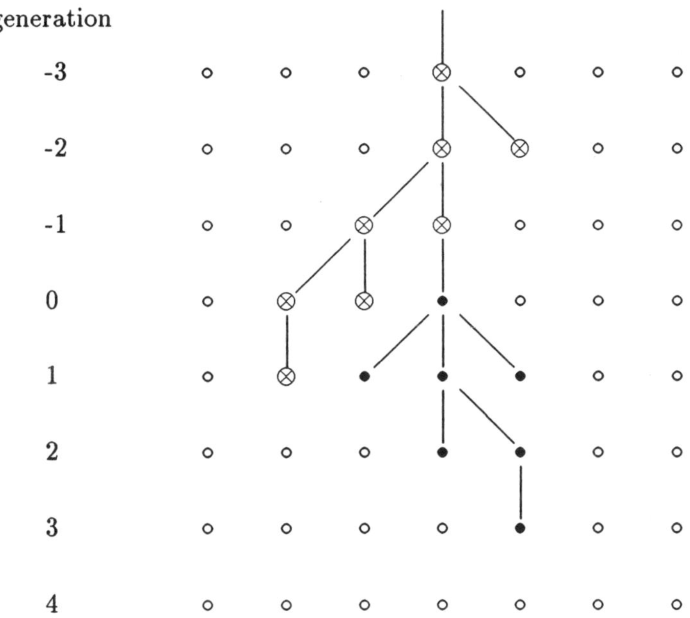

FIGURE 3. The Wright-Fisher model

The argument for $\mathcal{T}_n^* \xrightarrow{d}$ PGW$^\infty(1)$ is similar.

2.6. PGW$^\infty(1)$ as a limit within the uniform random n-tree. While the previous weak convergence results (Lemmas 2.2 and 2.3) were intuitively obvious to probabilists, the next may not be. It is due to Grimmett [10], who used it to explain earlier combinatorial results. Cayley's formula says there are n^{n-2} trees on vertex set $\{1, 2, \ldots, n\}$ rooted at vertex 1. Let $\mathcal{S}^n$ be a uniform random such tree.

LEMMA 2.4. *$\mathcal{S}^n \xrightarrow{d} PGW^\infty(1)$.*

PROOF. As in previous cases, the proof rests on an explicit formula. Fix h and $\mathbf{t} \in \mathbf{T}^h$ with $k = n(h, \mathbf{t}) \geq 1$. We claim that, regarding $r_h(\mathcal{S}^n)$ as a u-labeled tree, it has density

$$\frac{(n-1)(n-2)\ldots(n-|\mathbf{t}|+1)k(n-|\mathbf{t}|+k)^{n-|\mathbf{t}|-1}}{n^{n-2}}. \tag{2.10}$$

Then as $n \to \infty$, this density tends to $n(h, \mathbf{t})\exp(n(h, \mathbf{t}) - |\mathbf{t}|)$, and this limit is the density $f_h^*(\mathbf{t})$ of the restriction of PGW$^\infty(1)$ by (2.7).

To derive (2.10), recall another form of Cayley's formula. The number of forests on vertices $\{1, 2, \ldots, m\}$ with exactly k tree-components rooted at k specified vertices equals $c(m, k) = km^{m-k-1}$. Write $A_n(\mathbf{t})$ for the set of trees obtainable by giving the vertices of $\mathbf{t}$ labels from $\{1, 2, \ldots, n\}$. Then the number of trees on $\{1, 2, \ldots, n\}$ rooted at 1 whose restriction to height h is in $A_n(\mathbf{t})$ equals

$$(n-1)(n-2)\ldots(n-|\mathbf{t}|+1) \times c(n-|\mathbf{t}|+k, k)$$

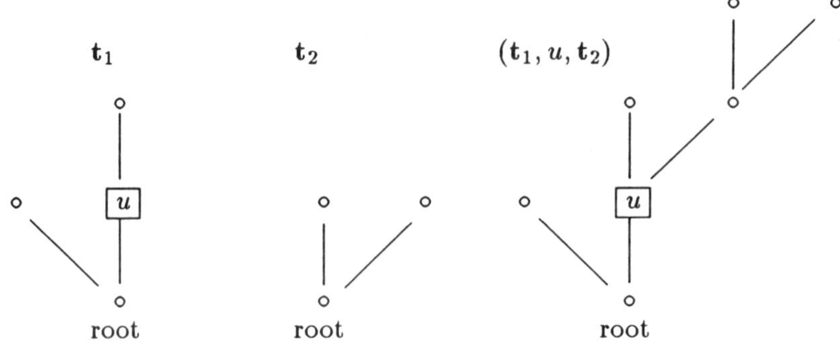

FIGURE 4. Joining

where the first term counts labelings of $\mathbf{t}$ and the second counts extensions above height h. So, before u-labeling,

$$P(r_h(\mathcal{S}_n) \in A_n(\mathbf{t})) = \text{ formula in (2.10).}$$

But u-labeling a tree in $A_n(\mathbf{t})$ has chance $du_1 \ldots du_{|\mathbf{t}|}$ to assign labels in $[u_i, u_i+du_i]$ for all i, and so we get the same formula for the density after u-labeling. □

2.7. Other limit representations of $\mathbf{PGW}^{\infty}(1)$. Lemma 2.4 seems less mysterious if one knows that the uniform labeled n-tree is distributed as the PGW(1) tree conditioned on having size equal to n. It is well known (see **[4]** Lemma 23) that several different formalizations of "PGW(1) conditioned to be large" lead to the $\text{PGW}^{\infty}(1)$ limit.

In a different direction, the uniform random tree $\mathcal{S}^n$ in Lemma 2.4 may be viewed as a special case (the complete graph) of the uniform random spanning tree of an given size-n graph. Lemma 2.4 can be extended to spanning trees of certain other graphs **[2, 18]**.

3. Markov chains with Poisson-Galton-Watson marginals

3.1. Notation for tree-valued chains. Our tree-processes evolve by "joining" and "pruning" operations. If u is a vertex-label of $\mathbf{t}_1$, then let $(\mathbf{t}_1, u, \mathbf{t}_2)$ denote the tree obtained by joining $\mathbf{t}_1$ and $\mathbf{t}_2$ via a new edge from u to the root of $\mathbf{t}_2$, and let the root of the new tree be the root of $\mathbf{t}_1$ (Figure 4). A transition $\mathbf{t}_1 \to (\mathbf{t}_1, u, \mathbf{t}_2)$ is a "joining" operation; the "pruning" operation of cutting an edge and retaining the subtree containing the root is a transition of the reverse form $(\mathbf{t}_1, u, \mathbf{t}_2) \to \mathbf{t}_1$.

We illustrate this notation with a simple example (somewhat different in spirit from our subsequent examples). Here is a Metropolis scheme for constructing a tree-valued discrete-time chain whose stationary distribution is (by design) PGW(λ), for fixed $\lambda \leq 1$, and where the only transitions allowed are to add or delete a leaf. In words, a step of the chain is

> Toss a fair coin. If Heads, pick a uniform random vertex v, and with chance $\lambda e^{-\lambda}$ append a leaf to v. If Tails, pick (if possible) a uniform random vertex v other than the root, and if v is a leaf, delete it.

Symbolically, writing $\bullet$ for the single-vertex tree, the transition probabilities are

$$\mathbf{t} \to (\mathbf{t}, u, \bullet) : \text{ probability } \frac{\lambda e^{-\lambda}}{2|\mathbf{t}|}$$

$$(\mathbf{t}, u, \bullet) \to \mathbf{t} : \text{ probability } \frac{1}{2|\mathbf{t}|}$$

for each vertex u of $\mathbf{t}$. To check that the stationary distribution is indeed PGW(λ) we need to confirm the detailed balance equation

$$f_\lambda(\mathbf{t}) \times \frac{\lambda e^{-\lambda}}{2|\mathbf{t}|} = f_\lambda(\mathbf{t}, u, \bullet) \times \frac{1}{2|\mathbf{t}|},$$

and indeed by (2.2) both sides equal $\lambda^{|\mathbf{t}|} e^{-\lambda|\mathbf{t}|-\lambda}/(2|\mathbf{t}|)$.

3.2. Caveat. In following sections we describe continuous-time tree-valued Markov chains by specifying their transition rates. Conceptually, the rates can be described in words, not depending on any choice of notation for trees; but our density notation is useful for discussing marginal distributions.

There is a technical issue – not pursued in this paper – of showing that rates do indeed specify a unique chain. In Aldous-Pitman [**4**] we give a rigorous constructive treatment of two of the chains (those in sections 3.4 and 3.5) without explicitly using rates or the density notation. It is not hard to make explicit constructions of the other chains, except for the "exotic" chain in section 4.3. But the present paper is intended as a more informal and intuitive presentation.

The definitions of the chains are mostly motivated as limits of finite-state chains. Writing out rigorous proofs of these weak limit results would be technically complicated, so we shall present them as "weak limit assertions" with only brief explanations. In only two places (sections 3.7 and 4.3) do there arise issues of substance rather than technique, so we write those as explicit conjectures.

We think of our chains as a "family" because their transition rates involve qualitatively similar pruning and joining rules. Note that some of the chains are (time)-homogeneous while some are non-homogeneous; and some are (time)-reversible while others are non-reversible.

3.3. The stationary PGW(λ) chain. Fix $0 < \lambda \leq 1$. Consider the homogeneous $\mathbf{T}$-valued chain $(\mathcal{H}_s, s \geq 0)$ with transition rates

$$\mathbf{t}_1 \to (\mathbf{t}_1, u, \mathbf{t}_2) \quad : \quad \text{rate } \lambda f_\lambda(\mathbf{t}_2) \tag{3.1}$$

$$(\mathbf{t}_1, u, \mathbf{t}_2) \to \mathbf{t}_1 \quad : \quad \text{rate } 1 \tag{3.2}$$

(these rates apply to each possible transition, i.e. each possible combination of $\mathbf{t}_1$, u and $\mathbf{t}_2$). In words,

> Each edge is pruned away at rate 1; at each vertex, at rate λ there is adjoined a PGW(λ)-distributed tree.

We claim this chain is reversible with stationary distribution PGW(λ). We need only verify the detailed balance equation

$$f_\lambda(\mathbf{t}_1) \times \lambda f_\lambda(\mathbf{t}_2) = f_\lambda(\mathbf{t}_1, u, \mathbf{t}_2) \times 1 \tag{3.3}$$

and indeed, by (2.2) both sides equal $\lambda^{|\mathbf{t}_1|+|\mathbf{t}_2|-1} e^{-\lambda(|\mathbf{t}_1|+|\mathbf{t}_2|)}$.

As motivation for the definition, regard the random graph $G(n, \lambda/n)$ as the stationary distribution of the chain whose state space is the set of graphs on

$\{1, 2, \ldots, n\}$, in which each of the $\binom{n}{2}$ possible edges is born at rate λ/n and dies at rate $1 - \lambda/n$. Restrict attention to the component containing vertex 1, and call this process $(\mathcal{H}_s^n, 0 \leq s)$. Lemma 2.2 showed that $P(\mathcal{H}_s^n = \mathbf{t}) \to f_\lambda(\mathbf{t})$ as $n \to \infty$ for fixed s.

WEAK LIMIT ASSERTION 3.1. $(\mathcal{H}_s^n, 0 \leq s) \xrightarrow{d} (\mathcal{H}_s, 0 \leq s)$, where the limit is the stationary chain with transition rates (3.1,3.2).

In brief: each edge of $\mathcal{H}_s^n$ dies at rate $1 - \lambda/n$, which is asymptotic to (3.2). Since there are $(1 - o(1))n$ vertices outside $\mathcal{H}_s^n$, at each vertex of $\mathcal{H}_s^n$ a new edge to some vertex v is created at rate $\frac{\lambda}{n}(1 - o(1))n \to \lambda$, and the distribution of the joined component containing v will be approximately the distribution of $\mathcal{H}_s^n$, i.e. with density f_λ.

Curiously, this chain $(\mathcal{H}_s)$ has not been studied, though (regarding the random graph as a mean-field model of percolation) it is analogous to the recently-studied topic of *flickering percolation* [**11**]. For $\lambda < 1$, using the fact that the distribution (2.4) of $|\mathcal{H}_s|$ has geometrically decreasing tail it is not hard to show that

$$T_a := \inf\{s : |\mathcal{H}_s| \geq a\} \xrightarrow{p} \infty \text{ as } a \to \infty$$

so that the chain stays in the state space $\mathbf{T}$ at all times. For $\lambda = 1$ this is more delicate. In the first draft, and at the Workshop, I asserted that $T_a < \infty$ a.s. when $\lambda = 1$. But this is false. More careful analysis, which may be presented elsewhere, shows that in fact T_a goes to infinity at rate $\log a$.

3.4. The PGW growth chain. Define $(\mathcal{G}_\lambda, 0 \leq \lambda \leq 1)$ to be the non-homogeneous $\mathbf{T}$-valued Markov chain with transition rate at time λ

$$\mathbf{t}_1 \to (\mathbf{t}_1, u, \mathbf{t}_2) : \text{ rate } f_\lambda(\mathbf{t}_2) \tag{3.4}$$

where $f_\lambda(\cdot)$ is the PGW(λ) distribution (2.2) and where $\mathcal{G}_0 = \bullet$ is the single-vertex tree. In words,

> At each vertex, at rate 1 there is adjoined a PGW(λ)-distributed tree.

It is intuitively clear (and proved in Aldous [**1**] in different language) that this process is the weak limit of the component containing vertex 1 in the usual random graph process $(G(n, \lambda/n), \lambda \geq 0)$. In particular, the marginal distribution at time λ is PGW(λ). We may verify this intrinsically by verifying the forwards equation

$$\frac{d}{d\lambda} f_\lambda(\mathbf{t}) = \sum_{(\mathbf{t}_1, u, \mathbf{t}_2) = \mathbf{t}} f_\lambda(\mathbf{t}_1) f_\lambda(\mathbf{t}_2) - f_\lambda(\mathbf{t}) \sum_{u \in \mathbf{t}, \mathbf{t}_2} f_\lambda(\mathbf{t}_2).$$

The right side becomes

$$\lambda^{-1} f_\lambda(\mathbf{t})(|\mathbf{t}| - 1) \; - \; f_\lambda(\mathbf{t})|\mathbf{t}|$$

and this is indeed the derivative of (2.2).

This process $(\mathcal{G}_\lambda)$ is studied in detail in Aldous-Pitman [**4**]. In particular, one can extend the time-interval from $0 \leq \lambda \leq 1$ to $0 \leq \lambda < A$, where $A > 1$ is the *ascension time*, indicating (in the random graph pre-limit) the time at which the component under study joins the giant component. Qualitatively, $\mathcal{G}_{A-}$ is a finite tree; at time A it acquires an edge linking it to an infinite tree. The distribution of A is specified by (3.9) below.

3.5. The process $(\mathcal{G}_s^*, 0 \le s \le 1)$. This process, and its extension to general Galton-Watson trees, is also studied rigorously in Aldous-Pitman [**4**], which contains elaborations of the results stated here (except for the "rate" result in Lemma 3.2(b)).

Let $\mathcal{G}_1^*$ have PGW$^\infty(1)$ distribution. Attach independent $U(0,1)$ random variables ξ_e to its edges; consider the subgraph consisting of edges e with $\xi_e \le s$, and let $\mathcal{G}_s^*$ be the component of that subgraph containing the root. So $\mathcal{G}_0^* = \bullet$, the single-vertex tree. Clearly the process $(\mathcal{G}_s^*, 0 \le s \le 1)$ is Markov when s is decreasing, so it is automatically (non-homogeneous) Markov when s is increasing. The first assertion of the lemma is that for fixed s the distribution of $\mathcal{G}_s^*$ is the PGW(s) distribution, size-biased by total population size.

LEMMA 3.2. *(a) For $0 \le s < 1$, the distribution g_s^* of $\mathcal{G}_s^*$ is*

$$g_s^*(\mathbf{t}) = (1-s)|\mathbf{t}|f_s(\mathbf{t}).$$

(b) $(\mathcal{G}_s^*)$ is the Markov chain with transition rates at time s

$$\mathbf{t}_1 \to (\mathbf{t}_1, u, \mathbf{t}_2) : \textit{ rate } \frac{|\mathbf{t}_1| + |\mathbf{t}_2|}{|\mathbf{t}_1|} f_s(\mathbf{t}_2) \tag{3.5}$$

for each vertex u of $\mathbf{t}_1$.

PROOF. (a) In constructing $\mathcal{G}_s^*$ from $\mathcal{G}_1^*$ we may distinguish a vertex u_h, the first vertex on the infinite path for which $\xi_{(u_h, u_{h+1})} > s$. Fix $\mathbf{t}$ and a height-h vertex u_h of $\mathbf{t}$. Consider the path (root $= u_0, u_1, u_2, \dots, u_h$), and write $(\mathbf{t}_0, \mathbf{t}_1, \dots, \mathbf{t}_h)$ for the trees obtained by breaking edges of that path. The chance that $\mathcal{G}_s^* = \mathbf{t}$ and u_h is the distinguished vertex equals $s^h(1-s)f_s(\mathbf{t}_0)f_s(\mathbf{t}_1)\dots f_s(\mathbf{t}_h)$. But this equals

$$s^h(1-s)\prod_{i=0}^{h} s^{|\mathbf{t}_i|-1}e^{-s|\mathbf{t}_i|} = (1-s)s^{|\mathbf{t}|-1}e^{-|\mathbf{t}|} = (1-s)f_s(\mathbf{t}).$$

So (a) follows by undistinguishing u_h.

For (b), as s decreases the transition rate at time s is (from the "pruning at uniform random times" construction)

$$(\mathbf{t}_1, u, \mathbf{t}_2) \to \mathbf{t}_1 : \text{ rate } 1/s. \tag{3.6}$$

So the forwards transition rate is $1/s \times g_s^*(\mathbf{t}_1, u, \mathbf{t}_2)/g_s^*(\mathbf{t}_1)$, and applying the formula in (a) this rate becomes

$$\frac{|\mathbf{t}_1| + |\mathbf{t}_2|}{|\mathbf{t}_1|} s^{-1} \frac{f_s(\mathbf{t}_1, u, \mathbf{t}_2)}{f_s(\mathbf{t}_1)}$$

which reduces to the formula in (b). □

Several relationships between $(\mathcal{G}_t^*)$ and $(\mathcal{G}_t)$ are explored in [**4**]. The first is that in reversed time they evolve in the same way. At ([**4**] equation (85)) this is formalized as: for fixed $\mathbf{t} \in \mathbf{T}$, $0 < \lambda < \infty$ and $0 < s < 1$,

$$\operatorname{dist}(\mathcal{G}_{t\lambda}, 0 \le t \le 1 | \mathcal{G}_\lambda = \mathbf{t}) = \operatorname{dist}(\mathcal{G}_{ts}^*, 0 \le t \le 1 | \mathcal{G}_s^* = \mathbf{t}). \tag{3.7}$$

Here is a reformulation in terms of transition rates.

LEMMA 3.3. *Each of the process $(\mathcal{G}_s, 0 \le s \le 1)$ and $(\mathcal{G}_s^*, 0 \le s \le 1)$ evolves, as s decreases, as the non-homogeneous Markov chain with transition rates at time s*

$$(\mathbf{t}_1, u, \mathbf{t}_2) \to \mathbf{t}_1 : \textit{ rate } 1/s. \tag{3.8}$$

PROOF. For $(\mathcal{G}_s^*)$ this is (3.6). For $(\mathcal{G}_s)$ it is intuitively clear from the random graph asymptotics: here is a formal verification. Since the forwards transition rates for $\mathcal{G}_s$ are (3.4)

$$\mathbf{t}_1 \to (\mathbf{t}_1, u, \mathbf{t}_2): \text{ rate } f_s(\mathbf{t}_2)$$

and the marginal distribution is f_s, the reversed transition rate is

$$\frac{f_s(\mathbf{t}_1)f_s(\mathbf{t}_2)}{f_s(\mathbf{t}_1, u, \mathbf{t}_2)} = 1/s.$$

□

A second relationship is that we can regard $(\mathcal{G}_s^*, 0 \leq s \leq 1)$ as having the conditional distribution of $(\mathcal{G}_s, 0 \leq s \leq 1)$ given $|\mathcal{G}_1| = \infty$. See [4] sections 3.3 and 4.4 for this and the analogous result for general Galton-Watson processes. Here is a third, more surprising relationship.

PROPOSITION 3.4 ([4] Proposition 26).

$$(\mathcal{G}_\lambda, 0 \leq \lambda < A) \stackrel{d}{=} \left(\mathcal{G}_{\lambda U}^*, 0 \leq \lambda < \frac{-\log U}{1-U}\right)$$

where U is $U(0,1)$, independent of $(\mathcal{G}_s^*, 0 \leq s \leq 1)$.

So in particular the distribution of A is

$$A \stackrel{d}{=} \frac{-\log U}{1-U}. \tag{3.9}$$

3.6. Stochastic models of coalescence. We digress to give an interpretation of two previous models and motivation for the next model. Given n and a kernel $K(i,j) \geq 0$, the *Marcus-Lushnikov* process takes values in the set of forests on $\{1, 2, \ldots, n\}$ and evolves according to the rule: trees $\mathbf{t}_1$ and $\mathbf{t}_2$ merge at rate $K(|\mathbf{t}_1|, |\mathbf{t}_2|)/n$, by adding a linking edge between uniform random vertices of $\mathbf{t}_1$ and $\mathbf{t}_2$. Initially there are 0 edges. If we suppose existence of the deterministic limit as $n \to \infty$

$$n^{-1}(\text{number of size-}i\text{ trees at time } s) \xrightarrow{p} c(i,s)$$

then these limits should satisfy the *Smoluchowski coagulation equation*

$$\frac{d}{ds}c(i,t) = \tfrac{1}{2}\sum_{j=1}^{i-1} K(j, i-j)c(j,s)c(i-j,s) \; - \; \sum_{j \geq 1} K(i,j)c(i,s)c(j,s). \tag{3.10}$$

See [3] for a survey of such models of stochastic coalescence. Now consider the process describing the tree component containing vertex 1 in the Marcus-Lushnikov process, considered as a rooted tree. As $n \to \infty$ we expect a limit tree-valued process $(\mathcal{T}_s, s \geq 0)$ such that

$$P(|\mathcal{T}_s| = i) = ic(i,s) \tag{3.11}$$

at least for s smaller than some "gelation time". Now change viewpoints, and consider seeking to define the candidate limit process directly. Combining the argument for (3.10) with the size-biasing relationship (3.11), one can see that the appropriate formula is as follows. Define $(\mathcal{T}_s)$ to be the $\mathbf{T}$-valued chain with $\mathcal{T}_0 = \bullet$ and with transition rate at time s

$$\mathbf{t}_1 \to (\mathbf{t}_1, u, \mathbf{t}_2): \quad \text{rate } \frac{K(|\mathbf{t}_1|, |\mathbf{t}_2|)}{|\mathbf{t}_1|\;|\mathbf{t}_2|}\; g_s(\mathbf{t}_2) \tag{3.12}$$

where g_s is the density of $\mathcal{T}_s$. Note (3.12) has the unusual feature that the formula for transition rates involves the marginal density of the process itself. Solving (3.12) for g_s is essentially equivalent to solving the Smoluchowski coagulation equation (3.10) for $c(i,s)$. But (3.4) says that the densities $(f_\lambda, 0 \le \lambda \le 1)$ of the PGWgrowth process $(\mathcal{G}_\lambda, 0 \le \lambda \le 1)$ solve (3.12) for $K(i,j) = ij$. And (3.5) says that the densities $(g_s^*, 0 \le s \le 1)$ of the process $(\mathcal{G}_s^*, 0 \le s \le 1)$ solve (3.12) for $K(i,j) = i+j$, except with an extra factor $(1-s)^{-1}$ on the right side. So after making the deterministic time-change to $(g^*_{1-e^{-s}}, 0 \le s < \infty)$ we do get a solution of (3.12) for $K(i,j) = i+j$.

Returning to the case $K(i,j) = ij$, for $\lambda > 1$ the PGW(λ) distribution has a sub-probability density f_λ on the set $\mathbf{T}$ of finite trees. But for $\lambda > 1$, (f_λ) is no longer a solution of (3.12) (with $\mathbf{t}_i$ finite trees), because in the random graph pre-limit there is coalescence between finite trees and the giant component which has no counterpart in (3.12). In the next section we construct a process $(\mathcal{G}_s^0)$ whose densities (h_s) do solve (3.12) on $\mathbf{T}$: see (3.15). For the weak limit interpretation see Conjecture 3.6.

3.7. The process $(\mathcal{G}_s^0, 0 \le s < Z)$. For $0 \le s < \infty$ let h_s be the density on $\mathbf{T}$ given by

$$\begin{aligned} h_s &= f_s, \ 0 \le s \le 1 \\ &= s^{-1} f_1, \ s > 1 \end{aligned}$$

where f_s is the PGW(s) density, so that h_s is a sub-probability density for $s > 1$.

PROPOSITION 3.5. *Let V have $U(0,1)$ distribution, independent of $(\mathcal{G}_s^*, 0 \le s \le 1)$. Define a $\mathbf{T}$-valued process $(\mathcal{G}_s^0, 0 \le s < Z)$ by*

$$\mathcal{G}_s^0 = \mathcal{G}_{sV}^*, \ 0 \le s < Z = 1/V. \tag{3.13}$$

Then

$$P(\mathcal{G}_s^0 = \mathbf{t}, Z > s) = h_s(\mathbf{t}). \tag{3.14}$$

And $(\mathcal{G}_s^0)$ is the non-homogeneous Markov chain with transition rates at time s

$$\mathbf{t}_1 \to (\mathbf{t}_1, u, \mathbf{t}_2) : \ \textit{rate } h_s(\mathbf{t}_2) \ , \textit{ for } \mathbf{t}_1, \mathbf{t}_2 \in \mathbf{T} \tag{3.15}$$

stopped at $Z = \inf\{s : |\mathcal{G}_s^0| = \infty\}$.

Note that the limit $\mathcal{G}_{Z-}^0 \stackrel{d}{=} \mathcal{G}_1^*$ has PGW$^\infty(1)$ distribution. And note that by Proposition 3.4

$$\mathcal{G}_V^* \stackrel{d}{=} \mathcal{G}_1 \text{ has density } f_1 \tag{3.16}$$

and that by definition (3.13)

$$P(Z > z) = 1/z, \ z \ge 1.$$

PROOF. For fixed $s \le 1$,

$$P(\mathcal{G}_s^0 = \mathbf{t}, Z > s) = P(\mathcal{G}_{sV}^* = \mathbf{t}) = f_s(\mathbf{t})$$

by Proposition 3.4. For fixed $s \geq 1$,

$$\begin{aligned} P(\mathcal{G}^0_s = \mathbf{t}, Z > s) &= P(\mathcal{G}^*_{sV} = \mathbf{t}, V < 1/s) = \\ &= s^{-1} P(\mathcal{G}^*_V = \mathbf{t}) \\ &\quad \text{because } P(V < 1/s) = 1/s \text{ and } \mathrm{dist}(sV|V < 1/s) = \mathrm{dist}(V) \\ &= s^{-1} f_1(\mathbf{t}) \text{ by (3.16).} \end{aligned}$$

From (3.7,3.8) the reversed transition rates are

$$(\mathbf{t}_1, u, \mathbf{t}_2) \to \mathbf{t}_1 : \text{ rate } 1/s$$

and so the forwards transition rate $\mathbf{t}_1 \to (\mathbf{t}_1, u, \mathbf{t}_2)$ is

$$s^{-1} h_s(\mathbf{t}_1, u, \mathbf{t}_2)/h_s(\mathbf{t}_1).$$

For $s \leq 1$ this is

$$s^{-1} f_s(\mathbf{t}_1, u, \mathbf{t}_2)/f_s(\mathbf{t}_1) = f_s(\mathbf{t}_2) = h_s(\mathbf{t}_2),$$

while for $s > 1$ it is

$$s^{-1} f_1(\mathbf{t}_1, u, \mathbf{t}_2)/f_1(\mathbf{t}_1) = s^{-1} f_1(\mathbf{t}_2) = h_s(\mathbf{t}_2)$$

establishing (3.15). □

Comparing Propositions 3.4 and 3.5 leads to an interesting paradox. Both processes $(\mathcal{G}_s)$ and $(\mathcal{G}^0_s)$ can be constructed as $(\mathcal{G}^*_{sV})$, stopped at times $\frac{-\log V}{1-V}$ and $1/V$ respectively. So we can say

> the process $(\mathcal{G}_s, 0 \leq s < A)$ is distributed as the process $(\mathcal{G}^0_s, 0 \leq s \leq Z)$ stopped at the random time $A = \frac{\log Z}{1-Z^{-1}}$.

But the transition rates (for $s > 1$) for the two processes are different. The resolution of the paradox is that, for a "stopped" process to have the same transition rates (before stopping, of course), the stopping time must be a (randomized) *stopping time* in the technical sense, and A here is not such a stopping time.

Motivation for study of $(\mathcal{G}^0_s)$ comes from the following limit considerations. Modify the random graph process $(G(n, s/n), s \geq 0)$ by specifying that components whose size exceeds a threshold $\phi(n)$ are forbidden to coalesce with other components (imagine a physical process of aggregation in a liquid, where sufficiently large aggregates precipitate out). Thus when the process is run to time infinity, instead of getting one component of size n we get a set of components, all of sizes between $\phi(n)$ and $2\phi(n)$ (except perhaps one smaller component).

CONJECTURE 3.6. *Let $\phi(n)$ be a function such that $\phi(n) \to \infty$ and $\phi(n)/n \to 0$ as $n \to \infty$. Consider the Marcus-Lushnikov process with*

$$K_n(x, y) = xy 1_{(\max(x,y) \leq \phi(n))}.$$

Let $\mathcal{Q}_n(t)$ be the component containing vertex 1 at time t, let $Q_n(t) = |\mathcal{Q}_n(t)|$ be its size, and let $Z_n = \min\{t : Q_n(t) > \phi(n)\}$. Then as $n \to \infty$,

$$(\mathcal{Q}_n(t), 0 \leq t < Z_n) \xrightarrow{d} (\mathcal{G}_t, 0 \leq t < Z)$$

and so in particular

$$P(Z_n > t) \to t^{-1},\ t > 1$$

$$P(Q_n(t) = q, Z_n > t) \to \frac{q^{q-1} e^{-q}}{q!} t^{-1},\ t > 1, q \geq 1.$$

$$Q_n(Z_n-) \xrightarrow{d} \infty.$$

This seems genuinely harder than our previous weak limit assertions, because we cannot write down useful exact expressions for the distribution of $Q_n(t)$.

4. Chains with boundary behavior

The chains in section 3 either stayed in the set $\mathbf{T}$ of finite trees for all time, or were stopped upon first becoming infinite. In this section we deal with chains which run for time $0 \le s < \infty$ but which sometimes momentarily become infinite (but are a.s. finite at fixed times).

Let us briefly mention theoretical background. Suppose we regard a continuous-time countable-state Markov chain as being specified by its finite-dimensional distributions. Then what can we say in general about its sample path behavior? The path must spend almost all (w.r.t. Lebesgue measure) time in the state space, but there are examples where in a natural topology the chain occasionally goes outside the original state space. An extensive general *compactification* theory was developed to handle such phenomena (see e.g. [**19**] Chapter III.6). But naturally-arising examples are rather scarce.

The specific chains we shall describe are novel, and indeed the notion of chains with such boundary behavior arising as weak limits within concrete discrete settings seems novel. The chain in sections 4.1 and 4.2 exhibits the qualitatively simplest boundary behavior, being infinite only at isolated times. A more exotic chain is outlined in section 4.3.

4.1. The backwards chain. This chain is motivated as a weak limit of a chain on spanning trees (weak limit assertion 4.3). It will also provide an alternative explanation (Corollary 4.6) of the striking "quasistationarity" property (3.14) of $(\mathcal{G}_s^0)$: that the conditional distribution of $\mathcal{G}_s^0$ given $Z > s$ is PGW(1) for all $s \ge 1$.

We study the $\mathbf{T}$-valued chain $(\overleftarrow{\mathcal{T}}_s)$ specified in words as

> Each edge is pruned away at rate 1; at rate 1 a PGW$^\infty$(1) infinite tree is appended at the root

and symbolically by transition rates

$$\mathbf{t}_1 \to (\mathbf{t}_1, \text{root}, \cdot) : \text{ rate } f_*(\cdot)$$

$$(\mathbf{t}_1, u, \mathbf{t}_2) \to \mathbf{t}_1 : \text{ rate } 1 \tag{4.1}$$

where f_* is the density of the PGW$^\infty$(1) infinite tree. Here $(\mathbf{t}_1, \text{root}, \mathbf{t}_2)$ indicates joining the root of $\mathbf{t}_2$ to the root of $\mathbf{t}_1$ by a new edge. The point is that an infinite branch is adjoined at the times at a Poisson (rate 1) process, thus taking the chain outside the set $\mathbf{T}$ of finite trees, but this new branch is immediately cut back to finite size, so the chain only spends isolated time-instants outside $\mathbf{T}$.

Proposition 4.1. *The chain $(\overleftarrow{\mathcal{T}}_s)$ has unique stationary distribution PGW(1).*

Proof. We consider restrictions to height h. Recall (2.6,2.7)

$$f^h(\mathbf{t}) = e^{n(h,\mathbf{t})-|\mathbf{t}|}, \ \mathbf{t} \in \mathbf{T}^h$$

$$f_*^h(\mathbf{t}) = n(h,\mathbf{t})e^{n(h,\mathbf{t})-|\mathbf{t}|}, \ \mathbf{t} \in \mathbf{T}^h$$

are the distributions of the restrictions of PGW(1) and of PGW$^\infty$(1). So we need to show that the chain on $\mathbf{T}^h$ with transition rates

$$\mathbf{t}_1 \to (\mathbf{t}_1, \text{root}, \mathbf{t}_2) : \text{ rate } f_*^{h-1}(\mathbf{t}_2) \text{ (for } \mathbf{t}_1 \in \mathbf{T}^h, \mathbf{t}_2 \in \mathbf{T}^{h-1})$$

$$(\mathbf{t}_1, u, \mathbf{t}_2) \to \mathbf{t}_1 : \text{ rate } 1 (\text{ for } (\mathbf{t}_1, u, \mathbf{t}_2) \in \mathbf{T}^h)$$

has stationary distribution f^h. So fix $\mathbf{t} \in \mathbf{T}^h$ and verify the balance equations for probability flow rates into and out of $\mathbf{t}$. We assert these flow rates are

out, by pruning	$(\lvert\mathbf{t}\rvert - 1)f^h(\mathbf{t})$
out, by joining	$f^h(\mathbf{t})$
in, by pruning	$(\lvert\mathbf{t}\rvert - n(\mathbf{t}, h))f^h(\mathbf{t})$
in, by joining	$n(\mathbf{t}, h)f^h(\mathbf{t})$

(these rates plainly do balance). The "out" rates are clear. To verify the "in, by pruning" rate, we need to show that for each vertex u of $\mathbf{t}$ at height less than h,

$$\sum_{\mathbf{t}_2 : (\mathbf{t}, u, \mathbf{t}_2) \in \mathbf{T}^h} \frac{f^h(\mathbf{t}, u, \mathbf{t}_2)}{f^h(\mathbf{t})} = 1.$$

But, if u is at distance j from the root, this reduces to

$$\sum_{\mathbf{t}_2 \in \mathbf{T}^{h-j-1}} f^{h-j-1}(\mathbf{t}_2) = 1$$

which holds because f^{h-j-1} is a probability distribution. To verify the "in, by joining" rate, for each edge e of $\mathbf{t}$ incident at the root, we may write $\mathbf{t} = (\mathbf{t}_1, \text{root}, \mathbf{t}_{h-1})$ where $\mathbf{t}_{h-1}$ is the subtree of height at most $h-1$ which is pruned away by cutting e. The flow rate from $\mathbf{t}_1$ to $\mathbf{t}$ equals

$$f^h(\mathbf{t}_1) f_*^{h-1}(\mathbf{t}_{h-1}) = m(e) f^h(\mathbf{t})$$

where $m(e)$ is the number of vertices at height h in $\mathbf{t}$ which lie in the branch through e. Summing over edges e gives the stated rate. □

The special case $\lambda = 1$ of Proposition 3.4 is the assertion that $\mathcal{G}_V^*$ has PGW(1) distribution. Here is an interesting independent proof of that fact, restated in words.

COROLLARY 4.2. *Take a realization from PGW$^\infty$(1), and independently let U have uniform distribution on $(0,1)$. Cut each edge of the PGW$^\infty$(1)-tree with probability U. Then the component containing the original root has distribution PGW(1).*

PROOF. By the PASTA (Poisson arrivals see time-averages) principle (e.g. [6]), observing the stationary process $(\overleftarrow{\mathcal{T}}_s)$ only at the instants immediately before an infinite branch is attached defines a discrete-time Markov chain, with PGW(1) stationary distribution, whose steps are specified by

> join a realization of PGW$^\infty$(1) at the root; prune each edge with probability $1 - e^{-V}$

where V (representing the time until the next infinite branch is attached) has exponential(1) distribution. But (from the definitions) joining a PGW$^\infty$(1) tree to a PGW(1) tree gives a PGW$^\infty$(1) tree, and $1 - e^{-V} \stackrel{d}{=} U$. □

Our motivation for the definition of $(\overleftarrow{\mathcal{T}}_s)$ came from the following weak limit assertion. Let (Q_i^n) be the discrete-time Markov chain taking values in the set of all n^{n-1} rooted trees on labeled vertices $\{1, \ldots, n\}$, which in one step goes from $\mathbf{t}$ to a tree $\mathbf{t}'$ as follows.

> Pick a uniform random vertex I. If $I = \mathrm{root}(\mathbf{t})$ then let $\mathbf{t}' = \mathbf{t}$. Otherwise, draw an edge from root($\mathbf{t}$) to I, creating a cycle, and delete the other edge in the cycle incident at root($\mathbf{t}$): let $\mathbf{t}'$ be the resulting tree, rooted at I.

It is easy to check that the stationary distribution is uniform. This chain may be regarded as the specialization (to the complete graph) of the now well-known *random walk algorithm* for constructing a uniform random spanning tree of a general finite undirected graph [**2**, **8**], which is part of the circle of ideas surrounding the Markov chain tree theorem [**18**, **15**]. Now define F_i^n to be the subtree of Q_i^n rooted at vertex 1 obtained by cutting the edge at vertex 1 which leads toward root(Q_i^n) (if root(Q_i^n) = 1 let $F_i^n = Q_i^n$). Regard $(F_i^n, i \geq 0)$ as a stationary process taking values in the set $\mathbf{T}$ of u-labeled trees. Lemma 2.4 implies that $F_0^n \xrightarrow{d}$ PGW(1).

Weak Limit Assertion 4.3. As $n \to \infty$,

$$(F_{\lfloor ns \rfloor}^n, s \geq 0) \xrightarrow{d} (\overleftarrow{\mathcal{T}}_s, s \geq 0).$$

Outline. If the vertex I chosen at step $i+1$ is not a vertex of F_i^n then $F_{i+1}^n = F_i^n$. If I is a non-root vertex u of F_i^n then F_{i+1}^n is obtained from F_i^n by pruning the edge from u to its parent. After time-rescaling, in $F_{\lfloor ns \rfloor}^n$ each edge is being pruned at rate asymptotic to 1. When I is vertex 1 (which happens at rescaled rate 1) the new tree F_{i+1}^n is obtained from F_i^n by attaching to the root an edge to a vertex v which is the root of a random tree on the remaining $n - |F_i^n|$ vertices. By Lemma 2.4 this attached tree has distribution asymptotic to PGW$^\infty$(1).

4.2. The forwards chain. The time-reversal $(\overrightarrow{\mathcal{T}}_s)$ of the stationary chain $(\overleftarrow{\mathcal{T}}_s)$ is automatically a Markov chain with the same PGW(1) stationary distribution.

Proposition 4.4. *$(\overrightarrow{\mathcal{T}}_s)$ is the homogeneous chain with transition rates*

$$\mathbf{t}_1 \to (\mathbf{t}_1, u, \mathbf{t}_2) : \textit{ rate } f_1(\mathbf{t}_2)$$

together with the rule:

at the instant the tree becomes infinite, the infinite branch is pruned away at the root.

Proof. To check that the transition rate is as asserted is to check

$$f_1(\mathbf{t}_1) \times f_1(\mathbf{t}_2) =^? f_1(\mathbf{t}_1, u, \mathbf{t}_2) \times 1$$

where the 1 is the reversed flow rate (4.1). But this equality is the $\lambda = 1$ case of (3.3). The final rule is just a sample path property inherited from the corresponding property of $(\overleftarrow{\mathcal{T}}_s)$. □

Remark. Our construction of $(\overrightarrow{\mathcal{T}}_s)$ via time-reversal finesses various difficulties which would arise if we tried to argue directly that the properties in Proposition 4.4 specify a positive-recurrent chain. First, it is not obvious (though in fact true, by the corresponding fact for $(\overleftarrow{\mathcal{T}}_s)$) that the restrictions to height h, i.e. the process

$(r_h(\vec{\mathcal{T}}_s), s \geq 0)$, is Markov. (The difficulty being that by only looking at the restriction we cannot see the infinite branch being grown and pruned). Thus we would have to deal directly with a compactification of $\mathbf{T}$ as state-space, and it is not elementary to write down the equations one must verify in order to check that PGW(1) is a stationary distribution.

Write $T_\infty > 0$ for the first time that $|\vec{\mathcal{T}}_s| = \infty$.

COROLLARY 4.5. *(a) For the stationary chain, T_∞ has exponential(1) distribution. Moreover, for each fixed $s > 0$ the distribution of $\vec{\mathcal{T}}_s$ is PGW(1) and is independent of the event $\{T_\infty > s\}$.*

(b)

$$P(T_\infty > s | \vec{\mathcal{T}}_0 = \mathbf{t}) = (H(s))^{|\mathbf{t}|},$$

where

$$H(s) = \exp(1 - s - e^{-s}).$$

PROOF. In the stationary $(\overleftarrow{\mathcal{T}}_s)$ chain, let C be the first time that an infinite branch is joined. Clearly $P(C > s) = e^{-s}$ and, for fixed s, the event $\{C > s\}$ is independent of $\overleftarrow{\mathcal{T}}_0$. The assertions of (a) follow immediately by time-reversal. For (b), from each vertex of $\mathbf{t}$ there grow independent and identically distributed processes of subtrees, so the first assertion holds for some $H(s)$. Then by (a)

$$e^{-s} = \Phi(H(s))$$

where $\Phi(z)$ is the generating function of the size of PGW(1), i.e. of the Borel(1) distribution (2.4). But it is elementary and well-known that $\Phi(z) = z\exp(\Phi(z) - 1)$, and hence that $\Phi^{-1}(u) = ue^{1-u}$. □

We now relate $(\vec{\mathcal{T}}_s)$ to $(\mathcal{G}^0_s)$.

COROLLARY 4.6.

$$(\mathcal{G}^0_s, 1 \leq s \leq Z) \stackrel{d}{=} (\vec{\mathcal{T}}_{\log s} : 1 \leq s \leq e^{T_\infty}).$$

PROOF. Corollary 4.5(a) shows that

$$P(\vec{\mathcal{T}}_{\log s} = \mathbf{t}, s < e^{T_\infty}) = s^{-1} f_1(\mathbf{t}) = P(\mathcal{G}^0_s = \mathbf{t}, s < Z).$$

Proposition 4.4 implies that the transition rates $\mathbf{t}_1 \to (\mathbf{t}_1, u, \mathbf{t}_2)$ for $(\vec{\mathcal{T}}_{\log s})$ are

$$\frac{d \log s}{ds} \times f_1(\mathbf{t}_2) = h_s(\mathbf{t}_2),$$

agreeing with the transition rates (3.15) for $(\mathcal{G}^0_s)$. □

Note that we could take Corollary 4.6 as a construction of $\mathcal{G}^0_s$, and then the quasistationarity property of $\mathcal{G}^0_s$ would be a consequence of the independence assertion for the stationary $\vec{\mathcal{T}}$-chain in Corollary 4.5(a).

Note also that Corollary 4.5(b) now translates to give distributional information for $(\mathcal{G}^0_s)$.

COROLLARY 4.7.

$$P(Z > z | \mathcal{G}_1^0 = \mathbf{t}) = (H_1(z))^{|\mathbf{t}|}, z \geq 1$$

where

$$H_1(z) = z^{-1}\exp(1 - z^{-1}).$$

4.3. An exotic process. Let us end by indicating a process with more exotic boundary behavior. Recall from section 4.1 the non-reversible discrete-time chain $(Q_i^n, i \geq 0)$ on the set of all rooted trees on $\{1, 2, \ldots, n\}$, with uniform stationary distribution. Here is a reversible chain with the same property. In one step, go from $\mathbf{t}$ to a tree $\mathbf{t}'$ as follows.

> Pick two distinct vertices v_1, v_2 uniformly at random. If (v_1, v_2) is an edge of $\mathbf{t}$, let $\mathbf{t}' = \mathbf{t}$. Otherwise, add an edge (v_1, v_2) thereby creating a cycle, and delete a uniformly-chosen other edge from that cycle to get $\mathbf{t}'$.

Call this chain (R_i^n), and regard the tree as rooted at vertex n. As in section 4.1, define F_i^n to be the subtree of R_i^n rooted at vertex 1 obtained by cutting the edge at vertex 1 which leads toward the root. Lemma 2.4 says that $F_0^n \xrightarrow{d}$ PGW(1). The arguments in the outline proof of weak limit assertion 4.3 suggest

$$(F_{\lfloor ns \rfloor}^n, s \geq 0) \xrightarrow{d} (\mathcal{Z}_s, s \geq 0),$$

where $(\mathcal{Z}_s)$ is a certain stationary reversible chain with stationary distribution PGW(1). For finite $\mathbf{t}$, the transition rates of $(\mathcal{Z}_s)$ are as follows.

> (i) At each vertex of $\mathbf{t}$ an infinite PGW$^\infty$(1)-distributed branch is attached at rate 1/2;
> (ii) at each vertex v of $\mathbf{t}$, at rate 1/2 the edge at the root leading toward v is cut, and simultaneously an infinite PGW$^\infty$(1)-distributed branch is attached to the root.

To complete a verbal specification of $(\mathcal{Z}_s)$ we need to define transitions which are qualitatively reverse to (i) when $\mathbf{t}$ is infinite. Obviously it is a challenge to formalize such a definition and show that it does specify a unique process.

Acknowledgement. This work has benefited from an ongoing collaboration with Jim Pitman. I also thank Vlada Limic for carefully reading a draft.

References

[1] D.J. Aldous. A random tree model associated with random graphs. *Random Structures Algorithms*, 1:383–402, 1990.

[2] D.J. Aldous. The random walk construction of uniform spanning trees and uniform labelled trees. *SIAM J. Discrete Math.*, 3:450–465, 1990.

[3] D.J. Aldous. Deterministic and stochastic models for coalescence: a review of the mean-field theory for probabilists. To appear in *Bernoulli*, 1997.

[4] D.J. Aldous and J. Pitman. Tree-valued Markov chains derived from Galton-Watson processes. Unpublished, 1997.

[5] N. Alon and J. H. Spencer. *The Probabilistic Method.* Wiley, 1992.

[6] S. Asmussen. *Applied Probability and Queues.* Wiley, 1987.

[7] B. Bollobás. *Random Graphs.* Academic Press, London, 1985.

[8] A. Broder. Generating random spanning trees. In *Proc. 30'th IEEE Symp. Found. Comp. Sci.*, pages 442–447, 1989.

[9] W. J. Ewens. *Mathematical Population Genetics*. Springer-Verlag, Berlin, 1979.
[10] G. R. Grimmett. Random labelled trees and their branching networks. *J. Austral. Math. Soc. (Ser. A)*, 30:229–237, 1980.
[11] O. Haggstrom. Dynamical percolation: early results and open problems. In this volume.
[12] H. Kesten. Subdiffusive behavior of random walk on a random cluster. *Ann. Inst. H. Poincaré Probab. Statist.*, 22:425–487, 1987.
[13] D.E. Knuth. *The Art of Computer Programming*, volume 3. Addison-Wesley, 1973.
[14] R. Lyons, R. Pemantle, and Y. Peres. Conceptual proof of $L \log L$ criteria for mean behavior of branching processes. *Ann. Probab.*, 23:1125–1138, 1995.
[15] R. Lyons and Y. Peres. Probability on Trees and Networks. Book in preparation, 1997.
[16] H. M. Mahmoud. *Evolution of Random Search Trees*. Wiley, 1992.
[17] J.A. Mann, B. Pittel, and W. A. Woycznski. Random tree-type partitions as a model for acyclic polymerization: Holtsmark (3/2 stable) distribution of the supercritical gel. *Ann. Probab.*, 18:319–341, 1990.
[18] R. Pemantle. Uniform random spanning trees. In J. Laurie Snell, editor, *Topics in Contemporary Probability*, pages 1–54, Boca Raton, FL, 1995. CRC Press.
[19] L.C.G. Rogers and D. Williams. *Diffusions, Markov Processes and Martingales*, volume 1,Foundations. Wiley, 1994. 2nd. edition.
[20] P. Whittle. *Systems in Stochastic Equilibrium*. Wiley, 1986.

Department of Statistics, University of California, 367 Evans Hall # 3860, Berkeley, CA 94720-3860

DIMACS Series in Discrete Mathematics
and Theoretical Computer Science
Volume **41**, 1998

On the central role of scale invariant Poisson processes on $(0, \infty)$

Richard Arratia

ABSTRACT. The scale invariant Poisson processes on $(0, \infty)$ play a central but mildly disguised role in number theory, combinatorics, and genetics. They give the continuous limits which underly and unify diverse discrete structures, including the prime factorization of a uniformly chosen integer, the factorization of polynomials over finite fields, the decomposition into cycles of random permutations, the decomposition into components of random mappings, and the Ewens sampling formula. They deserve attention as one of the fundamental and central objects of probability theory.

0. Introduction

The scale invariant Poisson processes on $(0, \infty)$ have almost always been overlooked, even though these processes play a fundamental role in combinatorics, number theory, and genetics. Quantities and relations which are most easily explained in terms of these Poisson processes have been studied directly without mentioning the Poisson connection. Examples in number theory (18.1) include Dickman's function ρ from [**17**] in 1930, Buchstab's function ω from [**15**] in 1937, and the "convolution powers of the Dickman function" from [**14, 29**]. The examples from combinatorics and genetics involve the Poisson-Dirichlet process and the GEM process from the 1970's, which are usually discussed in terms of products of independent, Beta-distributed random variables, or constructed by conditioning a Poisson process which is not scale invariant. We want to reveal the presence of the scale invariant Poisson processes, as well as their intrinsic beauty and simplicity, for an audience including probabilists, number theorists, geneticists, and others, without assuming background such as knowledge of point processes.

The sections correspond, more or less, to the transparancies from the lecture, except for the last two sections, which match discussions from the problem sessions, and have been substantially expanded.

1. Scale invariant Poisson processes on $(0, \infty)$

The only reference to "scale invariant Poisson processes" per se that we have found is Daley and Vere Jones [**16**], page 325, where the scale invariant Poisson processes on $\mathbb{R}$ are fully classified. We requested help from the audience in compiling a

1991 *Mathematics Subject Classification.* primary 60-02, secondary 60F17.

Work supported in part by NSF grant DMS 96-26412.

bibliography on scale invariant Poisson processes; C. Newman supplied a reference to [**1**]. Pitman and Yor [**45**] is a general study of spacings in scale invariant random closed subsets of $(0, \infty)$, including examples like the zeroes of Brownian motion, and the scale invariant Poisson processes.

The first subtle thing to realize about scale invariant Poisson process*es* is that the plural is essential! As with translation invariant Poisson processes, there is a nonnegative parameter θ which appears as a scalar factor for the intensity measure, but unlike the translation invariant case, varying θ leads to major qualitative changes. In particular, for certain natural aspects of qualitative behavior, there are phase transitions at $\theta = 1$ (see (16.2),) and at $\theta = 1/\log 2$ (see section 22.)

Our notation for scaling is $cX := \{cx : \ x \in X\}$, defined for any $X \subset R^d$ and $c > 0$. For a random set $\mathcal{X}$, scale invariance is the property that for all $c > 0$, $c\mathcal{X} =_d \mathcal{X}$, i.e. all scalings of the set have the same distribution.

In case $\mathcal{X} \subset (0, \infty)$, one can take logarithms: let $\mathcal{L} = \log(\mathcal{X}) := \{\log(x) : x \in \mathcal{X}\}$. In this situation, clearly, scale invariance for $\mathcal{X}$ is equivalent to translation invariance for $\mathcal{L}$.

2. Intensity

For us, a point process is a random set or multiset $\mathcal{X}$ such that for sets I in an appropriate class, $\mathcal{X} \cap I$ is a finite set or multiset, with finite mean size; the points of $\mathcal{X}$ are *not* labelled. The intensity measure μ is then defined by $\mu(I) = \mathbb{E}|\mathcal{X} \cap I|$, the expected number of points falling in I. For instance, if $\mathcal{X}$ is a translation invariant point process on $\mathbb{R}$, then its intensity is a translation invariant, nonnegative measure, and hence has the form θ times Lebesgue measure — we will say that the intensity is $\theta \ dx$ on $\mathbb{R}$. The interesting intensity measures for our purposes are, each with parameter $\theta > 0$:

$$\theta \ dx \text{ on } (-\infty, \infty) \qquad \text{translation invariant,}$$

$$(\theta/x) \ dx \text{ on } (0, \infty) \qquad \text{scale invariant,}$$

$$(\theta e^{-x}/x) \ dx \text{ on } (0, \infty) \qquad \text{“Gamma” or “Moran” subordinator.}$$

One can easily verify that $(\theta/u) \ du$ gives the intensity of scale invariant point processes $\mathcal{X}$ on $(0, \infty)$, starting from the translation invariance of $\mathcal{L} := \log(\mathcal{X})$, as follows. For $a = e^x < b = e^y$, $|\mathcal{X} \cap (a, b)| = |\mathcal{L} \cap (x, y)|$, with expectation $\int_x^y \theta \ dt = \theta(y - x) = \theta \log(b/a) = \int_a^b (\theta/u) \ du$. Another way to compute this is to start with the candidate for a scale invariant intensity, $\theta/x \ dx$, and the map $x \mapsto cx$, and to apply the change of variables formula correctly — an exercise which the author usually gets wrong at first! Note that scaling $(x \mapsto cx)$ the process with intensity $\theta e^{-x}/x \ dx$ yields the same θ but a different exponential decay, i.e. the new intensity is $\theta e^{-x/c}/x \ dx$, for $0 < c < \infty$.

3. Poisson processes, inversion invariance

The characterizing property of a Poisson process is that for disjoint $I_1, I_2, \ldots, I_k$, the number of points in I_j for $j = 1$ to k are independent, Poisson distributed random variables. This property is preserved by any mapping, not necessarily one-to-one; the relevant examples here are $x \mapsto x/\theta$ for $\theta > 0$, $x \mapsto -x$, $x \mapsto e^x$, $x \mapsto e^{-x}$, $x \mapsto \log x$, and $x \mapsto -\log x$.

Thus in checking that the following four statements about a point process $\mathcal{X} = \{X_i\}$ are equivalent, one ingredient is calculating intensities, and the other ingredient, mapping the *Poisson* properties, is automatic:

$$\{X_i\} \text{ is scale invariant Poisson } (\theta/x)\ dx \text{ on } (0,\infty),$$

$$\{\log X_i\} \text{ is translation invariant Poisson } (\theta\ dx) \text{ on } (-\infty,\infty),$$

$$\{-\log X_i\} \text{ is translation invariant Poisson } (\theta\ dx) \text{ on } (-\infty,\infty),$$

$$\{1/X_i\} \text{ is scale invariant Poisson } (\theta/x)\ dx \text{ on } (0,\infty).$$

The above shows how the reflection invariance of a translation invariant Poisson process on $\mathbb{R}$ is directly equivalent to the inversion invariance of a scale invariant Poisson process on $(0, \infty)$.

4. Favorite labellings

For the translation invariant Poisson process, label the points L_i for $i \in \mathbb{Z}$ with $L_i < L_{i+1}$ and $L_0 \le 0 < L_1$. This seems like the only decent choice.

$$(4.1) \qquad -\infty < \cdots < L_{-1} < L_0 \ \le 0 < \ L_1 < L_2 < \cdots < \infty\,.$$

For the scale invariant Poisson processes, there are two natural choices for labelling, and it does not seem reasonable to try to make one notation fit all situations.

If we focus from one down to zero, use $X_i := e^{-L_i}$, so that $X_{i+1} < X_i$, X_1 is the first point to the left of one, and $X_n \downarrow 0$ as $n \to \infty$.

$$(4.2) \qquad 0 < \cdots < X_3 < X_2 < X_1 < 1 \ \ \le X_0 < X_{-1} < X_{-2} < \cdots < \infty\,.$$

If instead we focus from one up to infinity, use $X_i := e^{L_i}$ so that $X_i < X_{i+1}$ and X_1 is the first point to the right of one:

$$(4.3) \qquad 0 < \cdots < X_{-1} < X_0 \ \ \le 1 < X_1 < X_2 < \cdots < \infty\,.$$

5. Spacings — for the translation invariant process

With the labelling of (4.1), the spacings of the points of the translation invariant process are $L_{i+1} - L_i$ for $i \in \mathbb{Z}$. Because of the exceptional behavior of $L_1 - L_0$, these spacings are *not* distributed the same as $\ldots, L_{-2} - L_{-1}, L_{-1} - L_0, 0 - L_0, L_1 - 0, L_2 - L_1, \ldots$, which are iid, each distributed like an exponentially distributed random variable W with mean $1/\theta$ and $\mathbb{P}(\theta W > t) = e^{-t}$ for $t > 0$. Note the familiar "waiting time for a bus" paradox: the interval that covers the origin has length $L_1 - L_0$ which can be expressed as the sum of two independent exponentials, $L_1 - 0$ and $0 - L_0$, and which is equal in distribution to the size biased spacing W^*, discussed in a different context in (11.1).

Write $W_1 := L_1, W_2 := L_2 - L_1, W_3 := L_3 - L_2$ etc. so that for $k = 1, 2, \ldots$, $L_k = W_1 + \cdots + W_k$. Recall that "the (negative) exponential of an exponential is uniform", at least for $\theta = 1$. To check this, note that $t > 0 \Leftrightarrow e^{-t} \in (0,1)$ and $\mathbb{P}(e^{-\theta W_i} < e^{-t}) = \mathbb{P}(\theta W_i > t) = e^{-t}$. Let

$$(5.1) \qquad U_i := e^{-W_i} = (e^{-\theta W_i})^{1/\theta} =_d (\text{UNIFORM})^{1/\theta}$$

where UNIFORM denotes a random variable uniformly distributed on [0,1].

In lecture, we offered the following closed book QUIZ: the distribution of U_i is either $\text{Beta}(1,\theta)$ or $\text{Beta}(\theta,1)$, but *which one?*

Use the notation (4.2) which focusses on the scale invariant process near zero, i.e. $X_i = \exp(-L_i)$. Then for $k = 1, 2, \ldots$ the expression

$$L_k = W_1 + W_2 + \cdots + W_k \tag{5.2}$$

for a sum of independent exponential mean $1/\theta$ random variables is immediately equivalent to

$$X_k = U_1 U_2 \cdots U_k$$

with a product of independent $(\text{UNIFORM})^{1/\theta}$ random variables.

6. Spacings — for the scale invariant process

The process of all spacings for the process on $(0, \infty)$ with the notation (4.2) is defined by

$$Y_k := X_{k-1} - X_k \in (0, \infty), \quad k \in \mathbb{Z}.$$

However, when considering the scale invariant process restricted to (0,1) or (0,1], the natural notation gives a different definition for the first spacing:

$$Y_k := X_{k-1} - X_k, \quad k = 2, 3, \ldots, \qquad \text{but } Y_1 := 1 - X_1. \tag{6.1}$$

In terms of the independent $U_1, U_2, \ldots =_d (\text{UNIFORM})^{1/\theta}$, (6.1) is $Y_1 = 1 - U_1$, $Y_2 = U_1 - U_1 U_2 = U_1(1 - U_2)$, and in general for $n \geq 1$,

$$Y_n = U_1 U_2 \cdots U_{n-1}(1 - U_n). \tag{6.2}$$

7. Residual allocation, GEM

Most geneticists have used the notation

$$Y_1 = U_1, Y_2 = (1 - U_1)U_2, Y_3 = (1 - U_1)(1 - U_2)U_3, \ldots$$

which is the opposite of (6.2). We wish to argue that the notation (6.2), which respects the sum (5.2), is preferable.

The product form above, with either notation, is referred to as a "residual allocation model", from Halmos 1944 [**27**]. The distribution of $(Y_1, Y_2, \ldots)$ in (6.1) is called the GEM with parameter θ, after Griffiths, Engen, and McCloskey; —[**39**] is the unpublished 1965 thesis by McCloskey; these historical notes are from chapter 41 by Ewens and Tavaré in [**33**].

The Beta function is defined, for $a, b > 0$, by $B(a, b) := \int_0^1 (1 - x)^{a-1} x^{b-1}\, dx$ and the corresponding distribution, Beta(a, b), has density $(1 - x)^{a-1} x^{b-1} / B(a, b)$ on (0,1). Since B is Beta(a, b) if and only if $1 - B$ is Beta(b, a), we see how Beta$(1,\theta)$ and Beta$(\theta,1)$ can easily be confused.

8. The Poisson-Dirichlet process

We write $(V_1, V_2, \ldots)$ for the Poisson-Dirichlet process with parameter θ. Survey references for the Poisson-Dirichlet process include [**46, 7**], and Chapter 41 in [**33**]. We do not assume the reader knows this process; we will give several characterizations of it, including one that relates it to the scale invariant Poisson process with parameter θ, following a review of its history. When the Poisson-Dirichlet process was first studied, none of its connections to the scale invariant Poisson process was explicitly noted.

One might view the Poisson-Dirichlet process with $\theta = 1$ as implicit in the 1930 paper of Dickman [**17**] describing the limit distribution of the largest prime factor of a randomly chosen integer. However if one insists on an explicit description of the process, it seems that the first appearance of the Poisson-Dirichlet is in Billingsley 1972 [**13**], which proves that

$$\left(\frac{\log P_1}{\log n}, \frac{\log P_2}{\log n}, \dots\right) \Rightarrow (V_1, V_2, \dots), \tag{8.1}$$

where the limit is the Poisson-Dirichlet process with $\theta = 1$. Here, P_i is the ith largest prime factor of an integer N chosen uniformly from 1 to n, using $P_i = 1$ when i is greater than the number $\Omega(N)$ of prime factors including multiplicities. (For example, if the random integer is 12, then $P_1 = 3$, $P_2 = P_3 = 2, 1 = P_4 = P_5 = \cdots$.) Billingsley specified the limit process in terms of the joint density of $(1/V_1, 1/V_2, \dots, 1/V_k)$. The second published proof of Billingsley's result, [**18**], is based on a size-biased permutation, and is especially robust, leading to a proof [**4**] of the analogous Poisson-Dirichlet convergence for general $\theta \neq 1$, in which the random integer is conditioned on the large deviation that the number of distinct prime factors is $\theta \log \log n$.

Ferguson [**25**] in 1973 described a class of processes that includes what we now know as the Poisson-Dirichlet; a recent survey of this is Pitman[**43**].

Kingman [**35**] in 1975 and Ignatov [**32**] in 1982 gave the first direct connection between the Poisson-Dirichlet process and the scale invariant Poisson process: they can be coupled, with $V_i =$ the ith largest of the spacings $Y_1, Y_2, \dots$ as in (6.1), i.e.

$$(V_1, V_2, \dots) =_d \text{RANK}(1 - X_1, X_1 - X_2, X_2 - X_3, \dots). \tag{8.2}$$

See [**45**] for generalizations.

Vershik and Schmidt [**53, 54**] in 1977 showed that

$$(\theta = 1) \qquad \left(\frac{L_1}{n}, \frac{L_2}{n}, \dots\right) \Rightarrow (V_1, V_2, \dots)$$

where L_i is the length of the ith longest cycle of a random permutation of n objects, choosing with all $n!$ possibilities equilikely.

Aldous in 1983 [**2**] gave the analogous Poisson-Dirichlet limit for the component sizes in a random mapping on n points, with all n^n maps equilikely; here $\theta = \frac{1}{2}$.

Hansen [**28**] gave a general treatment of decomposable combinatorial structures having a Poisson-Dirichlet limit; see also [**8**].

The following comes from the 1993 book [**36**], "Poisson processes," by Kingman. In 1968 Moran considered the Poisson process with intensity $\theta \exp(-x)/x \, dx$ on $(0, \infty)$. The points of this process can be labelled σ_i for $i = 1, 2, \dots$ and $0 < \cdots < \sigma_2 < \sigma_1$. The sum $\sigma := \sigma_1 + \sigma_2 + \cdots$ has a Gamma distribution with parameter θ, and is independent of the rescaled vector $(\sigma_1/\sigma, \sigma_2/\sigma, \dots)$, which has the Poisson-Dirichlet distribution. There is a somewhat similar model in physics, discussed in [**47**], 227-228, with a Poisson process on $(0, \infty)$ having intensity $cx^{-c-1} \, dx$ for a constant $c \in (0, 1)$, instead of $\theta \exp(-x)/x \, dx$. It is still the case that a.s. the points can be labelled σ_i for $i = 1, 2, \dots$ with $0 < \cdots < \sigma_2 < \sigma_1$, and $\sigma := \sigma_1 + \sigma_2 + \cdots < \infty$, and one may form the rescaled vector $(\sigma_1/\sigma, \sigma_2/\sigma, \dots)$. However, in this model, $\mathbb{E}\sigma = \infty$, in contrast to $\mathbb{E}\sigma = \theta$ for the Moran model. A unified treatment which includes the two models is [**44**].

9. Joint density for the Poisson-Dirichlet

The joint density f_k of the first k coordinates $(V_1, V_2, \dots, V_k)$ of the Poisson-Dirichlet distribution has support $\{(x_1, \dots, x_k) : \ 1 > x_1 > \cdots > x_k > 0$ and

$x_1 + \cdots + x_k < 1\}$. At such points, for the case $\theta = 1$,

$$f_k(x_1, \cdots, x_k) = \rho\left(\frac{1 - x_1 - \cdots - x_k}{x_k}\right)\frac{1}{x_1 x_2 \cdots x_k} \tag{9.1}$$

where ρ is Dickman's function [**17, 51**], characterized by $\rho = 0$ on $(-\infty, 0)$, $\rho = 1$ on [0,1], ρ continuous on $(0, \infty)$, and $\rho'(u) = -\rho(u-1)/u$ for $u > 1$.

The area under the graph of ρ is e^γ, where γ is Euler's constant, and $\rho \geq 0$ everywhere, so $g(u) := e^{-\gamma}\rho(u)$ is a probability density, which might naturally be called the Dickman distribution.

10. What is the "Dickman distribution"?

Dickman in 1930 showed that $\log P_1 / \log n \Rightarrow V_1$, where $\mathbb{P}(V_1 < 1/u) = \rho(u)$, i.e. the Dickman function ρ gives the tail probabilities for $1/V_1$. So if you hear someone refer to a random variable having "the Dickman distribution," you may be confident that the speaker refers to one of the following three, but which one?

- the random variable T with density $g(u) := e^{-\gamma}\rho(u)$ on $(0, \infty)$
- V_1, with density $\rho((1-x)/x)/x$ on $(0, 1)$
- $1/V_1$, with density $\rho(u-1)/u$ on $(1, \infty)$

The Dickman function decays superexponentially fast. Hildebrand 1990 [**30**] gives the following asymptotic expansion of the Dickman function. Write $L := \log u$ and $M := \log\log u$. As $u \to \infty$,

$$\frac{-\log \rho(u)}{u} = L + M - 1 + \frac{M}{L} - \frac{1}{L} - \frac{M^2}{2L^2} + \frac{M}{L^2} - \frac{2}{L^2} + O\left(\frac{M^2}{L^3}\right),$$

so that $\rho(u) = \exp(-u \log u - u \log\log u + u - \cdots)$.

11. A key random variable: T, the sum of the points in (0,1)

Let T be the sum of the locations of all points of the scale invariant Poisson process, with intensity $\theta/x\, dx$, restricted to $(0, 1)$. For the case $\theta = 1$, this is the random variable in the previous section, with density $e^{-\gamma}\rho(u)$. For any $\theta > 0$,

$$\begin{aligned} T &:= X_1 + X_2 + X_3 + \cdots \\ &= U_1 + U_1U_2 + U_1U_2U_3 + \cdots \\ &= U_1\ (1 + U_2 + U_2U_3 + \cdots) \\ &=_d U\ (1 + T') \end{aligned}$$

with T', U independent, $T' =_d T$, and $U =_d$ (UNIFORM $)^{1/\theta}$. Thus it is elementary to get an integral equation involving the density g of T, although this equation is not especially tractable.

Another approach is via size biasing; the following is taken from [**7**]. In general, if X is a non-negative random variable with mean $\mu \in (0, \infty)$, then the size biased random variable X^* is characterized by

$$\mathbb{E}h(X^*) = \mathbb{E}(Xh(X))/\mu; \tag{11.1}$$

if further X has density g then X^* has density $xg(x)/\mu$. If Z is a Poisson distributed random variable then $Z^* =_d 1 + Z$ and for any $x > 0$, $(xZ)^* =_d x + (xZ)$. If Z is a sum of independent nonnegative random variables, with $\mathbb{E}Z < \infty$, then Z^* is likewise a sum of independent random variables, using the same summands except that one summand is size biased, and the choice of which summand to bias is made with probability proportional to its contribution to μ. Thus if T is the sum of locations of points in a Poisson process on $(0,\infty)$ with intensity $f(x)\ dx$, such that $\mu := \mathbb{E}T = \int xf(x)\ dx \ < \infty$, then T^* is formed by choosing a location x with probability $xf(x)\ dx/\mu$ where the summand is size biased by deterministically adding in x. Hence the density of T^* at t is

$$\frac{tg(t)}{\mu} = \int_0^\infty \frac{xf(x)\ dx}{\mu}\ g(t-x)$$

For the case of the scale invariant Poisson process restricted to (0,1) we have $\mathbb{E}T = \int_0^1 x(\theta/x)\ dx \ = \theta$, and

$$\frac{tg(t)}{\theta} = \int_0^1 \frac{x(\theta/x)\ dx}{\theta}\ g(t-x) \ = \int_{t-1}^t g(u)du, \qquad t > 0.$$

Hence the differential-difference equation satisfied by g is

$$tg'(t) + (1-\theta)g(t) + \theta g(t-1) = 0, \qquad t > 0.$$

Another approach is that since T is the sum of locations of the Poisson process with intensity function $f(x) = \mathbb{1}(0 < x < 1)\theta/x$, it has Laplace transform

$$\mathbb{E}\exp(-sT) = \exp\left(-\int_0^\infty (1-e^{-sx})f(x)\ dx\right) = \exp\left(-\theta\int_0^1 \frac{1-e^{-sx}}{x}\ dx\right)$$

This leads to properties of the density g; see Vervaat 1972 [**55**], Watterson [**56**], and Hensley [**29**].

12. The scale invariant process of sums, and class L

For any fixed $\theta > 0$, our favorite random variable T, which is the sum of locations of points in (0,1). It satisfies $T =_d T_1 =_d (1/t)T_t$, where for $t > 0$, T_t is defined as the sum of locations of all points in $(0,t]$, for the scale invariant Poisson process with intensity $\theta/x\ dx$. Of course we take $T_0 \equiv 0$. As a process $(T_t)_{t\geq 0}$

- increases by jumps
- at time t, can only stay constant or jump up by t
- has a jump occurring in $(t, t+dt)$ with probability $(\theta/t)\ dt\ (1+o(1))$
- has independent increments
- is self-similar with index 1: for $t > 0$, $(T_{ts})_{s\geq 0} =_d t^1(T_s)_{s\geq 0}$

In particular, writing g_t for the density of T_t and g for the density of T, from $T_t =_d tT$ we have, for any $t > 0$,

$$g_t(x) = \frac{1}{t}\ g(\frac{x}{t}).$$

From Feller II [**24**], the Lévy class L of infinitely divisible distributions consists of those which are limit distributions of $\{S_n^*\}$ where $S_n := X_1 + \cdots + X_n$, with independent (not necessarily identically distributed) $X_1, X_2, \ldots$, and $S_n^* = (S_n - b_n)/\ a_n$ for constants $a_1, a_2, \ldots$ and $b_1, b_2, \ldots$ such that $a_n \to \infty$, $a_{n+1}/a_n \to 1$. To see that for each $\theta > 0$ the random variable T is in class L, take $X_n := T_n - T_{n-1}$, so that $S_n = T_n$, and take $b_n = 0, a_n = n$ so that $S_n^* := T_n/n =_d T$ for all n. We thank Larry Shepp for an enjoyable conversation on this topic.

We cannot understand a technical/aesthetic issue involving the process (T_t): why does the natural random Stieltjes measure $dT_{(\cdot)}$ always lose out to the competing counting measure? That is, to encode the scale invariant Poisson process as a random measure, the usual choice is the counting measure $\sum_{i\in\mathbb{Z}} \delta_{X_i}(\cdot)$. This has to be taken as a random measure on $(0,\infty)$ with the point at zero not in the underlying space; there is infinite mass in every neighborhood of zero. Another sensible choice would be $dT_{(\cdot)} \equiv \sum_{i\in\mathbb{Z}} X_i\, \delta_{X_i}(\cdot)$, which gives a sigma-finite random measure on $[0,\infty)$, with finite mass near zero.

13. Conditioning on $T = s$ in general

Consider *any* Poisson process on (0,1], having intensity $f(x)\,dx$, such that T_t, defined to be the sum of the locations of all points in $(0,t)$, has a density g_t. Write $T := T_1$ and $g := g_1$. Assume that g is strictly positive on $(0,\infty)$. Label the points in (0,1) so that $1 > X_1 > X_2 > \cdots > 0$.

The joint density of $X_1,\ldots,X_k$ is

$$f(x_1)f(x_2)\cdots f(x_k)\exp\left(-\int_{x_k}^1 f(u)\,du\right),$$

supported by points in $(0,1)^k$ where $x_1 > x_2\cdots > x_k$. The first k factors correspond to requiring points at $x_1,\ldots,x_k$, and the last factor corresponds to demanding no other points in $(x_k,1)$. Thus for any $s > 0$ the joint density of $X_1,\ldots,X_k$ conditional on $T = s$ is

$$\frac{g_{x_k}(s-x_1-\cdots-x_k)}{g(s)}\, f(x_1)f(x_2)\cdots f(x_k)\exp\left(-\int_{x_k}^1 f(u)\,du\right), \tag{13.1}$$

supported at $1 > x_1 > \cdots > x_k > 0$ with $x_1+\cdots+x_k < s$. For the special case of the scale invariant Poisson process with $f(x) = \theta/x$, this is

$$\frac{1}{x_k}\,\frac{g\left((s-x_1-\cdots-x_k)/x_k\right)}{g(s)}\,\frac{\theta}{x_1}\,\frac{\theta}{x_2}\cdots\frac{\theta}{x_k}e^{-\theta\log(1/x_k)}$$

14. Conditioning on $T = 1$ for the scale invariant process

For the special case $s = 1$ the conditional joint density of $X_1,\ldots,X_k$ given $T = 1$ simplifies to

$$\begin{aligned}&\frac{g\left((1-x_1-\cdots-x_k)/x_k\right)}{g(1)}\,\frac{\theta^k}{x_1x_2\cdots x_k}x_k^{\theta-1}\\ &= g\left(\frac{1-x_1-\cdots-x_k}{x_k}\right)\frac{e^{\gamma\theta}\,\theta^k\,\Gamma(\theta)\,x_k^{\theta-1}}{x_1x_2\cdots x_k},\end{aligned} \tag{14.1}$$

which is the joint density of the Poisson-Dirichlet; the final equality is just giving an explicit formula for the normalizing constant $g(1)$, which for $\theta \neq 1$ comes from [**56**]. In the special case $\theta = 1$, (14.1) simplifies to (9.1).

In summary, if one starts from the joint density of the Poisson-Dirichlet, with a differential-difference equation or Laplace transform to characterize the difficult factor, then it is very easy to prove that the Poisson-Dirichlet process is the scale-invariant Poisson process, restricted to (0,1), and conditioned on the event $T = 1$: for any $\theta > 0$

$$(V_1,V_2,\ldots) =_d (\ (X_1,X_2,\ldots) \mid T = 1\ ), \quad \text{i.e. } \mathrm{PD}(\theta) =_d (\mathrm{PP}(\theta) \mid \mathrm{T} = 1\ ). \tag{14.2}$$

The above result first was written in the 1996 version of [**7**]. One motivation for considering it is the analogy to the result that

$$(C_1(n),\ldots,C_n(n)) =_d (\ (Z_1,\ldots,Z_n)\ |T_n = n\ ), \tag{14.3}$$

relating the joint distribution of cycle counts for a uniformly distributed random permutation, to the joint distribution of $Z_1, Z_2, \ldots$, a sequence of independent Poisson random variables with $\mathbb{E}Z_i = 1/i$, conditioned on $T_n = n$, where $T_n := Z_1 + 2Z_2 + \cdots + nZ_n$.

Just as (14.2) is a continuum analog of (14.3), the Moran process representation implies

$$(V_1, V_2, \ldots) =_d ((\sigma_1, \sigma_2, \ldots) \mid \sigma = 1), \tag{14.4}$$

which is a continuum analog of the relation exploited by Shepp and Lloyd [**48**] in 1966 to analyze random permutations,

$$(C_1(n), \ldots, C_n(n), 0, 0, \ldots) =_d ((Z_1, Z_2, \ldots) \mid T_\infty = n). \tag{14.5}$$

Here $T_\infty = Z_1 + 2Z_2 + \cdots$, and the Z_i are independent Poisson with $\mathbb{E}Z_i = z^i/i$; this is valid for any $z \in (0,1)$ but especially useful with $z = 1 - 1/n$. Furthermore, (14.2) is to (14.4) as (14.3) is to (14.5). Loosely speaking, (14.4) and (14.5) both avoid the divergence of $\int^\infty x^{-1}$ by throwing in an exponentially decaying factor which does not affect the conditional distribution, while (14.2) and (14.3) avoid the divergence of $\int^\infty x^{-1}$ by not going out to infinity.

Another elementary proof of (14.2) can be given by comparison to the Moran representation; this is done in [**9**]. The general conditioning formula (13.1) is also used in [**40**].

15. More discrete analogs

The role of the scale invariant Poisson process in number theory may be found explicitly in DeKoninck and Galambos [**37**], with a theorem showing that the process of logarithms of the intermediate prime divisors of a random integer chosen uniformly from 1 to n converges in distribution to the scale invariant Poisson process, with $\theta = 1$. An attempt to metrize this result [**3**] leads to the following, which shows how the scale invariant Poisson process with $\theta = 1$ is very close to the discrete limit process that arises in number theory, with the number Z_p of occurrences of each prime p being independent, with $\mathbb{P}(Z_p = k) = (1 - 1/p)p^{-k}$. Theorem: it is possible to couple random values Q_i for $i \in \mathbb{Z}$, and the scale invariant Poisson process, so that

$$\mathbb{E} \sum_{i \in \mathbb{Z}} |X_i - \log Q_i| < \infty$$

with each Q_i either one or prime, and for each p, $Z_p = \sum_i \mathbb{1}(Q_i = p)$.

It is not obvious how the usual model of number theory, picking an integer uniformly from 1 to n, is related to the independent, geometrically distributed Z_p by something like conditioning; this is explained in [**6**]. The overall theme there is that discrete "logarithmic" combinatorial structures, such as random permutations, random mappings, and random polynomials over finite fields, together with prime factorizations of uniformly chosen random integers, have a dependent component size counting process, that is derived from an independent process by conditioning or something analogous. The discrete dependent processes, rescaled, are close to the Poisson-Dirichlet processes. The discrete independent processes, rescaled, are close to the scale invariant Poisson processes. For both the discrete and continuous situations, the dependent processes are obtained from the independent processes by conditioning, or something like conditioning — this is the point of studying (14.2).

16. Dependent versus independent: total variation distance

Recall, the dependent process is the Poisson-Dirichlet $(V_1, V_2, \dots)$ with $V_1 + V_2 + \cdots = 1$, and the related independent process is the scale invariant Poisson, restricted to (0,1), with $X_1 + X_2 + \cdots = T$, a random variable having a strictly positive density on $(0, \infty)$. For $\beta \in [0,1]$ let

$$H_\theta(\beta) := d_{TV}(\ \{V_1, V_2, \dots\} \cap [0,\beta],\ \{X_1, X_2, \dots\} \cap [0,\beta]\ ).$$

Note that $H_\theta(0) = d_{TV}(\emptyset, \emptyset) = 0$, and

$$H_\theta(1) = d_{TV}(\ (V_1, V_2, \dots), (X_1, X_2, \dots)\ ) = 1$$

because $1 = \mathbb{P}(V_1 + V_2 + \cdots = 1)$ and $0 = \mathbb{P}(X_1 + X_2 + \cdots = 1)$. For the case $\theta = 1$

$$H_1(1/1.9) = .4968\dots\ , H_1(1/2) = .4454\dots\ , H_1(1/3) = .1114\dots\ ,$$
$$H_1(1/3.5) = .0471\dots\ , H_1(1/4) = .0184\dots\ .$$

Informally, $H_1(.25) = .0184\dots$ means that if you are shown one sample of a process which is either the Poisson-Dirichlet ($\theta = 1$) or else the scale invariant Poisson ($\theta = 1$), but you only get to observe the set of components of size at most .25, then your edge over the house is at most 1.84 percent: if the examiner picks his distribution using a fair coin, then you have at most a 50.92 percent chance of answering correctly.

There is a phase transition in the qualitative behavior of $H_\theta(0+)$, with linear decay for $\theta \neq 1$, and superexponential decay for $\theta = 1$. In detail, from [**11**], p. 1368, combined with [**9, 50, 10**], as $\beta \to 0+$,

$$\frac{1}{\beta} H_\theta(\beta) \quad \to \quad |1-\theta|\, \frac{e^{-\gamma\theta}}{\Gamma(\theta)}\, \frac{\theta^\theta}{1+\theta} > 0 \text{ for } \theta \neq 1, \tag{16.1}$$

$$-\beta \log H_1(\theta) \quad \sim \quad \log(1/\beta) \to \infty. \tag{16.2}$$

17. Invariance principle for total variation distance

The function $H_\theta(\beta)$ gives the total variation distance between the dependent and independent processes, i.e. the Poisson-Dirichlet process with parameter θ, and the scale invariant Poisson process with parameter θ, when observing components of size at most β. These functions $H_\theta(\cdot)$, which can be most easily defined in terms of the scale invariant Poisson, showed up first in combinatorics. Consider random permutations of n objects, observing cycles of length at most βn: write $C_i(n)$ for the number of cycles of length i, and Z_i for a random variable which is Poisson, mean $1/i$, with $Z_1, Z_2, \dots$ independent. Then [**12, 50**] for any fixed $\beta \in [0,1]$, as $n \to \infty$,

$$d_{TV}(\ (C_1(n), C_2(n), \dots, C_{\lfloor \beta n \rfloor}(n)),\ (Z_1, Z_2, \dots, Z_{\lfloor \beta n \rfloor})\ ) \ \to H_1(\beta).$$

Similarly [**50**] with $\theta = \frac{1}{2}$, for the component counts for a random mapping, comparing to the independent limit process, and observing components of size at most βn,

$$d_{TV} \to H_{\frac{1}{2}}(\beta),$$

and also similarly with $\theta = 1$ for the factorization of a random polynomial of degree n over a finite field, observing factors of degree at most βn.

For the factorization of an integer chosen uniformly from 1 to n the analogous result holds [**10, 52**], again with $\theta = 1$. Here, we observe prime factors, with multiplicity, jointly for all primes $p \le n^\beta$. The independent comparison process has Z_p which are geometrically distributed with $\mathbb{P}(Z_p \ge k) = p^{-k}$.

18. Buchstab and the explicit formula for d_{TV}

The possiblity of numerically evaluating the limiting total variation distance $H_\theta(\beta)$ arises thanks to relation (14.2), which makes it possible to equate $H_\theta(\beta)$, defined as the total variation distance between processes, with the total variation distance between two random variables. Recall that the process $(T_s)_{s\ge 0}$ has independent increments, so that the $T \equiv T_1$ in conditioning on $T=1$ can be expressed as the sum $T = T_\beta + (T - T_\beta)$ with two independent summands. This leads to

$$H_\theta(\beta) := d_{TV} \text{ for processes}$$

$$= d_{TV} \text{ for random variables } \equiv d_{TV}(T_\beta, (T_\beta \mid T = 1\,))$$

Manipulation of the densities of the independent summands leads to the explicit expression for $H_\theta(\beta)$, which for $\theta = 1$ can be expressed as:

$$\begin{aligned} 2H_1(\beta) &= e^\gamma \mathbb{E}|\omega(u-T) - e^{-\gamma}| \;+ \rho(u) \\ &= \int_{t>0} |\omega(u-t) - e^{-\gamma}|\rho(t)\, dt \;\; + \rho(u). \end{aligned}$$

Here, ρ is Dickman's function, and and ω is Buchstab's function, with ω continuous on $(1,\infty)$, with $(u\omega(u))' = \omega(u-1)$ for $u > 2$ and $u\omega(u) = 1$ for $1 \le u \le 2$; see e.g. [**51**]. Both functions were described some sixty years ago in number theory, and each can also be described in terms of the scale invariant Poisson process for $\theta = 1$: ρ is e^γ times the density function of T, and for $u > 1$ with $\beta := 1/u$, $\omega(u)$ equals the density of $T - T_\beta$, evaluated at $1-$, i.e. $\omega(u) = \lim_{\Delta t \to 0+} \mathbb{P}(T - T_{1/u} \in (1-\Delta t, 1)\,)/\Delta t$. To state the connection with number theory, $\Psi(x,y)$ counts positive integers less than or equal to x whose largest prime factor is less than or equal to y, $\Phi(x,y)$ counts positive integers less than or equal to x with no prime factor less than or equal to y, and for $u > 1$ as $n \to \infty$,

$$\frac{1}{n}\,\Psi\left(n, n^{1/u}\right) \to \rho(u), \quad \frac{\log n}{u}\,\frac{1}{n}\,\Phi\left(n, n^{1/u}\right) \to \omega(u). \tag{18.1}$$

For every θ, H_θ is continuous and strictly monotone, mapping [0,1] into itself, with $H_\theta(0) = 0$ and $H_\theta(1) = 1$, but only for $\theta = 1$ is it the case that $H_\theta(\beta) = o(\beta)$ as $\beta \downarrow 0$.

From $H_\theta(\beta) \to 0$ as $\beta \downarrow 0$ it follows that the Poisson-Dirichlet process, "blown up", coverges in distribution to the scale invariant Poisson. That is, with $\mathcal{V} := \{V_1, V_2, \dots\} \subset (0,1)$, and $\mathcal{X} := \{X_i : i \in \mathbb{Z}\} \subset (0,\infty)$, as $v \to \infty$, $v\mathcal{V} \Rightarrow \mathcal{X}$. For a proof: for any fixed $x > 0$, for $v \ge x$ we have

$$d_{TV}(\, v\mathcal{V} \cap (0,x), \mathcal{X} \cap (0,x)\,) = d_{TV}(\, \mathcal{V} \cap (0,x/v), (v^{-1}\mathcal{X}) \cap (0,x/v)\,) \tag{18.2}$$

$$= H_\theta(x/v)$$

where we use the scale invariance of $\mathcal{X}$ to see that last equality. Thus for fixed x, $d_{TV}(v\mathcal{V} \cap (0,x), \mathcal{X} \cap (0,x)\,) \to 0$ as $v \to \infty$, which is quite a bit stronger than distributional convergence.

19. Insertion—deletion distance

That $H_\theta(1) = 1$ says that by looking at a single sample one can a.s. tell the random set $\mathcal{V} := \{V_1, V_2, \dots\}$ from $\mathcal{Y} := \mathcal{X} \cap (0,1) = \{X_1, X_2, \dots\}$. Could the two random sets nevertheless be close? Consider [3] the Wasserstein metric d_W using counts of insertions and deletions needed to convert one set to the other:

$$d_W := \min \; \mathbb{E}|\mathcal{V} \triangle \mathcal{Y}|$$

$$= \min(\; \mathbb{E}|\mathcal{V} \setminus \mathcal{Y}| + \mathbb{E}|\mathcal{Y} \setminus \mathcal{V}| \;).$$

The minimum is taken over all couplings of the Poisson-Dirichlet process and the scale invariant process, both with parameter θ.

For $\theta = 1$, this can be fully understood: $d_W = 2$, and there is a coupling such that

$$\mathcal{V} \setminus \{V_J\} = \mathcal{Y} \setminus \{X_{I_1}, \dots, X_{I_D}\}$$

where the number D of points of the scale invariant Poisson process restricted to (0,1) that need to be deleted is random, with $\mathbb{E}D = 1$.

That $d_W \geq 2$ for $\theta = 1$ starts with a proof that a.s. at least one deletion from $\mathcal{V}$ is needed, because with probability one no subsum of the points of $\mathcal{X}$ has the value $V_1 + V_2 + \cdots = 1$. This argument requires only that $\theta \leq 1/\log 2$; see problem 22.1. For the other half of the lower bound, $\theta = 1$ is needed, so that $\mathcal{X}$ and $\mathcal{V}$ have the same intensity, and hence the one required deletion from $\mathcal{V}$ must be matched by deletions from $\mathcal{X}$, averaging one in number.

The coupling showing that $d_W \leq 2$ for $\theta = 1$ can be given explicitly. In this coupling, the deleted compenent V_J is simply the first pick from $\mathcal{V}$ in a size biased permutation, i.e. given the value of $(V_1, V_2, \dots)$, the conditional probability that $V_J = V_k$ is V_k. The entire coupling is the continuum analog of the Feller coupling for random permutations [5], and depends on the "scale invariant spacing lemma".

20. Scale invariant spacing lemma

Label the scale invariant Poisson process as in (4.3), with $X_i < X_{i+1}$ for $i \in \mathbb{Z}$, and let $Y_i := X_{i+1} - X_i$.

LEMMA. For any $\theta > 0$, $\{X_i : i \in \mathbb{Z}\} =_d \{Y_i : i \in \mathbb{Z}\}$

and a.s. all the Y_i are distinct.

This is the continuous analog of a property of the Feller coupling: if $\xi_1, \xi_2, \dots$ are independent Bernoulli with $\mathbb{P}(\xi_i = 1) = \theta/(\theta + i - 1)$, and Z_k is defined as the number of k-spacings between consecutive ones in $\xi_1, \xi_2, \dots$, then the $Z_1, Z_2, \dots$ are independent, Poisson, with $\mathbb{E}Z_k = \theta/k$. In both the discrete and continuous cases, one starts with an independent process, and the surprise is that the spacings also form an independent process. In the discrete case, while each of the processes $\xi_1, \xi_2, \dots$ and $Z_1, Z_2, \dots$ has independent coordinates, the two processes are not the same.

The proof [3] of the scale invariant spacing lemma is based on the Poisson process on $(0,\infty)^2$ with intensity $\theta e^{-wy}\, dw\, dy$, whose projections on each coordinate axis give realizations of the scale invariant Poisson process. Labelling the points in decreasing order of their w coordinates gives the y coordinates in a permutation such that $X_n := \sum_{i \leq n} Y_i$ forms a simple point process $\mathcal{X} = \{X_n : n \in \mathbb{Z}\}$ on $(0,\infty)$ whose spacings, by construction, are the points Y_i of the scale invariant Poisson process. It is then a calculation that the distribution of $\mathcal{X}$ is also scale invariant Poisson. Since the joint density function of the (W_i, Y_i) can be written θe^{-wy} $= (\theta/y)\ (ye^{-wy})\ (= f_Y(y)\ f_{W|Y}(w),)$ which says that the conditional distribution of the w coordinate, given y, is exponential with mean $1/y$, our construction may be viewed as using a size-biased permutation of $\{Y_i\}$; see [**41**].

21. Problem session: what other processes are their own spacings?

PROBLEM 21.1. What other intensity functions $f(x)$ on $(0, \infty)$ beside those of the form $f(x) = \theta/x$ lead to Poisson processes whose spacings are again Poisson processes?

Tom Kurtz, upon being asked the above question, which is still open, immediately countered with a very different question:

PROBLEM 21.2. What point processes on $(0, \infty)$ are equal in distribution to their own spacings? Formally, what point processes with points X_i for $i \in \mathbb{Z}$ with $X_i < X_{i+1}$ for all i have

(21.1) $Y_i := X_{i+1} - X_i$ are all distinct, a.s., and $\{X_i :\ i \in \mathbb{Z}\} =_d \{Y_i :\ i \in \mathbb{Z}\}$.

The above question is reminiscent of an intriguing, still open 1978 question from Liggett [**38**], about invariant random measures for independent particle systems, a conjecture that $MP = M$ in distribution implies the existence of a nontrivial solution of $MP = M$ almost surely.

The answer to problem 21.2 involves mixtures of *distributional* solutions, such as the scale invariant Poisson processes and perhaps some others, and *deterministic* solutions. During a problem session at the DIMACS workshop, I asked what are all simple deterministic solutions, i.e. doubly infinite sequences, such as $x_i := 2^i$, with $0 < \cdots < x_{-1} < x_0 < x_1 < x_2 < \cdots < \infty$ such that

$$y_i := x_{i+1} - x_i \text{ are all distinct, and } \{y_i :\ i \in \mathbb{Z}\} = \{x_i :\ i \in \mathbb{Z}\}. \tag{21.2}$$

Gábor Tardos and László Lovász from the audience quickly contributed a partial solution: for any fixed $k \geq 0$ take

$$x_i := b^i, \quad \text{where } b > 1 \text{ solves } b^{k+1} - b^k = 1, \tag{21.3}$$

so that b is 2 when $k = 0$, and b is the golden ratio when $k = 1$. They observed that $y_i := x_{i+1} - x_i < x_{i+1}$ implies $y_i = x_{i+1} - x_i \leq x_i$, i.e. $x_{i+1} \leq 2x_i$, so that the example with $x_i = 2^i$ is extreme. They also showed how "entrance solutions" may be extended: working from left to right, the next spacing is chosen from the previously defined x_k, excluding those already used as spacings, i.e. $y_i \in \{\dots, x_{i-2}, x_{i-1}, x_i\} \setminus \{\dots, y_{i-2}, y_{i-1}\}$. For example, $b = (1 + \sqrt{5})/2$ and $x_i = b^i$ for $i \leq 0$ is an entrance solution with $y_i = x_{i-1}$ for $i < 0$, which may be extended with an infinite series of two-way choices: first $x_1 = x_0 + y_0$ where $y_0 \in \{x_0, x_{-1}\}$, then $x_2 = x_1 + y_1$ where $y_1 \in \{x_1, x_0, x_{-1}\} \setminus \{y_0\}$, and so on. To describe their idea in more detail, focus on the implied permutation, as follows.

Any solution of (21.2) determines a permutation π of $\mathbb{Z}$, such that $\forall i \in \mathbb{Z}$, $y_i = x_{\pi(i)}$, i.e. $\forall i \in \mathbb{Z}$

$$x_{i+1} = x_i + x_{\pi(i)} \in (0, \infty). \tag{21.4}$$

Note that always $\pi(i) \le i$ since $y_i \le x_i$. The geometric solutions in (21.3) correspond to permutations π with $\pi(i) = i - k$. In constructing deterministic solutions, for any given π with

$$\pi \text{ permutes } \mathbb{Z}, \quad \text{and} \quad \pi(i) \le i \quad \forall i \in \mathbb{Z}, \tag{21.5}$$

if there is an "entrance solution" from zero, then the solution can be uniquely extended out to infinity. That is, if for some k there are $0 < \cdots < x_{k-2} < x_{k-1} < x_k$ satisfying (21.4) for all $i < k$, then recursively defining $x_{k+1}, x_{k+2}, \ldots$ by (21.4) for $i = k, k+1, \ldots$ involves no problem — the solution cannot explode to plus infinity in finite time. Running this argument in the opposite direction is not so easy: extending backwards from $x_k, x_{k+1}, \ldots$ involves taking differences, which can produce values ≤ 0. We can show that entrance solutions exist by using a compactness argument on the sequence of ratios between successive points:

THEOREM 21.3. *If π satisfies (21.5), then there exists at least one solution of (21.4).*

Proof. Given $\mathbf{x} \equiv (x_i)_{i\in\mathbb{Z}} \in (0,\infty)^{\mathbb{Z}}$, let $V_i := (x_{i-1}/x_i, x_{i-2}/x_{i-1}, x_{i-3}/x_{i-2}, \ldots) \in (0,\infty)^{\mathbb{N}}$. Note that if the sequence $\mathbf{x}$ satisfies (21.4), then for all $i \in \mathbb{Z}$, $V_i \in [\frac{1}{2}, 1)^{\mathbb{N}} \subset K := [\frac{1}{2}, 1]^{\mathbb{N}}$.

If $V_i = (r_1, r_2, \ldots)$ then, factoring out $c := x_i$, we have $(x_i, x_{i-1}, x_{i-2}, \ldots) = c(1, r_1, r_1 r_2, \ldots)$. If $x_{i+1} = x_i + x_{i-d}$ then $x_{i+1} = c(1 + \prod_{1\le j\le d} r_j) =: c/r_0$; this defines r_0, such that $V_{i+1} = (r_0, r_1, r_2, \ldots)$. [In case $d = 0$ the product is empty and has value 1, and $x_{i+1} = 2x_i, r_0 = \frac{1}{2}$.]

This motivates us to define, for each $d \ge 0$, a function $f^{(d)}$: $(0,\infty)^{\mathbb{N}} \to (0,\infty)^{\mathbb{N}}$ by $(r_1, r_2, r_3 \ldots) \mapsto ((1 + \prod_{1\le j\le d} r_j)^{-1}, r_1, r_2, \ldots)$. Note that these functions are continuous and map K into itself. Given π satisfying (21.5), define $d(i) := i - \pi(i)$, and for $n = 1, 2, \ldots$ define a map $T_{-n,0} := f^{(d(-1))} \circ f^{(d(-2))} \cdots \circ f^{(d(1-n))} \circ f^{(d(-n))}$. Note that if $\mathbf{x}$ satisfies (21.4) then $T_{-n,0}(V_{-n}) = V_0$. Conversely, given *any* $\ldots, x_{-n-2}, x_{-n-1}, x_{-n} \in (0,\infty)$, (not necessarily satisfying (21.4),) the map $T_{-n,0}$ gives a recipe for extending the sequence with values $x_{-(n-1)}, \ldots, x_{-1}, x_0$ such that $x_{i+1} = x_i + x_{\pi(i)} \in (0,\infty)$ for $i = -n, \ldots, -2, -1$. For $n = 1, 2, \ldots$ let S_n be the image of the compact set K under $T_{-n,0}$, so $\cdots \subset S_2 \subset S_1 \subset K$. By the finite intersection property, $\cap_{n\ge1} S_n \ne \emptyset$. Any point $(r_1, r_2, \ldots) \in \cap_{n\ge1} S_n$ yields a sequence $x_0 = 1, x_{-1} = r_1, x_{-2} = r_1 r_2, \ldots$ satisfying $x_{i+1} = x_i + x_{\pi(i)}$ for $i = -1, -2, \ldots$, and this entrance solution can be extended to a full solution of (21.4). ∎

Clearly if a sequence (x_i) satisfies (21.4) then so does any scalar multiple (cx_i) for any $c > 0$. Are solutions uniquely determined by π, up to such scalar multiples? At least from the point of view taken in the proof of the following theorem, the situation resembles time inhomogeneous renewal chains; and we can only handle the bounded case. Do transient chains somehow correspond to counterexamples to uniqueness?

THEOREM 21.4. *Assume π satisfies (21.5) and $k := \sup\{i - \pi(i) : i \le 0\} < \infty$. Then solutions of (21.4) are unique up to a scalar multiple.*

Proof. We consider the effect of using $x_{i+1} = x_i + x_{\pi(i)}$ for $i = 0, -1, \ldots, -n$ to express x_1 and x_0 as linear combinations, with nonnegative integer weights, of $x_{-n}, x_{-n-1}, \ldots, x_{-n-k}$, viewed as indeterminates. The idea is to show that for large n, the weights in the combination for x_1 are close to being a multiple of the weights for x_0, so that the ratio x_1/x_0 is close to being determined by π.

The Hilbert projective metric ρ on $(0,\infty)^{k+1}$ (modulo scalar multiples) is defined by $\rho(u,v) := \max_i \log(u_i/v_i) - \min_i \log(u_i/v_i)$. If $u = (x_{-n}, x_{-n-1}, \dots, x_{-n-k})$ comes from a solution of (21.4) and $v = (1,1,\dots,1)$, then the property $x_i < x_{i+1} \le 2x_i$ implies that $\rho(u,v) \le k \log 2$.

Consider $k+1$ by $k+1$ matrices with nonnegative integer coefficients, indexed by $0 \le i,j \le k$, as follows. Write $E^{(i,j)}$ for the matrix having all zeroes, except for a single one in row i, column j. Write $B = \sum_{0<i\le k} E^{(i,i-1)}$ for the matrix with ones below the diagonal, and let $C = B + E^{(0,0)}$. For $0 \le d \le k$ let $A^{(d)} = C + E^{(0,d)}$. For $n \in \mathbb{Z}$ let $d(n) = n - \pi(n)$, $F^{(n)} = A^{(d(n))}$, and $M^{(n)} = F^{(-1)}F^{(-2)}\cdots F^{(-n)}$. Row i of $M^{(n)}$ gives the coefficients of x_{-i} as a linear combination of $x_{-n}, x_{-n-1}, \dots, x_{-n-k}$, and row 0 plus row $d(0)$ gives the coefficients of x_1.

Note that C^k is all ones in column zero, and zeroes elsewhere. Any product M of k or more factors of the form $A^{(d(i))}$ has all entries in column zero strictly positive, and any product with $2k$ or more factors has the property that every column is either all zeroes, or else has all entries strictly positive. The maximum entry in a product M with exactly $2k$ factors is at most $s = 2^{2k}$, and the least entry in a non-zero column is at least one; hence if $r = \rho(u,v)$, then

$$\rho(Mu, Mv) \le \log\left(\frac{1+se^r}{1+s}\right).$$

It follows that using $r = k\log 2$ and given $\epsilon > 0$, we can pick n so that for all u, v with $\rho(u,v) \le r$ we have $\rho(M^{(n)}u, M^{(n)}v) < \epsilon$. Using $v = (1,1,\dots,1)$ and the remarks at the ends of the previous two paragraphs, the ratio $a := (M^{(n)}v)_{d(0)}/(M^{(n)}v)_0)$ satisfies $|\log a - \log((x_1 - x_0)/x_0)| < \epsilon$, for any solution of (21.4). This show that any two solutions of (21.4) have exactly the same ratio x_0/x_1. The same argument shows that the permutation π determines the ratio x_i/x_{i+1} for each $i \in \mathbb{Z}$. ∎

Problem 21.2 also includes solutions placing probability $1/m$ on each point in an orbit of period $m > 1$ of the spacing transformation. The simple deterministic solutions discussed above are the case $m = 1$, fixed points. For the general case, there are m distinct deterministic sequences; the random process picks each of these with probability $1/m$ each; and the spacings of the kth sequence are all distinct and as a set give the points of the $(k+1)^{st}$ sequence mod m.

For example, for $m > 1$ there are solutions of the form $\mathbb{P}(X_i = x_i) = 1/m$, where the deterministic sequence is given by $x_i := b^i$, such that $\Delta x_i := x_{i+1} - x_i = (b-1)b^i, \dots, \Delta^m x_i = (b-1)^m x_i$ give m distinct sets of points, and the last of these sets is the same as $\{x^i : i \in \mathbb{Z}\}$. This is possible iff $(b-1)^m = b^k$ for some $k \in \mathbb{Z}, b \in (1,\infty)$. For example, with $m = 2$ and $k = 1$, $b = (3+\sqrt{5})/2 \doteq 2.618$ and with $m = 2$ and $k = -1$, $b \doteq 1.75487$. These examples with geometric sequences are misleading in that in general, the $y_i := x_{i+1} - x_i$ are not in increasing order, and the iterated transformation on sequences does *not* correspond to Δ^m, but rather to $(\text{RANK} \circ \Delta)^m$.

To compare simple deterministic solutions with the scale invariant Poisson processes, write σ for the inverse of π. Note that for all i, $i \le \sigma(i)$ and $y_{\sigma(i)} = x_i < x_{i+1} = y_{\sigma(i+1)}$. For the random solutions (21.1) there is a random permutation σ of $\mathbb{Z}$ such that for all i

$$Y_{\sigma(i)} < Y_{\sigma(i+1)},$$

with σ determined only up to translation. If we write $\sigma(i) = i + C(i)$, then it is easy to show, for the scale invariant Poisson process, that a.s. $\limsup C_i = \infty$ and $\liminf C_i = -\infty$, a qualitative property possessed by no mixture of simple deterministic solutions. Large deviations for the permutation σ are studied in [**57**].

To summarize the above discussion: the solution to problem 21.2 includes a) the scale invariant Poisson processes, b) simple deterministic solutions, i.e. fixed points of the spacing transformation, c) deterministic orbits of length $m > 1$, and d) other extreme points of the set of distributional solutions. For b), it remains to resolve the question of uniqueness relative to permutations satisfying (21.5), but with unbounded displacements; and for c) and d), everything is open.

Peter Baxendale recently asked, for simple deterministic solutions, what are the possible values of the ratio $r := x_0/x_1$ of two adjacent points? Is $\frac{2}{3}$ achievable? More generally, writing $r_i := x_{i-1}/x_i$, what are the possible configurations of $k+1$ consecutive points $(x_0, x_1, \ldots, x_k)$, i.e. what points in $[\frac{1}{2}, 1)^k$ are realizable as the value of $(r_1, r_2 \ldots, r_k)$, for $k = 1, 2, \ldots$? Two other ways to generalize the question about r are to ask which finite sets $A \subset (0, \infty)$ can satisfy $A \subset \{x_i : i \in \mathbb{Z}\}$ for some solution of (21.2), and similarly which finite sets $B \subset [\frac{1}{2}, 1)$ can satisfy $B \subset \{r_i : i \in \mathbb{Z}\}$?

22. Problem session: a phase transition at $\theta = 1/\log 2$

For any $\theta > 0$, starting from a realization $\{X_i : i \in \mathbb{Z}\}$ of the scale invariant Poisson process with intensity $\theta/x\, dx$, let $A \equiv A(\theta)$ be the random closed set which is the closure of the countable set whose points are $\sum_{i \in I} X_i$ for finite $I \subset \mathbb{Z}$. From the scale invariance of the underlying $\{X_i\}$ it follows easily that A is also scale invariant: for any $c > 0$ and for each $\theta > 0$

$$cA(\theta) =_d A(\theta).$$

The process $(A(\theta))_{\theta > 0}$ has stationary, independent increments in the following sense. The Minkowski sum of two sets is $A \oplus B := \{a + b : a \in A, b \in B\}$. With $\mathcal{X}(\theta)$ to denote the set of points in the scale invariant Poisson process with intensity $\theta/x\, dx$, the process $(\mathcal{X}(\theta))_{\theta>0}$ and hence also $(A(\theta))_{\theta>0}$ can be constructed with increasing sample paths: $\theta_1 < \theta_2$ implies $\mathcal{X}(\theta_1) \subset \mathcal{X}(\theta_2)$, and hence $A(\theta_1) \subset A(\theta_2)$. The process $\mathcal{X}$ has stationary, independent increments in the strong sense that $\mathcal{X}(\theta_2) \setminus \mathcal{X}(\theta_1)$ is independent of $\mathcal{X}(\theta_1)$ and equal in distribution to $\mathcal{X}(\theta_2 - \theta_1)$, but the process $A(\theta)_{\theta>0}$ has stationary independent increments in a weaker sense. If $\theta_1 < \theta_2$, and if $A'(\theta_2 - \theta_1)$ is independent of $A(\theta_1)$ and equal in distribution to $A(\theta_2 - \theta_1)$, then $A(\theta_1) \oplus A'(\theta_2 - \theta_1) =_d A(\theta_2)$.

Some of the structure of the set of divisors of a random integer, as in [**26**], is captured by the scale invariant closed set $A \subset [0, \infty)$. First, for any $\theta > 0$ define a random closed set $B(\theta) \subset [0, 1]$ by $B := \{\sum_{i \in I} V_i : I \subset \mathbb{N}\}$. From the Poisson-Dirichlet convergence in (8.1) and its extension to the large deviation case for $\theta \neq 1$, it follows easily that the random finite set

$$\mathcal{D}_n := \{\log d / \log n \} \subset [0, 1], \tag{22.1}$$

where d runs over the divisors of our random integer, has $\mathcal{D}_n \Rightarrow B(\theta)$. (See Theorem 22.2 for an extension.) In fact, with the Hausdorff metric for closed subsets of $[0, 1]$, and the l_1 metric on $\mathbb{R}^\infty$, the map which produces $\mathcal{D}_n$ from $(\log P_1, \log P_2, \ldots)$ and $B(\theta)$ from $(V_1, V_2, \ldots)$ is a contraction. Second, just like (18.2) it is easy to see that as $v \to \infty$, $vB(\theta) \Rightarrow A(\theta)$. In fact, for any $0 < x \le v$,

$$(22.2)\qquad \begin{aligned} d_{TV}(vB \cap (0,x), A\cap(0,x)\,) &= d_{TV}(B\cap(0,x/v),(v^{-1}A)\cap(0,x/v)\,) \\ &\le d_{TV}(\mathcal{V}\cap(0,x/v),(v^{-1}\mathcal{X})\cap(0,x/v)\,) = H_\theta(x/v). \end{aligned}$$

Thus for fixed x, $d_{TV}(vB\cap(0,x), A\cap(0,x)\,) \to 0$ as $v \to \infty$, which is even stronger than the distributional convergence $vB \Rightarrow A$.

Let $f(\theta) := \mathbb{P}(1 \in A(\theta)\,)$. Using scale invariance, $\forall x > 0, \mathbb{P}(x \in A(\theta)\,) = f(\theta)$. Hence with $m(A)$ to denote Lebesgue measure, for $0 \le a < b < \infty$, $\mathbb{E}m(A \ \cap (a,b)\,) = (b-a)f(\theta)$. It is easy to show that for $\theta \le 1/\log 2$, $f(\theta) = 0$. Note that for any θ, $f(\theta) \le \mathbb{P}(T < \ 1) < 1$. In lecture, we asked a semi-open question:

PROBLEM 22.1. Prove the conjecture that if $\theta > 1/\log 2$, then $f(\theta) > 0$.

At the time of lecture, the conjecture was semi-open in the following sense: Tenenbaum (private communication) had proved a related property about the set of divisors of a random integer from 1 to n with $\theta \log\log n$ distinct prime divisors. From the early version of this number theory result, it was not yet possible to deduce the continuum result as a corollary, so the conjecture was indeed open. However, it was highly plausible to guess that the underlying principles from Tenenbaum's proof would also work directly on the continuum problem. Subsequent to the workshop, the conjecture has indeed been proved this way. Further, Tenenbaum gave a sharper version of the number theoretic result, from which the continuum result follows as a corollary. This amounts to using a more complicated process, prime divisors of a random integer, to approximate a simpler process, the scale invariant Poisson!

The event that $A(\theta)$ has positive length is a tail event with respect to the independent exponentials W_i in (5.1), and hence by the Kolmogorov zero-one law and the above, for every $x > 0$ and any $\theta > 1/\ \log 2$, $1 = \mathbb{P}(m(A\cap(0,x)\,) > 0)$.

The random sets above are closely related to the theory of Bernoulli convolutions; we learned of the connection thanks to Jim Pitman. For Bernoulli convolutions, the usual setup is to start with a deterministic sequence $r_1, r_2, \ldots > 0$ with $\sum r_n < \infty$ and to consider the random variable $Y = \sum_{i\ge1} S_i r_i$, where the S_i have values 1, -1, independently with probability 1/2 each. The closed support of the distribution of Y is the set K of all points of the form $\sum s_i r_i$, where the s_i are arbitrarily chosen from $\{1,-1\}$. There is a straightfoward translation to the situation where $s_i \in \{0,1\}$, which is natural for the application to number theory.

Erdős focussed on the case $r_n := \lambda^n$ for a fixed $0 < \lambda < 1$. Here it is obvious that K has zero length for $0 < \lambda < \frac{1}{2}$, and that K is an interval for $\frac{1}{2} \le \lambda < 1$. Erdős asked, when is the distribution μ_λ of Y absolutely continuous with respect to Lebesgue measure? For $\lambda < \frac{1}{2}$, since the support K has zero length, μ_λ is singular, but for $\lambda > \frac{1}{2}$, even though the support K is an interval, the absolute continuity question is subtle. In 1939 Erdős [**21**] showed that there are values in $(\frac{1}{2},1)$, such as $\lambda = 2/(1+\sqrt{5})$, for which μ_λ is singular, and in 1940 [**22**] he showed that there exists some $t < 1$ such that for almost every $\lambda \in (t,1)$, μ_λ is absolutely continuous. In 1995, Solomyak [**49**] showed that t in the previous sentence can be taken to be 1/2. Peres and Solomyak [**42**] give a simple proof of this. In 1958 Kahane and Salem [**34**] discussed a variant of the problem where the summands r_n are random, of the form $r_n = U_1U_2\cdots U_n$ where the U_i are independent, and uniformly chosen from intervals $[a_i,b_i] \subset [\frac{1}{2},1]$.

Consider the random closed set $C := \{\sum_{i\ge1} c_i X_i \ : \ c_i \in \{0,1\}\ \}$, with the scale invariant Poisson process labelled as in (4.2), so that $C \subset [0,T]$. This is close to A in that always $A \cap [0,1) = C \cap [0,1)$, and related to B: as a corollary of (14.2), for each $\theta > 0$, $B =_d (C \mid T = 1\ )$. Let $J_1, J_2, \dots$ be fair coins with values in $\{0,1\}$, independent of each other and everything else. The random series $Y^* := \sum_{i \geq 1} J_i X_i$ has values in the random set C, and the random series $Y := \sum_{i \geq 1} J_i V_i$ has values in the random set B. Consider the conditional distribution $\mu_{Y^*|X}$ of Y^* given $X_1, X_2, \dots$, and the conditional distribution $\mu_{Y|V}$ of Y given the Poisson-Dirichlet process. These are *random* probability measures on $[0,\infty)$ and $[0,1]$, respectively, and correspond to the fixed probability distribution μ_λ from the situation where only the signs are random. The method of [**42**] can be adapted to show that for $\theta > 1/\log 2$, a.s. the distributions $\mu_{Y^*|X}$ and $\mu_{Y|V}$ are absolutely continuous with respect to Lebesgue measure, with density in L_2 (Yuval Peres, private communication).

The closed support in $[0,\infty)$ of $\mu_{Y^*|X}$ is C, and that of $\mu_{Y|V}$ is B. The distribution of Y itself, without conditioning on the value of the Poisson-Dirichlet process, is the probability measure $\mathbb{E}\,\mu_{Y|V}$, averaging over the values of $(V_1, V_2, \dots)$. From Hirth [**31**] and Donnelly-Tavaré [**19**], the distribution of Y is simply Beta$(\frac{\theta}{2}, \frac{\theta}{2})$, which for $\theta = 1$ is the arc-sine distribution. It gives the limit of the distribution of $\log d/\log n$, and the limit distribution of $\log d/\log N$, where a random integer N is chosen uniformly from 1 to n, and then d is chosen uniformly from the $\tau(N)$ divisors of N; see [**51**], II.6.2. All this background suggests looking at the distribution of (the rescaled log of) a randomly chosen divisor of uniformly chosen random integer, as a random probability distribution. We need some notation.

Let $\mathcal{P}$ denote the space of probability measures on $\mathbb{R}$, with the topology of weak convegence, as given by the Lévy metric ρ; see e.g. [**20**]. For $m \geq 1$ let $\mu_m := \left(\sum_{d|m} \delta_{\log d/\log m}\right) / \left(\sum_{d|m} 1\right) \in \mathcal{P}$ be the probability measure on [0,1] that puts mass $1/\tau(m)$ on each point of the form $\log d/\log m$, where d runs over the divisors of m. (For $m = 1$, interpret $0/0$ as 1, so that $\mu_1 := \delta_1$.)

For $n \geq 1$ let $N \equiv N(n)$ be chosen uniformly from 1 to n, i.e. $\mathbb{P}_n(N = m) = 1/n$ for $m = 1, 2, \dots, n$. Let μ_N denote the random measure whose value is μ_m on the event $\{N = m\}$, so that the law of μ_N as a random element of $\mathcal{P}$ is $(1/n)\sum_{1 \leq m \leq n} \delta_{\mu_m}$.

THEOREM 22.2. *As $n \to \infty$,*

$$\mu_N \Rightarrow \mu_{Y|V},$$

i.e. as random a element of $\mathcal{P}$, $\mu_{N(n)}$ converges in distribution to $\mu_{Y|V}$, using the Poisson-Dirichlet process with $\theta = 1$.

Proof. As noted in [**31**], if $m = p_1 p_2 \cdots p_k$ is squarefree, then μ_m, supported at 2^k distinct points, is a Bernoulli convolution, and the essential difficulty is to deal with m such that $p^2|m$ for some prime. Write $P_i(m)$ for the ith largest prime factor of m, with $P_i = 1$ if i is greater than the number $\Omega(m)$ of prime factors of m, including multiplicities. Let J_i be $\{0,1\}$-valued fair coins, independent of everything else. For $k = 1, 2, \dots$ define $\mu_m^{(k)} \in \mathcal{P}$ to be the distribution of $\sum_1^k J_i\,(\log P_i(m)/\log m)$, so that $\mu_m^{(k)} = \mu_m$ iff m is squarefree and has at most k prime factors. Similarly, define $\mu_{Y|V}^{(k)} \in \mathcal{P}$ to be the conditional distribution of $\sum_1^k J_i V_i$ given $(V_1, V_2, \dots)$. Write $\mathcal{L}(S)$ for the law of S. The Lévy metric ρ on $\mathcal{P}$ has the property that for random variables S, T, $1 = \mathbb{P}(|S - T| \leq \epsilon)$ implies that $\rho(\mathcal{L}(S), \mathcal{L}(T)) \leq \epsilon$. Hence with

$S = \sum_1^k J_i v_i$ and $T = \sum_1^\infty J_i v_i$, it follows that $\rho(\mu_{Y|V}^{(k)}, \mu_{Y|V}) \leq V_{k+1} + V_{k+2} + \cdots = 1 - (V_1 + \cdots + V_k)$.

Given $\epsilon > 0$, pick k so that $\mathbb{P}(V_1 + \cdots + V_k < 1 - \epsilon) < \epsilon$. This implies $\mathbb{P}(\rho(\mu_{Y|V}^{(k)}, \mu_{Y|V}) > \epsilon) < \epsilon$. For any m write $r = P_1(m) \cdots P_k(m)$ and consider the map g_m with $g_m(d) = (d, r)$, the greatest common divisor. This map, applied to the set of divisors of m, is exactly $\tau(m)2^{-k}$ to one, *provided that* m has at least k prime divisors and the k largest prime divisors are distinct. For m satisfying this condition, g_k applied to a divisor chosen randomly from the $\tau(m)$ divisors of m yields a uniform pick from the 2^k divisors of r, and hence $\rho(\mu_m^{(k)}, \mu_m) \leq \log(m/r)/\log m$.

We note that Billingsley [**13**] actually proved both (8.1) and the slightly different form we need here, that $(\log P_i(N)/\log N)_{i \geq 1} \Rightarrow (V_i)_{i \geq 1}$. Thus we can pick n_1 so that for $n \geq n_1$, $\mathbb{P}_n(P_1(N) \cdots P_k(N) \geq N^{1-\epsilon}$ and $P_1(N) > \cdots > P_k(N) > 1) > 1 - \epsilon$. This yields, for $n \geq n_1$, that $\mathbb{P}_n(\rho(\mu_N^{(k)}, \mu_N) > \epsilon) < \epsilon$.

The map $f_k : (x_1, \ldots, x_k) \mapsto \mathcal{L}(\sum_1^k J_i x_i)$, from $(\mathbb{R}^k, l_1)$ to $(\mathcal{P}, \rho)$ is continuous; in fact it is a contraction. Now $\mu_{Y|V}^{(k)} = f_k(V_1, \ldots, V_k)$, and $\mu_N^{(k)} = f_k(\log_N P_1(N), \ldots, \log_N P_k(N))$, so using Billingsley's result again, as $n \to \infty$, $\mu_N^{(k)} \Rightarrow \mu_{Y|V}^{(k)}$. By Skorohod embedding [**20**] for the complete separable metric space $\mathcal{P}$, there exists a coupling in which $\mu_{N(n)}^{(k)} \to \mu_{Y|V}^{(k)}$ almost surely. This coupling can be extended to a coupling of $N(1), N(2), \ldots$ and the Poisson-Dirichlet process, so that for one fixed k, $\mu_{N(n)}^{(k)}, \mu_N(n), \mu_{Y|V}^{(k)}$, and $\mu_{Y|V}$ are all defined on the same probability space, and a.s. $\mu_{N(n)}^{(k)} \to \mu_{Y|V}^{(k)}$. Pick n_2 such that for $n \geq n_2$, $\mathbb{P}(\rho(\mu_{N(n)}^{(k)}, \mu_{Y|V}^{(k)}) > \epsilon) < \epsilon$. The triangle inequality yields, for $n \geq \max(n_1, n_2)$, that $\mathbb{P}(\rho(\mu_N(n), \mu_{Y|V}) > 3\epsilon) < 3\epsilon$. ■

References

[1] Aizenman, M., and Newman, C. (1986) Discontinuity of the percolation density in one-dimensional $1/|x-y|^2$ percolation models. Comm. Math. Phys. **107**, 611-647.

[2] Aldous, D. J. (1983) Exchangeability and related topics. Springer, Lecture Notes in Mathematics, vol. 1117.

[3] Arratia, R. (1996) Independence of small prime factors: total variation and Wasserstein metrics, insertions and deletions, and the Poisson-Dirichlet process. Draft, 73 pages, available from `rarratia@math.usc.edu`

[4] Arratia, R. (1996) Large and small prime factors conditional on a large deviation for the number of prime factors of a random integer. Draft, 17 pages, available from `rarratia@math.usc.edu`

[5] Arratia, R., Barbour, A. D., and Tavaré, S. (1992) Poisson process approximation for the Ewens Sampling Formula. Ann. Appl. Probab. **2**, 519-535.

[6] Arratia, R., Barbour, A. D., and Tavaré, S. (1997) Random combinatorial structures and prime factorizations. AMS Notices **44**, 903-910.

[7] Arratia, R., Barbour, A. D., and Tavaré, S. (1997) Logarithmic combinatorial structures. Monograph, 187 pages, in preparation.

[8] Arratia, R., Barbour, A. D., and Tavaré, S. (1997) On Poisson-Dirichlet limits for random decomposable combinatorial structures. To appear in Comb., Prob., and Comp.

[9] Arratia, R., Barbour, A.D., and Tavaré, S. (1997) The Poisson-Dirichlet distribution and the scale invariant Poisson process. To appear in Comb., Prob., and Comp.

[10] Arratia, R., and Stark, D. (1997) A total variation distance invariance principle for primes, permutations and Poisson-Dirichlet. Preprint.

[11] Arratia, R., Stark, D., and Tavaré, S. (1995). Total variation asymptotics for Poisson process approximations of logarithmic combinatorial assemblies. Ann. Probab. **23**, 1347-1388.

[12] Arratia, R., and Tavaré, S. (1992) The cycle structure of random permutations. Ann. Probab. **20**, 1567-1591.

[13] Billingsley, P. (1972) On the distribution of large prime factors. Period. Math. Hungar. **2**, 283-289.

[14] de Bruijn, N. G., and van Lint, J. H. (1964) Incomplete sums of multiplicative functions, I, II. Nederl. Akad. Wetensch. Proc. Ser. A bf 67 339-347, 348-359.

[15] Buchstab, A. A. (1937) An asymptotic estimation of a general number-theoretic function. Mat. Sbornik (2) **44**, 1239-1246.

[16] Daley, D.J., and Vere-Jones, D. (1988) An introduction to the theory of point processes. Springer-Verlag, Springer series in statistics.

[17] Dickman, K. (1930) On the frequency of numbers containing prime factors of a certain relative magnitude. Ark. Math. Astr. Fys. **22**, 1-14.

[18] Donnelly, P., and Grimmett, G. (1993) On the asymptotic distribution of large prime factors. J. London Math. Soc. (2) **47**, 395-404.

[19] Donnelly, P., and Tavaré, S. (1987) The population genealogy of the infinitely-many neutral alleles model. J. Math. Biol. **25**, 381-391.

[20] Dudley, R. M. (1989) Real Analysis and Probability. Wadsworth and Brooks/Cole.

[21] Erdős P. (1939). On a family of symmetric Bernoulli convolutions. Am. J. Math. **61**, 974-975.

[22] Erdős P. (1940). On the smoothness properties of Bernoulli convolutions. Am. J. Math. **62**, 180-186.

[23] Ewens, W. J. (1972) The sampling theory of selectively neutral alleles. Theoret. Population Biol. **3**, 87-112.

[24] Feller, W. (1966) An introduction to probability and its applications, volume II. Wiley.

[25] Ferguson, T. S. (1973) A Bayesian analysis of some nonparametric problems. Ann. Stat. **1**, 209-230.

[26] Hall, R.R., and Tenenbaum, G. (1988) Divisors. Cambridge Tracts in Mathematics **90**, Cambridge.

[27] Halmos, P. R. (1944) Random Alms. Ann. Math. Stat. **15**, 182-189.

[28] Hansen, J. C. (1994). Order statistics for decomposable combinatorial structures. Random Structures and Algorithms **5**, 517-533.

[29] Hensley, D. (1986). The convolution powers of the Dickman function. J. London Math. Soc. (2) **33**, 395-406.

[30] Hildebrand, A. (1990). The asymptotic behavior of the solutions of a class of differentiual-difference equations. J. London Math. Soc. (2) **42**, 11-31

[31] Hirth, U. M. (1997) Probabilistic number theory, the GEM/Poisson-Dirichlet distribution and the Arc-sine law. Comb., Prob., and Comp. **6**, 57-77.

[32] Ignatov, T. (1982) On a constant arising in the asymptotic theory of symmetric groups, and on Poisson-Dirichlet measures. Theor. Prob. Appl. **27**, 136-147.

[33] Johnson, N. S. and Kotz, S., and Balakrishnan, N. (1997). Discrete Multivariate Distributions. Wiley, New York.

[34] Kahane, J. P., and Salem, R. (1958). Sur la convolution d'une infinité de distributions de Bernoulli. Colloquium Mathematicum **6**, 193-202.

[35] Kingman, J.F.C. (1975). Random discrete distributions. J. Royal Stat. Soc. **37**, 1-22.

[36] Kingman, J.F.C. (1993) Poisson Processes. Oxford Science Publications

[37] De Koninck, J.-M., and Galambos, J. (1987) The intermediate prime divisors of integers. Proc. Amer. Math. Soc. **101**, 213-216.

[38] Liggett, T. (1978) Random invariant measures for Markov chains, and independent particle systems. Z. Wahrsch. verw. Gebiete **45**, 297-313.

[39] McCloskey, J. W. (1965) A model for the distribution of individuals by species in an environment, Michigan State University, unpublished Ph.D. thesis.

[40] Perman, M. (1993). Order statistics for jumps of normalized subordinators. Stoch. Pr. Appl. **46**, 267-281.

[41] Perman, M., Pitman, J., and Yor, M. (1992). Size-biased sampling of Poisson point processes and excursions. Prob. Th. Rel. Fields **92**, 21-39.

[42] Peres, Y., and Solomyak, B. (1996). Absolute continuity of Bernoulli convolutions, a simple proof. Math. Res. Lett. **3**, 231-239.

[43] Pitman, J. (1996). Some developments of the Blackwell-MacQueen urn scheme. In T.S. Ferguson et al., editor, Statistics, Probability and Game Theory; Papers in honor of David

Blackwell, Lecture Notes-Monograph Series **30**, 245–267. Institute of Mathematical Statistics, Hayward, California, 1996.

[44] Pitman, J. (1997). Coalescents with multiple collisions. Technical report No. 495, Department of Statistics, Berkeley.

[45] Pitman, J., and Yor, M. (1996) Random discrete distributions derived from self-similar random sets. Electron. J. Probab. **1**, no. 4.

[46] Pitman, J., and Yor, M. (1997) The two-parameter Poisson-Dirichlet distribution derived from a stable subordinator. Ann. Probab. **25**, 855-900.

[47] Ruelle, D. (1987) A mathematical reformulation of Derrida's REM and GREM. Commun. Math. Phys. **108**, 225-239.

[48] Shepp, L. A., and Lloyd, S. P. (1966) Ordered cycle lengths in a random permutation. Trans. Amer. Math. Soc. **121**, 340-357.

[49] Solomyak, B. (1995). On the random series $\sum \pm\lambda^n$ (An Erdős problem). Ann. Math **142**, 611-625.

[50] Stark, D. (1997) Explicit non-zero limits of total variation distance in independent Poisson approximations of logarithmic combinatorial assemblies. Comb., Prob., and Comp. **6**, 87-106.

[51] Tenenbaum, G. (1995) Introduction to analytic and probabilistic number theory. Cambridge studies in advanced mathematics, **46**. Cambridge University Press.

[52] Tenenbaum, G. (1997). Crible d'Ératosthène et modèle de Kubilius. Preprint.

[53] Vershik, A.M., and Shmidt, A.A. (1977) Limit measures arising in the theory of groups, I, Theory Probab. Appl. **22**, 79- 85.

[54] Vershik, A.M., and Shmidt, A.A. (1978) Limit measures arising in the theory of groups, II, Theory Probab. Appl. **23**, 36- 49.

[55] Vervaat, W. (1972) Success Epochs in Bernoulli Trials with Applications in Number Theory, Mathematical Center Tracts, vol. 42, Mathematisch Centrum, Amsterdam.

[56] Watterson, G. A. (1976). The stationary distribution of the infinitely-many alleles diffusion model. J. Appl. Probab. **13**, 639-651.

[57] Zeitouni, O. (1997) Superexponential decay for the GEM process. To appear, J. Appl. Prob.

Current address: University of Southern California, Department of Mathematics, DRB 155, Los Angeles CA 90089-1113

E-mail address: `rarratia@math.usc.edu`

DIMACS Series in Discrete Mathematics
and Theoretical Computer Science
Volume 41, 1998

Beyond the Method of Bounded Differences

Anant P. Godbole[†]
Paweł Hitczenko[‡]

ABSTRACT. It is well known that any objective function $Y = Y_n = Y_n(X_1, \ldots, X_n)$ that depends on an underlying sequence (X_n) of random variables may be represented as a martingale in the canonical fashion introduced by J. L. Doob. Inequalities such as the Azuma-Hoeffding inequality may then be used to bound the probabilities $\mathbf{P}(|Y - \mathbf{E}Y| > \lambda)$ in a meaningful way provided, e.g., that the corresponding martingale differences are uniformly bounded by appropriately small quantities. In many applications, however, this last condition is not satisfied; the purpose of this note is to survey and compare results that enable one to give a useful bound on the corresponding tail probability in such cases. Applications to random proximity graphs will be presented.

1. Introduction

Consider the following generic set up: $(\Omega, \mathcal{F}, \mathbf{P})$ is a probability space and (X_n) a sequence of random variables on $(\Omega, \mathcal{F}, \mathbf{P})$. For our purposes, we will always suppose that they are i.i.d., though they may not, in general, even be independent. Let $Y = Y_n = Y(X_1, \ldots, Y_n)$ be an objective function, and consider the filtration $(\mathcal{F}_n)$, $\mathcal{F}_0 = \{\emptyset, \Omega\}$, $\mathcal{F}_k = \sigma(X_1, \ldots, X_k)$. Let $\mathbf{E}_k Z$ denote the conditional expectation of Z with respect to $\mathcal{F}_k$ (with $\mathbf{E} = \mathbf{E}_0$) and set $d_k = \mathbf{E}_k Y - \mathbf{E}_{k-1} Y$. Then it is well known that $(d_k, \mathcal{F}_k)$ is a martingale difference sequence, and that we have $Y - \mathbf{E}Y = \sum_{k=1}^{n} d_k$. In applications one is often interested in the degree of concentration of Y around $\mathbf{E}Y$, while concentration of measure questions have played a fundamental role in theoretical probability for quite some time. In a landmark paper, Talagrand (1996) asserts that "the idea of concentration of measure (which was discovered by V. Milman) is arguably one of the great ideas of analysis of our times," while Steele (1997), in a monograph that moves easily from the theoretical to the concrete and back, states that "The number and depth of new applications continues to build, and the concentration of measure phenomenon is emerging as one of the central ideas of probability - or perhaps of analysis."

[†] Supported in part by NSF grant DMS 9619889
[‡] Supported in part by NSF grant DMS 9401345
AMS 1991 Subject Classification: 60C05, 60D05, 60E15, 05C80, 60G42

One of the first (and key) methods used towards gaining an understanding of the concentration of Y was the method of bounded differences, also known as the Azuma (1967)-Hoeffding (1963) inequality:

Theorem 1.1. *(Azuma-Hoeffding) For all $\lambda > 0$,*

$$\mathbf{P}(|Y - \mathbf{E}Y| \geq \lambda) \leq 2\exp\left\{-\frac{\lambda^2}{2\sum \|d_i\|_\infty^2}\right\}. \tag{1.1}$$

See, e.g. Steele (1997) pp. 4 - 5 for a development that only requires that the d_i's satisfy a product identity:

$$\mathbf{E}\prod_{i=1}^{n}(a_i + b_i d_i) = \prod_{i=1}^{n} a_i.$$

In any event, it is worth noting that the proof of the Azuma-Hoeffding inequality is quite uncompromising about the fact that one needs to use the ℓ_∞ (i.e. the worst - case scenario, in applications) bound in (1.1). The guiding principle behind (1.1) is simple: note first that letting (X_i^*) be an independent copy of (X_i), we have

$$\mathbf{E}_{i-1}Y(X_1, \ldots, X_n) = \mathbf{E}_i Y(X_1, \ldots, X_{i-1}, X_i^*, X_{i+1} \ldots, X_n),$$

so that d_i can be written as a single conditional expectation as follows:

$$d_i = \mathbf{E}_i\big(Y(X_1, \ldots, X_n) - Y(X_1, \ldots, X_{i-1}, X_i^*, X_{i+1} \ldots, X_n)\big),$$

and thus

$$\begin{aligned}\|d_i\|_\infty &= \|\mathbf{E}_i\{Y(X_1, \ldots, X_n) - Y(X_1, \ldots, X_{i-1}, X_i^*, X_{i+1} \ldots, X_n)\}\|_\infty \\ &\leq \|Y(X_1, \ldots, X_n) - Y(X_1, \ldots, X_{i-1}, X_i^*, X_{i+1} \ldots, X_n)\|_\infty.\end{aligned}$$

The philosophy behind the method of bounded differences is that the change in the value of Y resulting from a change in a single point should be small. More specifically, a useful concentration statement can be made about Y whenever, say, $\|d_i\|_\infty = o(\mathbf{E}Y/\sqrt{n})$ as $n \to \infty$, for all i. For example, in the case when $\mathbf{E}Y \approx c \cdot n$, (1.1) would imply

$$\mathbf{P}(|Y - cn| \geq \lambda) \leq 2\exp\frac{-\lambda^2\phi(n)}{2n^2c^2},$$

for some $\phi(n) \to \infty$ so that one would have Y concentrated in an interval of length $o(n)$ around $\mathbf{E}Y$. The method of bounded differences is thus somewhat of a misnomer since (1.1) may make a useful statement even when these differences tend to ∞ with n. Surveys of the method of bounded differences from a combinatorial viewpoint may be found in McDiarmid (1989), Bollobás (1987, 1990), and Alon and Spencer (1992), while Steele's monograph (1997) provides the reader with a comprehensive view of concentration phenomena with a range of examples drawn mainly from combinatorial optimization.

It is important to note, however, that the $\|d_i\|_\infty$ may grow too rapidly for the Azuma-Hoeffding inequality to provide any meaningful information: we give three examples of this phenomenon below, and remark that such situations abound in the fields of random graph theory and probabilistic combinatorics (among others).

Example 1.2. *Random nearest neighbor graphs.* (See Bass and Godbole (1997) for references.) Let $X_1, \ldots, X_n$ be uniformly distributed on $[0,1]^d$; for simplicity, we consider the case $d = 2$. The X_i's constitute the vertices of the graph, while its edge set consists of undirected edges (X_i, X_j) where

$$\tau(X_i, X_j) = \min\{\tau(X_i, X_k) : \ k \neq i\},$$

or

$$\tau(X_i, X_j) = \min\{\tau(X_j, X_k) : \ k \neq j\},$$

(τ is Euclidean distance.) Now let

$$Y = \sum_{\{\text{edges } (X_i, X_j)\}} \tau(X_i, X_j).$$

It was proved in Bass and Godbole (1997) that

$$\mathbf{E}Y \approx c \cdot \sqrt{n},$$

for a well specified constant (similar results were proved for all d and for random kth nearest neighbor graphs, $k \geq 2$.) In any event, it is fairly evident that for $d = 2$ and $k = 1$ the degree of any vertex is at most six, so that a relocation of any point can lead to a change in Y of some constant magnitude. The Azuma-Hoeffding inequality thus yields, for some constant A,

$$\mathbf{P}(|Y - c \cdot \sqrt{n}| \geq \lambda) \leq 2e^{-\frac{\lambda^2}{An}},$$

which does not even yield a strong law for Y. This example illustrates, moreover, that the inequalities we survey in this paper will be of use even when the martingale differences *are* bounded by a constant.

Example 1.3. *Random sphere of influence graphs.* (See Chalker et al. (1997) for references.) Here again, we confine ourselves to the case $d = 2$. For (X_i) as in Example 1.2, the random sphere of influence graph is constructed as follows: for each i, let B_i be a ball around X_i with radius equal to $\min\{\tau(X_i, X_j) : i \neq j\}$ (i.e. the distance from X_i to its closest neighbor). Draw an edge between X_j and X_k if and only if the balls B_j and B_k have non-empty intersection. The quantity of interest, Y, is the total number of edges in the graph. It is known (see e.g. Michael and Quint (1994)) that we always have $c \cdot n \leq Y \leq C \cdot n$ for some absolute constants c and C. Chalker et al. (1997) showed that (in the case $d = 2$),

$$\mathbf{E}Y \approx (1 + \pi/4)n.$$

Now, a relocation of one point can lead to a radical change in the number of edges – up to $\alpha \cdot n$, which is comparable to the maximal size of the graph. (This is seen by considering the following situation: $n - 1$ points are evenly distributed on a circle and the remaing point is in the center of that circle (so that the degree of the center is $n-1$ while the degree of each point on the circle is 5.) Relocating the central point to the boundary of the circle will cause a change of order n.) Therefore, (1.1) yields only a pointless estimate

$$\mathbf{P}(|Y - c\cdot n| \geq \lambda) \leq 2e^{-\frac{\lambda^2}{An^3}}.$$

Example 1.4. *Wiener index of a random graph.* (See Godbole et al. (1997) for references.) Given a connected graph G, we define its Wiener index $W = W(G)$ as the sum $\sum_{i<j} d(i,j)$ of all the distances between pairs of vertices (i,j). If G is not connected then $W(G)$ is not finite, so, in an effort to bypass this difficulty, we set (for example) $d(i,j) = n$ for any (i,j) between which no path exists. Next, consider the random graph $G(n,p)$ with this modified definition of distance between pairs of vertices. We shall assume that $p \geq (1+o(1))\log n/n$ so as to guarantee that G is connected with high probability. It was shown in Godbole et al. (1997) that

$$\begin{aligned}\mathbf{E}(W) &\approx \binom{n}{2}\left[J + e^{-(np)^J/n}\right]\\ &\approx \binom{n}{2}\left[J + e^{-1}\right]\end{aligned}$$

where $J = \lceil \frac{\log n}{\log np} \rceil$; the proofs used the Janson and extended Janson inequalities (Alon and Spencer (1992)). Now consider the following scenario: Assume without loss of generality that $n \equiv 0(\text{mod} 3)$. Imagine that the the random graph aligns itself into two $K_{n/3}$'s, with the only possible connections between two vertices a and b (one in each of the complete components) being through (i) a path of length 2, of the form $a-c-b$ or (ii) a path of length $(n/3)+1$ of the form $a-d_1-d_2-\ldots-d_{n/3}-b$. Let us "resample" the edge $i=(a,c)$, finding that it is now absent. It is now clear that $W(G)$ increases by a quantity of order n^3. Thus (1.1) yields

$$\mathbf{P}(|W - \mathbf{E}(W)| > \lambda) \leq 2e^{-\lambda^2/An^8};$$

since $\mathbf{E}(W) \leq n^3/2$, this fact is of little value.

Thus, the Azuma-Hoeffding inequality does not get us too far in any of the above cases. A natural thought, however, is that if the most pessimistic (i.e. ℓ_∞) bound is not too adequate, and our d_k's are much smaller "most of the time," then we ought to be able to find much better bounds on the tail probabilities than those obtained by a direct application of the Azuma-Hoeffding inequality. This idea is not new, of course. In the context of combinatorics it was used for the first time, it seems, by Shamir and Spencer (1987) in their study of the chromatic number of sparse random graphs. (The key new ingredient in the Shamir-Spencer argument was that their d_k were, roughly speaking, bounded by the degree of the kth vertex in a random graph $G(n,p)$. The latter, being a binomial random variable, is tightly concentrated around its mean; a quantity *much* smaller than the ℓ_∞ bound, if p is small.) We find it a bit ironic, therefore, that their bound on the deviation of the chromatic number of a sparse graph has been quickly improved upon by Alan Frieze whose proof was based on the method of bounded (!) differences (see e.g., McDiarmid (1989)).

As the number of situations that do not submit to the method of bounded differences has grown considerably in recent years, in this paper we attempt to survey some of the fairly new inequalities that go beyond that method. Here is a typical result, taken from Chalker et al. (1997):

Theorem 1.5. *Let (d_i) be a martingale difference sequence. Then for each $\lambda > 0$, all sequences of positive numbers (w_i), and any $0 < \varepsilon < 1$,*

$$\mathbf{P}\left(\left|\sum_{i=1}^{n} d_i\right| > \lambda\right) \le 2\exp\left\{-\varepsilon^2\lambda^2 / \left(8\sum_{i=1}^{n} w_i^2\right)\right\} + \left(1 + \frac{\|d^*\|_\infty}{\lambda(1-\varepsilon)}\right)\sum_{i=1}^{n}\mathbf{P}(|d_i| > w_i), \tag{1.2}$$

where $\|d^*\|_\infty = \sup_i \|d_i\|_\infty$.

The main appeal of (1.2) is the simplicity of its statement and proof, its generality, and the ease of application it enjoys. In the next section we will give an elementary proof of (1.2). In Section 3 we will show how (1.2) can lead to improved concentration results. Since (1.2) is just one of a number of inequalities of this type, in Section 4 we will survey some of these results and see how Theorem 1.5 compares with other inequalities of the same genre. In particular, specific comparisons will be made with the inequalities of Kim (1995); Shamir and Spencer (1987); and McKay and Wormald (1997), while the related results of Pinelis; Kesten; de la Peña; Burkholder and Gundy; and others will be surveyed. We will also briefly discuss Talagrand's isoperimetric inequalities.

2. A method of "controllable" martingale differences

Proof of Theorem 1.5. For any λ and (w_k) as above we have

$$\mathbf{P}\left(\left|\sum d_k\right| \ge \lambda\right) \le \mathbf{P}\left(\bigcup\{|d_k| > w_k\}\right) + \mathbf{P}\left(\left|\sum d_k I(|d_k| \le w_k)\right| \ge \lambda\right).$$

By adding and subtracting the conditional expectation $\mathbf{E}_{k-1} d_k I(|d_k| \le w_k)$ inside the second probability, we see that the second term on the right hand side above is no larger than

$$\mathbf{P}\left(\left|\sum d_k I(|d_k| \le w_k) - \mathbf{E}_{k-1} d_k I(|d_k| \le w_k)\right| \ge \varepsilon\lambda\right) + \mathbf{P}\left(\left|\sum \mathbf{E}_{k-1} d_k I(|d_k| \le w_k)\right| \ge (1-\varepsilon)\lambda\right). \tag{2.1}$$

Note that a summand appearing in the first probability is a martingale difference bounded by $2w_k$. Therefore, applying (1.1) to the first term and Chebyshev's inequality to the second, we get that (2.1) is bounded by

$$2\exp\left\{-\frac{\varepsilon^2\lambda^2}{8\sum w_k^2}\right\} + \frac{1}{(1-\varepsilon)\lambda}\mathbf{E}\left|\sum \mathbf{E}_{k-1} d_k I(|d_k| \le w_k)\right|. \tag{2.2}$$

The crucial observation is the following obvious fact: since the d_k's are martingale differences, $\mathbf{E}_{k-1}d_kI(|d_k| \le w_k) = -\mathbf{E}_{k-1}d_kI(|d_k| > w_k)$, and therefore,

$$|\mathbf{E}_{k-1}d_kI(|d_k| \le w_k)| = |\mathbf{E}_{k-1}d_kI(|d_k| > w_k)| \le \|d_k\|_\infty \mathbf{E}_{k-1}I(|d_k| > w_k).$$

Consequently, we can further upper bound the second term in (2.2) by

$$\frac{1}{(1-\varepsilon)\lambda}\mathbf{E}\sum|\mathbf{E}_{k-1}d_kI(|d_k| > w_k)| \le \frac{\|d^*\|_\infty}{(1-\varepsilon)\lambda}\sum \mathbf{P}(|d_k| > w_k).$$

This proves (1.2).

Remarks.

(i) The first proof of Theorem 1.5 due to the authors involved the notion of decoupling (see e.g. Kwapień and Woyczyński (1992) and the references therein for more information.) Iosif Pinelis (personal communication) used similar methods to give a shorter proof. The elementary proof presented above is shorter than either of these proofs and avoids appealing to decoupling techniques.

(ii) In the case when the martingale differences are conditionally symmetric (i.e. d_k and $-d_k$ have the same conditional distributions, given $\mathcal{F}_{k-1}$), it is easy to see that we get the sharper estimate

$$\mathbf{P}\left(|\sum d_k| \ge \lambda\right) \le \mathbf{P}\left(\bigcup |d_k| > w_k\right) + 2\exp\left\{-\frac{\lambda^2}{2\sum w_k^2}\right\}.$$

(One just notices that if the (d_k) are conditionally symmetric, then $(d_kI(|d_k| \le w_k))$ are martingale differences, so centering is not needed.) In general, however, the extra weight $1 + (\|d^*\|_\infty/(1-\varepsilon)\lambda)$ is necessary in (1.2), as the following example shows: Let $X_1, \ldots, X_n$ be i.i.d. random variables with

$$\mathbf{P}(X_1 = -\sqrt{n}) = \frac{n^3}{\sqrt{n}+n^3} = 1 - \mathbf{P}(X_1 = n^3).$$

Set $\lambda = n^{3/2}-1$, and $w_k = n/\sqrt{\log n}$. Then it is easy to verify that $\mathbf{P}(|\sum X_k| \ge \lambda) = 1$, $\exp\{-\varepsilon^2\lambda^2/8\sum w_k^2\} = n^{-\alpha}$ for some $\alpha > 0$, and $\sum \mathbf{P}(|X_k| > w_k) \approx n \cdot (1/n^{5/2}) = n^{-3/2}$. Finally, $\|d^*\|_\infty/((1-\varepsilon)\lambda) \approx n^{3/2}$ for ε close to zero.

(iii) In a recent paper, Pinelis (1997, Theorem 5.4) proved the following sharpening of Azuma's inequality: if (d_k) is a martingale difference sequence with $|d_k| \le a_k$, then for every positive λ,

$$\mathbf{P}(|\sum d_k| \ge \lambda) \le \frac{4e^3}{9}\left(1 - \Phi(\lambda/A)\right),$$

where Φ is the distribution of a standard Gaussian random variable, and $A = (\sum a_k^2)^{1/2}$. Accordingly, the first term on the right hand side in (1.2) can be replaced by $(4e^3/9)\left(1 - \Phi\left(\varepsilon\lambda/(\sum 4w_k^2)^{1/2}\right)\right)$. Note that the ratio of this last quantity to the first term on the right hand side in (1.2) goes to zero, as $\varepsilon\lambda/(\sum 4w_k^2)^{1/2} \to \infty$.

3. Some applications

In this section we will briefly illustrate the power of (1.2) with reference to Examples 1.2 and 1.3 discussed above. More applications to combinatorial and combinatorial optimization problems can be found in the references corresponding to most of the inequalities discussed below. In order to avoid getting into too many technicalities we will just extract key ideas from the corresponding papers, rather than reprove those results here. We refer the reader to Shamir and Spencer (1987), Kim (1995), and McKay and Wormald (1997) for more details. Let us begin by returning to Example 1.3:

Example 1.3. Let r_i be the radius of the "sphere of influence" of X_i, and denote by $B_r(x)$ the ball of radius r centered at x. The crucial observation here is that the change in the size of the graph resulting from relocating just one point X_i is, roughly speaking, no more than a constant multiple of the number of remaining points that fall into $B_{2r_i}(X_i)$ (see Chalker et al. for precise statements). Since with overwhelming probability all radii are bounded by a constant times $\sqrt{\log n/n}$ (in two dimensions), we see that our d_i is bounded by a multiple of a $\text{Bin}(n,p)$ random variable, where p is proportional to $c\log n/n$ (the volume of a ball with radius $\sqrt{\log n/n}$. Therefore, although $\|d_i\|_\infty \approx n$, d_i is heavily concentrated around $c \cdot \log n$. Applying (1.2) with $w_k = O(\log n)$, $\lambda = O(\sqrt{n\log^3 n})$, and using standard bounds for tails of binomial random variables, we obtain the following bound: for every $\alpha > 0$, there exists $C = C_\alpha$ such that for every $n \geq 1$ we have:

$$\mathbf{P}(|Y - \mathbf{E}Y| \geq C\sqrt{n}\log^{3/2} n) \leq \frac{1}{n^\alpha}. \tag{3.1}$$

There is no reason to believe that this bound is optimal, but it is the best that is known. Next consider Example 1.2:

Example 1.2. The benefits of (1.2) are even more readily apparent here: We have $\|d_i\|_\infty \leq 6C_i$ as noted above, where C_i is the maximal distance to X_i's nearest neighbour(s). Now C_i can be as large as $\sqrt{2}$, say, but with overwhelming probability (as seen with sphere of influence graphs), C_i does not exceed a constant times $\sqrt{\log n/n}$ (once again in 2 dimensions.) It follows that for each $\alpha > 0$, there exist A, B such that (1.2) yields (on choosing $w_i = A\sqrt{\log n/n}$ and $\lambda = B\log n$)

$$\mathbf{P}(|Y - \mathbf{E}(Y)| > B\log n) \leq \frac{1}{n^\alpha}. \tag{3.2}$$

The concentration here is far tighter than in Example 1.3; this benefit decreases gradually with the dimension, however. See Bass and Godbole (1997) for details. As we will see in Section 5 below, (3.2) can be improved upon by using Talagrand's (1994) method.

Example 1.4 (Wiener indices of random graphs) can be handled in a similar fashion, but the analysis gets far more complicated. See Godbole et al. (1997) for detailed results on concentration in this case, where it is proved, e.g., that W is in an interval of length $n\sqrt{\log n}J^2(np)^{J-1}$ around $\mathbf{E}(W)$ with high probability.

4. Some comparisons and related inequalities

In this section we survey some of the related inequalities of a similar type. Let us begin with the closest one – due to Shamir and Spencer (1987, Theorem 7). More precisely, we present a slight variation of the original Shamir–Spencer inequality that can be found in McKay and Wormald (1997, Lemma 3.1) and was used in their study of the joint distribution of the degrees of vertices in random graphs.

Theorem 4.1. *Let (d_k) be a martingale difference sequence such that, for all i, $|d_i| \le c$ with probability at least $1 - r$, and $|d_i| \le K$ always. Then, for any $t \ge 0$, $n \ge 1$ and $0 < p < 1$ we have*

$$\mathbf{P}\left(|\sum_{k=1}^{n} d_k| > t(c + Kp)\sqrt{n} + nKp\right) \le nr(1 + 1/p) + 2e^{-t^2/2}. \tag{4.1}$$

We claim that (1.2) is stronger than (4.1). Reparametrizing slightly and letting $\varepsilon = 1/2$, (1.2) can be rewritten, for any $t > 0$, as

$$\mathbf{P}\left(|\sum d_k| > 4tc\sqrt{n}\right) \le 2e^{-t^2/2} + nr\left(1 + \frac{K}{2tc\sqrt{n}}\right);$$

this provides for a good comparison between (1.2) and (4.1) since the exponential terms are now the same in both cases. Now, if we are to make a reasonable comparison, we must have

$$t(c + Kp)\sqrt{n} + nKp = 4tc\sqrt{n},$$

or

$$p = \frac{3tc}{tK + \sqrt{n}K}.$$

Now this choice reduces (4.1) to

$$\mathbf{P}\left(|\sum d_k| > 4tc\sqrt{n}\right) \le 2e^{-t^2/2} + nr\left(1 + \frac{tK + \sqrt{n}K}{3tc}\right);$$

it is elementary to check that

$$\frac{tK + \sqrt{n}K}{3tc} > \frac{K}{2tc\sqrt{n}},$$

verifying our claim.

Another interesting result of the above kind is an inequality due to Kim (1995, Lemma 5.2)

Theorem 4.2. *Let $(d_k, \mathcal{F}_k)$ be a martingale difference sequence and suppose that for each $k \ge 1$ there exists $A_k \in \mathcal{F}_k$ such that*

$$I_{A_{k-1}^c}\mathbf{E}_{k-1}e^{wd_k} \le C_k, \tag{4.2}$$

for some $C_k \geq 1$. Then

$$\mathbf{P}\left(|\sum_{k=1}^{n} d_k| \geq \lambda\right) \leq 2\left\{e^{-w\lambda}\prod_{k=1}^{n} C_k + \mathbf{P}(\bigcup_{k=0}^{n-1} A_k)\right\}. \tag{4.3}$$

Kim's proof is based on Doob's optional sampling theorem (the stopping time being the first time of entering $\bigcup_{k=0}^{n-1} A_k$.) In most applications, the C_k's would be exponential functions, and thus the first term in (4.3) would give some sort of exponential bound, by following the usual steps. This inequality has potentially a very wide range of applications, for it does not require *any* bound on the ℓ_∞ – norm of differences, just a bound on their conditional moment generating functions. The price to pay is that (4.2) involves information on the *conditional* distributions of the d_k's. This typically requires some information on the joint distributions of the d_k's, which often is harder to get. However, *if* (4.2) can be checked, then (4.3) will usually give a better bound than (1.2). For example, suppose, as in Theorem 1.5 that $\mathbf{P}(|d_k| > w_k)$ is small. Now, *suppose* that $A_{k-1} \in \mathcal{F}_{k-1}$ can be found such that $\{|d_k| > w_k\} \subset A_{k-1}$, and that its measure is comparable to that of $\{|d_k| > w_k\}$, say $\mathbf{P}(A_{k-1}) \leq c\mathbf{P}(|d_k| > w_k)$. Then,

$$I_{A_{k-1}^c}\mathbf{E}_{k-1}e^{wd_k} \leq e^{w^2 w_k^2/2},$$

and Kim's result gives

$$\mathbf{P}(|\sum d_k| \geq \lambda) \leq 2\left\{c\sum \mathbf{P}(|d_k| > w_k) + \exp\left\{-\frac{\lambda^2}{2\sum w_k^2}\right\}\right\},$$

which is better than (1.2). In the applications considered by Kim, d_k assumed only two values: c_k on B_k and c_k' on its complement for a certain B_k independent of $\mathcal{F}_{k-1}$. In that case, (4.3) gives the following

Corollary. *(Kim (1995, Lemma 5.3)). Suppose that there is a set $I \subset \{1\dots,n\}$ such that*

$$\begin{aligned} |d_k|I_{A_{k-1}^c} &\leq c_k I_{B_k} + c_k\mathbf{P}(B_k), \qquad &\forall k \in I,\\ d_k I_{A_{k-1}^c} &\leq c_k, \qquad &k \in I^c, \end{aligned} \tag{4.4}$$

for some constants (c_k), $A_k \in \mathcal{F}_k$ and B_k independent of $\mathcal{F}_{k-1}$. Then for each $w > 0$ such that $wc^* = w\max_{k\in I} c_k \leq 1/6$,

$$\begin{aligned} P\left(\sum d_k \geq \lambda\right) \leq \mathbf{P}&\left(\bigcup_{k=0}^{n-1} A_k\right)\\ &+ \exp\left\{-w\left(\lambda - \sum_{k\in I^c} c_k\right) + 3w^2\sum_{k\in I} c_k^2\mathbf{P}(B_k)\right\}. \end{aligned} \tag{4.5}$$

The main drawback seems to be that one needs to control d_k on a *predictable* (i.e. $\mathcal{F}_{k-1}$ -measurable) set A_{k-1}^c. This may not always be easy to accomplish. In case it can be done, one can write

$$\mathbf{P}\left(|\sum d_k| > \lambda\right) \le \mathbf{P}\left(|\sum d_k I_{A_{k-1}}| > 0\right) + \mathbf{P}\left(|\sum d_k I_{A_{k-1}^c}| > \lambda\right)$$
$$\le \mathbf{P}\left(\bigcup_{k=0}^{n-1} A_k\right) + \mathbf{P}\left(|\sum d_k I_{A_{k-1}^c}| > \lambda\right).$$

Under the assumptions of the corollary, $(d_k I_{A_{k-1}^c})$ is a martingale difference sequence, and the second probability can be bounded in various ways. For example, assuming for simplicity that $I^c = \emptyset$ since we are using trivial bounds on I^c anyway, and that $\lambda \le (1/c^*) \sum c_k^2 \mathbf{P}(B_k)$, Kim's bound gives, by minimizing over $w \le 1/(6c^*)$ in (4.5)

$$\mathbf{P}\left(\bigcup_{k=0}^{n-1} A_k\right) + 2\exp\left\{-\frac{\lambda^2}{12\sum c_k^2 \mathbf{P}(B_k)}\right\}, \tag{4.6}$$

while (1.2) used with $w_k = c_k \mathbf{P}(B_k)$, $\|d^*\|_\infty = 2c^*$, and $\varepsilon = 1/2$ yields (the usually weaker) bound

$$\mathbf{P}\left(\bigcup_{k=0}^{n-1} A_k\right) + \left(1 + \frac{4c^*}{\lambda}\right) \sum \mathbf{P}(B_k) + 2\exp\left\{-\frac{\lambda^2}{32\sum c_k^2 \mathbf{P}^2(B_k)}\right\},$$

but with no restrictions on λ.

Still more possibilities are provided by other inequalities given below, utilizing the fact that the assumption (4.4) gives the bound

$$\mathbf{E}_{k-1}(d_k I_{A_{k-1}^c})^2 \le c_k^2 \mathbf{E}_{k-1}(I_{B_k} + \mathbf{P}(B_k))^2 \le 4c_k^2 \mathbf{P}(B_k), \tag{4.7}$$

by independence. Controlling the conditional variance opens several possibilities. For example, using an inequality

$$\mathbf{P}\left(|\sum d_k| \ge \lambda\right) \le 2\exp\left\{-\frac{\lambda}{2\|d^*\|_\infty} \operatorname{arcsinh}\left(\frac{\lambda \|d^*\|_\infty}{2\sum \|\mathbf{E}_{k-1} d_k^2\|_\infty}\right)\right\}, \tag{4.8}$$

proved by Johnson, Schechtman and Zinn (1985), we get the upper bound

$$\mathbf{P}\left(\bigcup_{k=0}^{n-1} A_k\right) + 2\exp\left\{-\frac{\lambda}{4c^*} \operatorname{arcsinh}\left(\frac{\lambda c^*}{4\sum c_k^2 \mathbf{P}(B_k)}\right)\right\}.$$

While this last bound looks weaker than (4.6) this is not really the case; it does not impose any restrictions on λ and is comparable to (4.6) for the range of λ's specified there. Inequality (4.8) is a martingale extension of Prokhorov's inequality for sums of independent random variables (see e.g. Stout (1974)). It was subsequently strengthened by Hitczenko (1990a) and independently by Leventhal (1989), who showed that $\sum \|\mathbf{E}_{k-1} d_k^2\|_\infty$ could be replaced by the smaller quantity $\|\sum \mathbf{E}_{k-1} d_k^2\|_\infty$. Although this improvement was of some significance (see the above two papers for examples), in practical applications controlling the latter quantity is typically harder than the former so (4.8) seems to be easier to use. It is perhaps worth noticing that the inequality (4.8) has the convenience of giving a subgaussian bound when $\|d^*\|_\infty$ is much smaller than $\sum \|\mathbf{E}_{k-1} d_k^2\|_\infty$ and the "Poissonian" bound $\exp\{-\lambda \log(1+\lambda)\}$ when they are comparable (this just follows from the fact that $\operatorname{arcsinh}(x) \approx x$ when x is small and $\approx \log(1+x)$ when x is large.)

Another way of controlling $\mathbf{E}_{k-1} d_k^2$ was proposed by Kesten (1993). His result is as follows, where the constants K and C are "generic" and may vary from line to line:

Theorem 4.3. *Suppose that (d_k) satisfy $d_k \le c$ for some constant c and*

$$\mathbf{E}_{k-1} d_k^2 \le \mathbf{E}_{k-1} U_k,$$

for a sequence (U_k) of $\mathcal{F}$ – measurable random variables. Suppose further that for $\lambda_0 \ge e^2 c^2$,

$$\mathbf{P}\left(\sum U_k \ge \lambda\right) \le C \exp\{-C\lambda\}, \qquad \lambda \ge \lambda_0. \tag{4.9}$$

Then,

$$\mathbf{P}\left(\left|\sum d_k\right| \ge \lambda\right) \le C(1+1/\lambda_0) \exp\left\{-K \frac{\lambda}{\sqrt{\lambda_0} + \lambda^{1/3}}\right\},$$

so that for $\lambda \le C\lambda_0^{3/2}$

$$\mathbf{P}\left(\left|\sum d_k\right| \ge \lambda\right) \le C(1+1/\lambda_0) \exp\left\{-K \frac{\lambda}{\sqrt{\lambda_0}}\right\}.$$

Here again, one needs to know something about the joint distribution of the U_k's to check (4.9). But, under Kim's assumption, for example, (4.7) and the independence of $\mathcal{F}_{k-1}$ and B_k imply that U_k may be taken to be $4c_k^2 I_{B_k}$, and that the (U_k) are independent. Therefore, one can try to control the left hand side of (4.9) through the Hoeffding inequality.

Similar in spirit are inequalities known as good λ inequalities. They were introduced by Burkholder and Gundy (1970) as a tool for proving moment inequalities for martingales. We refer to Burkholder (1973) for a detailed account. Basically, a good λ inequality aims at bounding the probability that a martingale exceeds a certain level, and at the same time that a certain functional of the martingale (typically, the square function or a conditional square function) does not exceed another level. Although most of the bounds given in Burkholder' paper were too weak to give good exponential bounds, it was realized by various researchers (see e.g. Hitczenko (1990b) for more specific references) that the method could be used to yield fairly good results. For example, one has:

Theorem 4.4. *Let $\beta > 1$, $\delta > 0$, and let (d_k) be a martingale difference sequence such that $|d_k| \le u_k$. (It is enough to assume that u_k is $\mathcal{F}_{k-1}$ measurable.) Then for any $\lambda > 0$ we have*

$$\begin{aligned}&\mathbf{P}\left(\left|\sum d_k\right| \ge \beta\lambda,\ \left(\sum \mathbf{E}_{k-1} d_k^2\right)^{1/2} \vee \max u_k \le \delta\lambda\right) \\ &\le 2\exp\left\{-\frac{\beta-1-\delta}{2\delta} \log\left(1+\frac{\beta-1-\delta}{2\delta}\right)\right\} \mathbf{P}\left(\left|\sum d_k\right| \ge \lambda\right),\end{aligned}$$

implying that

$$\mathbf{P}\left(|\sum d_k| \geq \beta\lambda\right) \leq \mathbf{P}\left(\max u_k \geq \delta\lambda\right) + \mathbf{P}\left(\sum \mathbf{E}_{k-1} d_k^2 \geq \delta^2\lambda^2\right)$$
$$+ 2\exp\left\{-\frac{\beta-1-\delta}{2\delta}\log\left(1+\frac{\beta-1-\delta}{2\delta}\right)\right\}\mathbf{P}\left(|\sum d_k| \geq \lambda\right).$$

Results along similar lines were obtained by Freedman (1975) whose Theorem (1.6) implies, assuming that $|d_k| \leq c$,

$$\mathbf{P}(|\sum d_k| \geq \beta) \leq \mathbf{P}(\sum \mathbf{E}_{k-1} d_k^2 \geq \delta^2)$$
$$+ \exp\left\{-\frac{\delta^2}{c^2}\left((1+\frac{c\beta}{\delta^2})\log(1+\frac{c\beta}{\delta^2}) - \frac{c\beta}{\delta^2}\right)\right\}.$$

A recent paper by de la Peña (1997) contains more inequalities of similar type, and we refer the reader there.

It is perhaps worth noting that many classical exponential inequalities may be improved upon by using techniques associated with decoupling. We would like to avoid getting into details, and we will give just one result obtained by Pinelis (1994), referring the reader to that paper for an illustration of this method. We give a formulation for real valued variables:

Theorem 4.5. *(Extension of the Fuk-Nagaev inequality). Let (d_k) be a martingale difference sequence, and let $p \geq 2$. Then for some absolute constants C, δ, K and M, and any $\lambda > 0$ we have*

$$\mathbf{P}\left(|\sum d_k| \geq \lambda\right) \leq C\left\{\mathbf{P}(d^* \geq K\lambda) + \mathbf{E}\exp\left\{-\frac{\delta\lambda^2}{\sum \mathbf{E}_{k-1} d_k^2}\right\} + M\frac{\sum \mathbf{E}|d_k|^p}{\lambda^p}\right\}.$$

Here, of course, the improvement is in the middle term, which has expectation "outside", but one would need to find a way of controlling this expectation. As in Kim's inequality, no bound on the ℓ_∞ norm is needed, only a bound on the tails of the maximum of increments.

Finally, we mention three papers that contain results in the general spirit of this survey: Khoshnevisan (1996) and Dembo (1996) study inequalities for continuous martingales and those with bounded jumps respectively, while Alon et al. (1997) provide a game-theoretic version of a martingale inequality with applications to near-perfect matchings in regular simple hypergraphs.

5. Talagrand's concentration of measure

This paper focuses on martingale inequalities that go one step beyond the method of bounded differences. As is by now well recognized, in many cases where this method can be applied, Talagrand's isoperimetric methods (1994, 1996) do at least as well and in most of these instances perform significantly better. Talagrand's methods apply to functions $Y(X_1, \ldots, X_n)$ that depend "smoothly" on the coordinates (which is precisely not the case in our considerations). Yet, in each case, outside a small exceptional set our differences are reasonably well behaved. It is, therefore, quite possible that Talagrand's methods could be adapted to situations discussed in this paper, and, in fact it has been done in some situations like that (see e.g. Talagrand (1994, sec. 11.4)). Perhaps the best general method that would allow one to decide whether concentration of measure can be applied successfully, is to check one of the regularity conditions discussed e.g. in Talagrand (1994). However, this may not always be so easy. For example, it is not clear, at this time, what kind of regularity conditions sphere of influence graphs satisfy. On the other hand, it is easy to show that the length of the nearest neighbor graph has enough regularity to have subgaussian tails, thus improving upon (3.2).

Example 1.2 revisited. If Y is the length of the nearest neighbor graph, and $\mathbf{m}(Y)$ is a median of Y, then for any $\lambda > 0$,

$$\mathbf{P}(|Y - \mathbf{m}(Y)| > \lambda) \leq C\exp\{-\lambda^2/C\},$$

for some absolute constant C. It follows that the concentration inequality (3.2) can be improved to

$$\mathbf{P}(|Y - \mathbf{m}(Y)| > B\sqrt{\log n}) \leq \frac{1}{n^\alpha}.$$

This can be seen by following Talagrand's (1994) argument for a minimum spanning tree, so we will only give a brief sketch of a proof. It is convenient to change notation; for a finite set $F \subset [0,1]^2$ we will denote by $Y(F)$ the length of nearest neighbor graph built on vertices F, and we will let $|F|$ be the cardinality of F. Our intent is to use Theorem 11.3.2 in Talagrand (1994), and to this end it suffices to check that Y satisfies the following regularity property (Lemma 11.3.1 in the same paper.)

Lemma 5.1. *Let, for $k \geq 1$, $\mathcal{S}_k$ be a family of 2^{2k} subsquares of $[0,1]^2$ with vertices at the points $(j/2^k, m/2^k)$, $0 \leq j, m \leq 2^k$. Let $S \in \mathcal{S}_k$, and suppose that F and G are finite subsets of $[0,1]^2$ such that $G \subset S$, and $F \cap S' \neq \emptyset$ for every $S' \in \mathcal{S}_{k-1}$ which is within distance of 2^{-k+5} of S. Then*

$$|Y(F \cup G) - Y(F)| \leq C2^{-k}\sqrt{|G|},$$

for some absolute constant C.

Proof. The proof uses the same reasoning as for a minimum spanning tree, but is actually much simpler: first we note that Y is subadditive, i.e. $Y(F \cup G) \leq Y(F) + Y(G)$, and for such functions we always have a bound (see Rhee (1983)) $Y(G) \leq C\mathrm{diam}(G)\sqrt{|G|}$. This implies

$$Y(F \cup G) \leq Y(F) + C2^{-k}\sqrt{|G|}.$$

To show that

$$Y(F) \leq Y(F \cup G) + C2^{-k}\sqrt{|G|},$$

consider a nearest neighbor graph for $F \cup G$ and remove all edges incident to points in G. As a result, some points in F lose their nearest neighbor and will have to find a new one. But there are at most $6|G|$ such points (since degree of any vertex in the nearest neighbor graph is no more than 6) and each of them is at distance at most 2^{-k+3} of S. (For otherwise, if for such a point x, $d(x,S) > 2^{-k+3}$, then, by the assumption that F meets every square in $\mathcal{S}_{k-1}$ at distance at most 2^{-k+5} of S, we would have $d(x, F \setminus \{x\}) \leq 2^{-k+3}$, and x would not be connected to G.) So, we have a set of at most $6|G|$ points with diameter at most $K2^{-k}$. By Rhee's result again, we can built a nearest neighbor graph on those points with total length no more than $K2^{-k}\sqrt{|G|}$, and clearly, if we combine the length of this nearest neighbor graph with the length of edges that were not removed, we get at least $Y(F)$.

We close with one more result of Talagrand that is very much in line with our considerations:

Theorem 5.2. *(Talagrand (1996, Theorem 6.6)). Let Y be a real function on $[-1,1]^n$ such that for every real a the set $\{Y \leq a\}$ is convex. Let B be a convex set such that Y is σ – Lipschitz on B i.e. $|Y(x) - Y(y)| \leq \sigma\tau(x,y)$ for all $x, y \in B$ (τ is the Euclidean distance.) If $X_1, \dots, X_n$ are independent random variables with values in $[0,1]$ and $b = \mathbf{P}((X_1, \dots, X_n) \notin B) < 1/2$, then*

$$\mathbf{P}(|Y - \mathbf{E}Y| \geq \lambda) \leq 4b + \frac{1}{1-2b} \exp\left\{-\frac{\lambda^2}{16\sigma^2}\right\}.$$

References

1. Alon, N., Kim. J., and Spencer, J. (1997). Nearly perfect matchings in regular simple hypergraphs, *Israel J. Math* **100**, 171–187.
2. Alon, N., and Spencer, J. (1992). *The Probabilistic Method*, Wiley Interscience Series in Discrete Mathematics and Optimization, Wiley and Sons, New York.
3. Azuma, K. (1967). Weighted sums of certain dependent random variables, *Tohoku Math. J.* **19**, 357–367.
4. Bass, T., and Godbole, A.P. (1997). Random nearest neighbour graphs, preprint.
5. Bollobás, B. (1987). Martingales, isoperimetric inequalities, and random graphs, *Colloq. Math. Soc. János Bolyai* **52**, 113–139.
6. Bollobás, B. (1990). Sharp concentration of measure phenomena in random graphs, In *Random Graphs*, Karoński, M., Jaworski, J., Ruciński, A., eds., 1–15. Ann. Discrete Math. Ser., Wiley, New York.
7. Burkholder, D.L. (1973). Distribution function inequalities for martingales, *Ann. Probab.* **1**, 19 – 42.
8. Burkholder, D.L., and Gundy, R.F. (1970). Extrapolation and interpolation of quasi linear operators on martingales *Acta Math.* **124**, 248 – 304.
9. Chalker, T.K., Godbole, A.P., Hitczenko, P., Radcliff, J., and Ruehr, O.G. (1997). On the size of a random sphere of influence graph, preprint.
10. de la Peña, V.H. (1997). A general class of exponential inequalities for martingales and ratios, Technical Report no. 378, Department of Theoretical Statistics, Aarhus University.

11. Dembo, A. (1996). Moderate deviations for martingales with bounded jumps, *Elect. Comm. in Probab.* **1**, 11–17.
12. Freedman, D. (1975). On tail probabilities for martingales, *Ann. Probab.* **3**, 100 – 118.
13. Godbole, A.P., Schneck, R., Sidman, J., and Taylor, I. (1997). On the Wiener index of a random graph, preprint.
14. Hitczenko, P. (1990a). Best constants in the martingale version of Rosenthal's inequality, *Ann. Probab.* **18**, 1656–1668.
15. Hitczenko, P. (1990b). Upper bounds for the L_p norms of martingales, *Probab. Theory Rel. Fields* **86**, 225–238.
16. Hoeffding, P. (1963). Probability inequalities for sums of bounded random variables, *J. Amer. Statist. Assoc.* **58**, 13–30.
17. Johnson, W.B., Schechtman, G., and Zinn, J. (1985). Best constants in moment inequalities for linear combinations of independent and exchangeable random variables, *Ann. Probab.* **13**, 234 – 253.
18. Kesten, H. (1993). On the speed of convergence in first – passage percolation, *Ann. Appl. Probab.* **3**, 296 – 338.
19. Khoshnevisan, D. (1996). Deviation inequalities for continuous martingales, *Stoch. Processes Appl.* **65**, 17–30.
20. Kim, J. H. (1995). On Brooks' theorem for sparse graphs, *Combin. Probab. Comput.* **4**, 97 – 132.
21. Kwapień, S. and Woyczyński, W. (1992). *Random Series and Stochastic Integrals. Single and Multiple*, Birkhäuser, Boston.
22. Leventhal, S. (1989). A uniform CLT for uniformly bounded families of martingale differences, *J. Theoret. Probab.* **2**, 271 – 281.
23. McDiarmid, C.J.H. (1989). On the method of bounded differences, in *Surveys in combinatorics: London Math. Soc. Lecture Notes* **141**, 148–188, Cambridge University Press, Cambridge.
24. McKay, B., and Wormald, N. (1997). The degree sequence of a random graph, I. The models, *Random Structures and Algorithms* **11**, 97 – 117.
25. Michael, T. S., and Quint, T. (1994). Sphere of influence graphs: a survey, *Congressus Numerantium* **105**, 153–160.
26. Pinelis, I. (1994). Sharp exponential inequalities for the martingales in the 2-smooth Banach spaces and applications to "scalarizing" decoupling, in Probability in Banach Spaces 9, *Progress in Probability* **35**, J. Hoffmann-Jørgensen et al. (eds.) Birkhäuser, Boston, pp. 55–72.
27. Pinelis, I. (1997). Optimal tail comparison based on comparison of moments, preprint.
28. Rhee, W.T. (1993) A matching problem and subadditive Euclidean functionals, *Ann. Appl. Probab.* **3**, 794 – 801.
29. Shamir, E., and Spencer, J. (1987). Sharp concentration of the chromatic number on random graphs $G_{n,p}$, *Combinatorica* **7**, 121–129.
30. Steele, J. M. (1997). *Probability Theory and Combinatorial Optimization*, NSF-CBMS Regional Research Conference Lecture Notes Series Vol. 69, Society for Industrial and Applied Mathematics, Philadelphia.
31. Stout, W. F. (1974). *Almost Sure Convergence*, Academic Press, New York.
32. Talagrand, M. (1994). Concentration of measure and isoperimetric inequalities in product spaces, *Publications Math. IHES* **81**, 73 – 205.
33. Talagrand, M. (1996). A new look at independence, *Ann. Probab.* **24**, 1 – 34.

DEPARTMENT OF MATHEMATICAL SCIENCES, MICHIGAN TECHNOLOGICAL UNIVERSITY, HOUGHTON, MI 49931.

E-mail address: anant@mtu.edu

DEPARTMENT OF MATHEMATICS, NORTH CAROLINA STATE UNIVERSITY, BOX 8205, RALEIGH, NC 27695.

E-mail address: pawel@math.ncsu.edu

DIMACS Series in Discrete Mathematics
and Theoretical Computer Science
Volume 41, 1998

Dynamical Percolation: Early Results and Open Problems

OLLE HÄGGSTRÖM

Abstract. Dynamical percolation is a way of imposing time dynamics on static percolation models such as the usual bond percolation setup. This paper contains a summary of what has been done so far in dynamical percolation. Some of the main open problems are also discussed.

1 Introduction

Percolation theory deals with connectivity properties of random media. In traditional setups, the medium is modelled by some stochastic process which is static in the sense that it varies only in space and not in time. The general philosophy of **dynamical percolation**, which was recently introduced by Häggström, Peres and Steif [15] and which is the topic of this survey paper, is to introduce some natural stationary time dynamics whose invariant distribution equals the distribution P of a given static process. By considering the state of the dynamical process at some fixed time t, one thus recovers the original static process. If $\mathcal{A}$ is some event for the static process which has P-probability 1, then Fubini's Theorem implies that a.s. $\mathcal{A}$ occurs for Lebesgue-a.e. t in the dynamical process. The type of questions mainly studied in dynamical percolation is whether with positive probability there are exceptional times at which $\mathcal{A}$ does not occur.

I do not claim that the models to be discussed here are of any immediate physical relevance. The following naive physical interpretation can nevertheless be good to keep in mind for motivational purposes. Suppose that we have a static percolation model for the microscopic behaviour of a random electrical medium, and that the event $\mathcal{C}$ that there exists an unbounded connected component in this percolation model corresponds to the macroscopic event that the medium has nonzero conductance. Thus, if $P(\mathcal{C}) = 0$, then

[1]Work supported by a grant from the Swedish Natural Science Research Council

[2]**1991 Mathematics Subject Classification:** Primary 60K35, Secondary 82C43, 60J25.

the medium has zero conductance a.s. In the corresponding dynamical percolation model (representing some equilibrium microscopic fluctuations), the medium will be nonconductive at a.e. time t, and the question is whether we can strengthen this to having zero conductance at *every* t. In fact, it was a similar question on electrical conductance, asked by Paul Malliavin at the Mittag-Leffler Institute in 1995, which was the original impetus for introducing dynamical percolation in [15].

Before entering a detailed discussion of dynamical percolation, it is worth mentioning that there is another way of incorporating time aspects into percolation theory, namely first-passage percolation (see e.g. [19] or [27]). The general flavour of first-passage percolation, which can be thought of as a model for the time it takes for a fluid to spread through a random medium, is, however, completely different from that of dynamical percolation.

The purpose of this paper is to summarize what has been done so far in the field of dynamical percolation, and to mention some interesting open problems. For simplicity and concreteness, I will not always state the results in their most general form. Most of the results quoted are from [15, 29, 7, 6, 13, 4] which to my knowledge are the only papers so far in this young field. As is evident from the reference list, several connections with classical percolation theory will also be discussed.

Here is a brief outline of the rest of the paper. In Section 2, I will discuss the basic dynamical percolation model on an infinite locally finite graph G, with particular focus on the event $\mathcal{C}$ that there is an infinite connected component; the static counterpart of this model is ordinary independent bond percolation. Section 3 specializes to the case $G = \mathbf{Z}^d$, the cubic lattice in d dimensions. The setup in Section 4 is the same, but focus is shifted to the event $\mathcal{C}^!$ that there is a *unique* infinite cluster (i.e. a unique maximal connected component). Section 5 treats the case where G is a tree, where some rather detailed results have been obtained. Finally, in Section 6, I will move on from discrete to continuum models, i.e. to models where the percolation process takes place on $\mathbf{R}^d$ rather than on a graph.

2 Dynamical percolation on a graph

Let G be an infinite locally finite connected graph with vertex set V and edge set E. In the standard bond percolation setup, one fixes a parameter $p \in [0,1]$ and considers the probability measure P_p on $\{0,1\}^E$ where each edge independently takes value 0 (absent, closed) or 1 (present, open) with respective probabilities $1-p$ and p. The event $\mathcal{C}$ that there exists an infinite cluster of open edges is a tail event, and therefore it has probability 0 or 1. A simple coupling argument shows that there exists a critical retention probability $p_c = p_c(G) \in [0,1]$ such that

$$P_p(\mathcal{C}) = \begin{cases} 0 & \text{if } p < p_c \\ 1 & \text{if } p > p_c. \end{cases}$$

Whether $P_{p_c}(\mathcal{C})$ equals 0 or 1 depends on the choice of G. The most important, and by far the most studied, example is when G equals the cubic lattice $\mathbf{Z}^d$, $d \geq 2$, with edges connecting nearest neighbours. The reader is well-advised to consult Grimmett [11] for an introduction and a detailed discussion of what is known on this model, with emphasis on the $\mathbf{Z}^d$ case.

In the corresponding dynamical percolation process, one defines a $\{0,1\}^E$-valued Markov process $\{\eta_t\}_{t\geq 0}$ as follows. The initial configuration η_0 is chosen according to P_p, and after that each edge flips (changes its value) independently at rate

$$\lambda(\eta_t, e) = \begin{cases} p & \text{if } \eta_t(e) = 0 \\ 1-p & \text{if } \eta_t(e) = 1. \end{cases}$$

Write Ψ_p for the underlying probability measure of this process. Clearly, P_p is the unique stationary distribution, and η_t has distribution P_p for each t.

For a Borel set $\mathcal{A} \subseteq \{0,1\}^E$, write $\mathcal{A}_t$ for the event that $\eta_t \in \mathcal{A}$. Thus, $\mathcal{C}_t$ is the event that the edge configuration at time t contains an infinite cluster. As indicated in the introduction, it is immediate from Fubini's Theorem that

$$\begin{cases} \Psi_p(\mathcal{C}_t \text{ occurs for Lebesgue-a.e. } t) & = & 1 & \text{if} & P_p(\mathcal{C}) = 1 \\ \Psi_p(\neg\mathcal{C}_t \text{ occurs for Lebesgue-a.e. } t) & = & 1 & \text{if} & P_p(\mathcal{C}) = 0 \end{cases}$$

The first result in [15] states that for $p \neq p_c$, there are a.s. no exceptional times.

Proposition 2.1: *For any graph G we have for each $p > p_c(G)$ that*

$$\Psi_p(\mathcal{C}_t \textit{ occurs for every } t) = 1,$$

whereas if $p < p_c(G)$, then

$$\Psi_p(\neg\mathcal{C}_t \textit{ occurs for every } t) = 1.$$

The proof is simple, and I encourage the reader to try to figure it out him/herself before consulting [15].

This leaves open the critical case $p = p_c$. Part of what makes dynamical percolation an interesting subject is that for certain choices of G, the process $\{\eta_t\}_{t\geq 0}$ is less well-behaved at criticality, as stated in the following result from [15].

Theorem 2.2: *There exist graphs G such that for $p = p_c(G)$ we have $P_p(\mathcal{C}) = 0$ while*

$$\Psi_p(\mathcal{C}_t \text{ occurs for some } t) = 1. \tag{1}$$

There also exist graphs G such that for $p = p_c(G)$ we have $P_p(\mathcal{C}) = 1$ while

$$\Psi_p(\neg\mathcal{C}_t \text{ occurs for some } t) = 1. \tag{2}$$

Note that when exceptional times exist, the set of such times has to be dense in $\mathbf{R}_+$, because for any $\tau, \delta > 0$ the event that there is an exceptional $t \in (\tau, \tau + \delta)$ has probability 0 or 1 by Kolmogorov's 0-1 law, and the probability is clearly not 0.

Theorem 2.2 may perhaps seem a bit surprising. Let me therefore sketch a constructive proof, following [15]. As a warm-up and a first step, I will show that if the switching rates are allowed to vary from edge to edge, then the behaviour in Theorem 2.2 is easy to obtain. More precisely, pick G and p in such a way that $P_p(\mathcal{C}) = 0$, let $(e_1, e_2, \ldots)$ be an enumeration of the edge set E, and for a sequence $\alpha = (\alpha_1, \alpha_2, \ldots)$ of positive real numbers define a $\{0,1\}^E$-valued Markov process $\{\xi_t\}_{t\geq 0}$ as follows. The initial state ξ_0 is picked according to P_p, and thereafter each edge e_i independently flips at rate

$$\lambda^\alpha(\xi_t, e_i) = \begin{cases} \alpha_i p & \text{if } \eta_t(e_i) = 0 \\ \alpha_i(1-p) & \text{if } \eta_t(e_i) = 1. \end{cases}$$

As before, the configuration at time t has distribution P_p for each t. Note also that taking $\alpha_i = 1$ for each i yields the original dynamical percolation model. The preliminary task is to pick $\alpha = (\alpha_1, \alpha_2, \ldots)$ in such a way that there will be times that $\mathcal{C}$ occurs. The following procedure for picking α_i rapidly increasing will yield not only this but also times at which *all* $e \in E$ are present. First pick a sequence $\varepsilon = (\varepsilon_1, \varepsilon_2, \ldots)$ such that $0 < \varepsilon_i < 1$ for each i and furthermore $\prod_{i=1}^\infty (1 - \varepsilon_i) > 0$. Pick α_1 so large that $\xi_t(e_1) = 1$ for some $t \in [0,1]$ with probability at least $1 - \varepsilon_1/2$. We can then find $\delta_1 > 0$ small enough so that with probability at least $1 - \varepsilon_1$ there is an interval of length δ_1 in $[0,1]$ throughout which $\xi_t(e_1) = 1$. Similarly, we can find an even larger α_2 and smaller δ_2 such that for any time interval $[\tau, \tau + \delta_1]$ of length δ_1 we can with probability at least $1 - \varepsilon_2$ find an interval of length δ_2 during which $\xi_t(e_2) = 1$ throughout. Continuing inductively in this way, we obtain a sequence α such that with probability at least $\prod_{i=1}^\infty (1 - \varepsilon_i)$ there is some $t \in [0,1]$ at which $\xi_t(e) = 1$ for all $e \in E$. The probability that this occurs for some $t \in [0, \infty)$ is therefore 1, by timewise ergodicity of $\{\xi_t\}_{t\geq 0}$.

Note that this shows that even the analogue of Proposition 2.1 fails in this wider class of processes. The above construction can be mimicked in the context of ordinary dynamical percolation if we take some graph G and instead of changing the flip rates we replace edges by so-called n-**aggregates**. An n-aggregate A_n consists of two endpoints between which there are 2^n

branches in parallel, each branch consisting of n edges in series; see Figure 1. Call A_n open if one can go from one endpoint to the other via open edges in A_n. The key to using n-aggregates for constructing graphs with interesting percolation behaviour is the following lemma (see [15] for a proof; (3) and (4) are easy while (5) requires some more work).

Lemma 2.3: *For n-aggregates A_n, we have*

$$\lim_{n\to\infty} P_p(A_n \text{ is open}) = \begin{cases} 0 & \text{if } p < \frac{1}{2} \\ 1 - e^{-1} & \text{if } p = \frac{1}{2} \\ 1 & \text{if } p > \frac{1}{2}, \end{cases} \tag{3}$$

and furthermore, for any $\delta > 0$,

$$\lim_{n\to\infty} \Psi_{\frac{1}{2}}(A_n \text{ is open for some } t \in [0,\delta]) = 1 \tag{4}$$

and

$$\lim_{n\to\infty} \Psi_{\frac{1}{2}}(A_n \text{ is closed for some } t \in [0,\delta]) = 1. \tag{5}$$

Parts (4) and (5) of the lemma show that at $p = 1/2$, an n-aggregate with n very large behaves somewhat like an edge with very high flip rate.

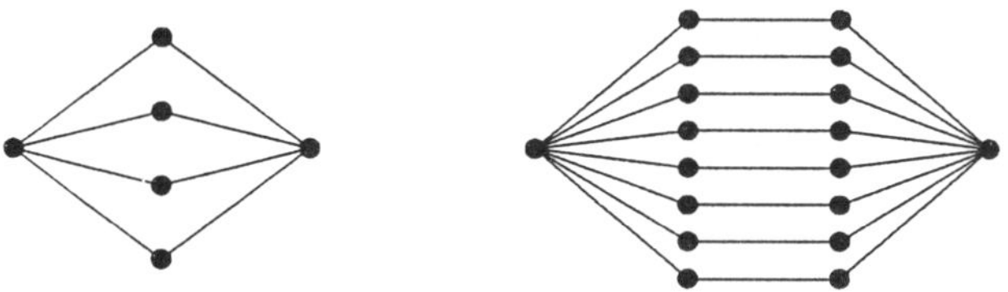

Figure 1: The n-aggregates A_2 and A_3.

Now let G^* be some graph with $p_c(G^*) > 1 - e^{-1}$ (for concreteness, one might pick G^* to be the honeycomb lattice; see [33]). Lemma 2.3 implies that by replacing the edges of G^* by n-aggregates with rapidly increasing n, similarly to the above sequential construction in the case of non-homogeneous flip rates, a graph G is obtained for which

(i) $p_c(G) = \frac{1}{2}$,
(ii) $P_{\frac{1}{2}}(\mathcal{C}) = 0$, while
(iii) $\Psi_{\frac{1}{2}}(\mathcal{C}_t \text{ occurs for some } t) = 1$.

This proves part (1) of Theorem 2.2. To get (2), essentially the same construction can be used, except that one should start with a graph G^* with $p_c(G^*) < 1 - e^{-1}$, such as e.g. the square lattice $\mathbf{Z}^2$.

One can of course argue that these examples are somewhat artificial. There are other examples of graphs exhibiting the behaviour indicated in Theorem 2.2; Section 5 contains some examples which are nicer than those presented here e.g. in the sense that they have bounded degree.

3 The $\mathbf{Z}^d$ case

Given the results in the previous section, an obvious next question is what happens at criticality on the cubic lattice $\mathbf{Z}^d$, $d \geq 2$. This problems remains to a large extent unsolved. However, the following general result from [15] takes care of the case where d is large.

Theorem 3.1: *Let G be an infinite locally finite connected graph, let v be a vertex of G, and define*

$$\theta(p) = P_p(v \text{ belongs to an infinite cluster}).$$

If there exists a constant $C < \infty$ with the property that for all $p \geq p_c$ we have

$$\theta(p) \leq C(p - p_c) \tag{6}$$

then

$$\Psi_{p_c}(\neg\mathcal{C}_t \text{ occurs for every } t) = 1.$$

Condition (6) is easily verified for regular trees, and it has also been shown by Hara and Slade [16] to hold for $\mathbf{Z}^d$ with $d \geq 19$. Regular trees and high-dimensional lattices thus behave tamely, in the sense that there are no exceptional times as far as the event $\mathcal{C}$ is concerned, even at criticality. It is generally believed that (6) should hold for $\mathbf{Z}^d$ down to $d = 7$ or possibly $d = 6$. However, for lower dimensions it is believed (and for $d = 2$ known [21]) that (6) fails. Partly for this reason, it is unclear what answer to Question 3.2 below should be conjectured. For all $d \geq 2$, it is expected (but only known for $d = 2$ and $d \geq 19$; see [18] and [16], respectively) that $P_p(\mathcal{C}) = 0$.

Question 3.2: *For $\mathbf{Z}^d$, $d = 2, \ldots, 5$, is it the case that*

$$\Psi_{p_c}(\neg\mathcal{C}_t \text{ occurs for every } t) = 1 \text{ ?}$$

The answer may of course depend on d. It seems to me that the case which is most likely to be solved before the turn of the century is $d = 2$, due to the rather powerful methods based on self-duality of the square lattice which are available (see [11]), and the rigorously known facts that $p_c = 1/2$ and $P_{p_c}(\mathcal{C}) = 0$. Moreover, some progress towards understanding critical dynamical percolation on $\mathbf{Z}^2$ has already been made by Benjamini, Kalai and Schramm [4]. They prove a result analogous to Lemma 2.3 in the situation where the n-aggregates are replaced by $(n+1)\times n$-sized rectangular portions of the square grid, and the rectangle is called open if there is a path of open edges connecting its two shorter sides to each other. Häggström and

Pemantle [13] made another attempt at shedding light on critical dynamical percolation on $\mathbf{Z}^2$; the results of that paper are quoted in Section 5 below.

One more result from [15] concerning $\mathbf{Z}^d$ says that if $P_{p_c}(\mathcal{C}) = 0$ and exceptional times at which $\mathcal{C}_t$ occurs exist, then a.s. for *all* times at which an infinite cluster exists the limiting spatial density of the set of vertices belonging to infinite clusters is 0. I leave it to the reader to decide whether this is substantial evidence for answering Question 3.2 by 'yes'.

It would certainly be more exciting if it turned out that Question 3.2 could be answered by 'no', i.e. if at criticality there are times at which η_t contains an infinite cluster. It might then be possible to connect dynamical percolation with the somewhat mysterious theory of "incipient infinite clusters". To explain this, let me for simplicity concentrate on the case $d = 2$. There exists a large physics literature in which the percolation systems studied are finite and very large, rather than infinite (see e.g. [31]). The role which the infinite cluster plays in infinite systems is for such models played by the *largest* cluster, which exhibits interesting fractal-like behaviour at the critical point $p = p_c$. For the infinite system $\mathbf{Z}^2$, it would therefore be interesting to study properties of the infinite cluster at criticality, were it not for the unfortunate fact that such a cluster is nowhere to be found, since $P_{p_c}(\mathcal{C}) = 0$. For this reason, several different proposals have been made on how to modify P_{p_c} in such a way that

(i) the infinite cluster is recovered, while

(ii) the modified measure is still, in some sense, as "close" as possible to the original measure P_{p_c}.

One way to do this is to let $p > p_c$, to condition on the event that the origin is in an infinite cluster, and to consider a weak limit of such conditional distributions as $p \searrow p_c$ (see [20]). Another approach is to have inhomogeneous retention probabilities decaying (at some carefully chosen rate) to p_c as one moves away from the origin [9], whereas yet another approach is based on invasion percolation [10]. If it turned out that there exist exceptional times (under Ψ_{p_c}) for the dynamical percolation process at which an infinite cluster exists, then the distribution of the configuration η_T with

$$T = \inf\{t : \text{ the origin is in an infinite cluster in } \eta_t\} \tag{7}$$

would provide a fourth candidate for the incipient infinite cluster. The next step would then be to compare the fractal dimension and other properties of this incipient infinite cluster to the other three candidates.

But then again, it may turn out that the answer to Question 3.2 is 'yes' so that no such exceptional times exist, or that the problem is simply too difficult. One could then try to modify or extend the dynamical percolation process in such a way that the existence of exceptional times can be proved. Here are two suggestions for how this might be done in the case $G = \mathbf{Z}^2$:

1. Associate with each edge $e_i \in E$ an independent stationary standard Ornstein–Uhlenbeck diffusion $\{\zeta_t^i\}_{t\geq 0}$, and let, for each t,

$$\eta_t(e_i) = \begin{cases} 1 & \text{if } \zeta_t^i \geq 0 \\ 0 & \text{otherwise.} \end{cases}$$

The process $\{\eta_t\}_{t\geq 0}$ will then no longer be a Markov process, but η_t will still have distribution $P_{1/2}$ for each t. The point of this modification is that the set of times that an edge switches will get a self-similar-like behaviour, so that for each t there will be plenty of edges that have very high flip rate in some neighbourhood of t. This might help to produce exceptional times at which $\mathcal{C}_t$ occurs; recall from Section 2 how varying flip rates can facilitate such a phenomenon.

2. Relativity theory has taught us that space and time are not as distinct as they may seem in everyday life. This train of thought can be brought into dynamical percolation as follows. Take $p = p_c$, define the dynamical percolation process $\{\eta_t\}_{t\geq 0}$ as usual, and extend the process to negative times in the obvious way. For each $e \in E$, write $x(e)$ and $y(e)$ for the x- and y-coordinates of the midpoint of e. For $a_x, a_y \in \mathbf{R}$, define the $\{0,1\}^E$-valued random object η_{t,a_x,a_y} by letting

$$\eta_{t,a_x,a_y}(e) = \eta_{t+a_x x(e)+a_y y(e)}(e)$$

for each $e \in E$, and think of this as the configuration at a "time" which is "tilted" in three-dimensional space-time. The distribution of η_{t,a_x,a_y} is P_{p_c} for every choice of (t, a_x, a_y), whence $\neg\mathcal{C}_{t,a_x,a_y}$ (with obvious notation) will occur for Lebesgue-a.e. (t, a_x, a_y). Will there be exceptional (t, a_x, a_y) for which $\mathcal{C}_{t,a_x,a_y}$ occurs? This question appears to be closely related to Question 6.4 in the final section of this paper.

4 Uniqueness of the infinite cluster

One of the basic questions in percolation theory is the following: If infinite clusters exist, then how many of them can there be? In the case of the cubic lattice $\mathbf{Z}^d$, it is by now well-known that if an infinite cluster exists, then it is a.s. unique. In other words, if $\mathcal{C}^!$ denotes the existence of *exactly one* infinite cluster, then $P_p(\mathcal{C}^!) = P_p(\mathcal{C})$. For the case $d = 2$, this was proved by Harris [17], and for general d by Aizenman *et al.* [2]; see also [8] for the "definitive" proof of this result, and [25, 5, 14] for some further aspects of the uniqueness problem.

For dynamical percolation, we can thus conclude in the usual way that if $P_p(\mathcal{C}) = 1$, then

$$\Psi_p(\mathcal{C}_t^! \text{ occurs for a.e. } t) = 1.$$

Can "a.e. t" be replaced by "every t" in this statement? For $p > p_c$, the answer is yes, as stated in the following theorem of Peres and Steif [29].

Theorem 4.1: *For dynamical percolation on $\mathbf{Z}^d$ with parameter $p > p_c$, we have a.s. uniqueness of the infinite cluster simultaneously for all t, i.e.*

$$\Psi_p(\mathcal{C}_t^! \text{ occurs for every } t) = 1.$$

This leaves open the critical case $p = p_c$. Although it is believed "beyond reasonable doubt" that $P_{p_c}(\mathcal{C}) = 0$ for all $d \geq 2$, it may still very well be the case that

$$\Psi_{p_c}(\mathcal{C}_t \text{ occurs for some } t) = 1$$

as discussed in the previous section. One would then expect that there is a unique infinite cluster for, in some sense, most of the exceptional times at which $\mathcal{C}_t$ occurs. For instance, it is not hard to prove that a.s. there will be a unique infinite cluster at the stopping time T, defined in (7). On the other hand, there might be "exceptional exceptional" times at which there are two or more infinite clusters. Readers who find such a scenario too bizarre to be taken seriously are advised to go on reading; Theorems 5.2 and 6.5 below show that similar phenomena do take place at criticality in certain other models.

Before closing this short section, it is worth mentioning that a result somewhat analogous to Theorem 4.1, in the context of monotone couplings of percolation processes, was proved a couple of years ago by Alexander [3]. A natural coupling of the measures P_p for all $p \in [0, 1]$ is obtained if one first assigns i.i.d. random variables $Y(e)$, uniformly distributed on $[0, 1]$, to all edges on $\mathbf{Z}^d$, and then for each p considers the random graph with edge set $\{e \in E : Y(e) \leq p\}$. Uniqueness of the infinite cluster for ordinary (static) percolation in conjunction with the usual Fubini argument shows that a.s. there is a unique infinite cluster for a.e. $p > p_c$. Alexander's result states that this a.s. holds simultaneously for *every* $p > p_c$. Recently, Häggström and Peres [14] obtained a (slightly more complicated) analogue of this result in the setting of general Cayley graphs.

5 The tree case

Quite often in probability theory, it turns out that a process taking place on the integer lattice $\mathbf{Z}^d$ is intractable for exact calculation. One way to proceed in such a situation is to study the corresponding process on a tree; in many cases, the simpler connectivity structure of the tree will be enough to make computations tractable. For the case of regular trees, this tradition dates back at least to the 1970's (see e.g. [30]). Later, it became apparent through the seminal work of Lyons [22, 23, 24] and others that several random processes can be analyzed with great precision in the considerably more general context of *arbitrary* trees. This is particularly true for (static) percolation, and as it turns out to a good extent also for dynamical percolation. The methods developed by Lyons have been successfully applied outside of the tree setting (e.g. by Peres [28] to obtain an improved understanding of intersection properties of Brownian paths) and it is not unlikely that the same will happen for the capacity estimation techniques used in [15] and [29] to study dynamical percolation on trees. Here, however, space will permit only a summary of the results, and not a discussion of the techniques.

Let Γ be an infinite locally finite tree with vertex set V, edge set E, and a distinguished vertex $\rho \in V$ called the root. For $v \in V$, write $|v|$ for the distance between v and ρ. If v and w are nearest neighbours with $|w| = |v| + 1$, then w is called a child of v. The vertex set $\{v \in V : |v| = n\}$ is called the n:th level of Γ, and is denoted Γ_n. Further, $|\Gamma_n|$ denotes the number of vertices on the n:th level. If two vertices at the same level always have the same number of children, then Γ is called spherically symmetric.

For static percolation, Lyons [23] computed p_c for any tree, in terms of a natural quantity called the branching number of the tree. In [24], he completed the picture by establishing the following precise condition on Γ for $P_p(\mathcal{C})$ to be strictly positive. Consider the electrical network $\mathcal{N}_\Gamma^{p^{-n}}$ obtained by assigning each edge between the $(n-1)$:st and the n:th level in Γ resistance p^{-n}. Then $P_p(\mathcal{C}) > 0$ if and only if the effective resistance between ρ and "infinity" is finite. In [15] an analogous condition for dynamical percolation on Γ was obtained, namely that there exist times at which $\mathcal{C}_t$ occurs if and only if the effective resistance between ρ and infinity is finite in the network $\mathcal{N}_\Gamma^{p^{-n}/n}$ where each edge between levels $n-1$ and n has resistance p^{-n}/n. For spherically symmetric trees, these criteria amount to the following.

Theorem 5.1: *For an infinite spherically symmetric tree Γ and $p \in (0,1)$, we have $P_p(\mathcal{C}) = 1$ if and only if*

$$\sum_{n=1}^{\infty} \frac{p^{-n}}{|\Gamma_n|} < \infty,$$

whereas we have $\Psi_p(\mathcal{C}_t$ occurs for some $t) = 1$ if and only if

$$\sum_{n=1}^{\infty} \frac{p^{-n}}{n|\Gamma_n|} < \infty.$$

Specializing further to the k-ary regular tree (i.e. the tree in which ρ has $k+1$ children and all other vertices have k children), we get $|\Gamma_n| = (k+1)k^{n-1}$ so that $p_c = 1/k$ and $\Psi_{p_c}(\neg\mathcal{C}_t$ occurs for all $t) = 1$. As mentioned in Section 3, this last conclusion can also be obtained from Theorem 3.1.

If we instead consider a spherically symmetric tree $\Gamma^{(\beta)}$ with $|\Gamma_n^{(\beta)}|$ being of the order $2^n n^\beta$ for some $\beta \in (0,1]$ (the reader will have no problem constructing such a tree), we get $p_c = 1/2$, $P_{p_c}(\mathcal{C}) = 0$ while at the same time $\Psi_p(\mathcal{C}_t$ occurs for some $t) = 1$.

In [13], it is shown, building on Theorem 5.1, that for spherically symmetric trees with critical value p_c and $\theta(p_c) = 0$, there are exceptional times at criticality with infinite clusters if and only if

$$\int_{p_c}^{1} (\theta(p))^{-1} dp < \infty. \tag{8}$$

An interesting aspect of this result is that (8) holds for $G = \mathbf{Z}^2$ (see [21]), so that this can be taken as heuristic evidence that there are exceptional times with infinite clusters in the critical dynamical percolation process on $\mathbf{Z}^2$. On the other hand, an indication that this need not be taken too seriously, is that [13] contains an example showing that the 'if' direction in the assertion involving (8) does *not* extend to general trees.

So what about the number of infinite clusters in the tree case? Whereas percolation models on $\mathbf{Z}^d$ tend to get 0 or 1 infinite cluster, tree-indexed percolation processes tend to get 0 or infinitely many infinite clusters (see e.g. [12]). A dynamical percolation version of this meta-theorem was proved by Peres and Steif [29], and says that for dynamical percolation on a tree Γ with $p \in (p_c(\Gamma), 1)$, the number of infinite clusters is a.s. infinite for all t.

At the critical value p_c the behavior is considerably more intricate, as evidenced by the following theorem from [29]. Write T^k for the set of times at which there are at least k infinite clusters.

Theorem 5.2: *Let Γ be an infinite spherically symmetric tree with $p_c(\Gamma) \in (0,1)$ and $P_{p_c}(\mathcal{C}) = 0$. For each $k \in \{1, 2, \ldots\}$, we have that $\Psi_{p_c}(T^k \neq \emptyset) = 1$ if and only if*

$$\int_0^1 \int_0^1 \left(\sum_{n=1}^{\infty} \frac{(1 + \frac{1-p_c}{p_c} e^{-|s-t|})^{n-1}}{|\Gamma_n|} \right)^k ds\, dt < \infty. \tag{9}$$

Moreover, if (9) *holds for every $k < \infty$, then $\Psi_{p_c}(T^\infty \neq \emptyset) = 1$.*

This means e.g. that if $p = 1/2$ and $\Gamma^{(\beta)}$ is as above with $\beta \in (0, 1/2]$, then there are times at which an infinite cluster exists, but there is never more than one infinite cluster. If instead $\beta \in (1/2, 2/3]$, then there are sometimes two but never three infinite clusters, and so on. If $\beta = 1$, then there are, for any $k \in \{0, 1, \ldots\} \cup \{\infty\}$, times at which exactly k infinite clusters exist.

Another result in [29] concerning spherically symmetric trees gives an explicit expression for the Hausdorff dimension of the random set T^k. For the case $k = 1$ this was done in [15]. For $\Gamma^{(\beta)}$ at criticality, the Hausdorff dimension of T^1 turns out to be exactly β. Similar questions seem natural to ask in any dynamical percolation model exhibiting exceptional times.

6 Continuum models

Continuum percolation theory is about percolation processes which take place on $\mathbf{R}^d$ without being confined to a lattice. Many such models have been studied, and in the past 10 years the continuum percolation literature has grown rapidly; see the monograph by Meester and Roy [26] for a summary of what has been done up to about 1995.

The only static continuum model I will discuss here is the so-called Poisson Boolean model, sometimes also known as the lilypad model. This model arises by taking a Poisson process in $\mathbf{R}^d$ with intensity λ (i.e. a random point configuration in $\mathbf{R}^d$ where the number of points in a set of volume V is a Poisson random variable with mean λV, and the number of points in disjoint sets are independent) and placing a closed sphere of radius 1 centered at each point. Write P_λ for the ditributions of this process. Also, write U for the union of the spheres, and let $\mathcal{C}$ be the event that U contains an unbounded connected component. The starting point of continuum percolation theory is the result that for $d \geq 2$, there exists a critical value $\lambda_c = \lambda_c(d) \in (0, \infty)$ such that

$$P_\lambda(\mathcal{C}) = \begin{cases} 0 & \text{for } \lambda < \lambda_c \\ 1 & \text{for } \lambda > \lambda_c. \end{cases}$$

At the critical value we have $P_{\lambda_c}(\mathcal{C}) \in \{0, 1\}$ by ergodicity, and it is expected that $P_{\lambda_c}(\mathcal{C}) = 0$ for each $d \geq 2$. However, as for the lattice case this is known only for $d = 2$ and for d sufficiently large (see [26] and [32]).

A dynamical variant of this model was recently introduced by van den Berg, Meester and White [7]. A special case of their model is as follows. At time 0, the spheres are distributed in $\mathbf{R}^d$ according to P_λ, and after that the spheres (or more precisely, the centers of the spheres) move according to independent d-dimensional standard Brownian motions. It is intuitively clear (and can be proved) that this dynamics preserves P_λ. Hence, writing Ψ_λ for the probability measure underlying the dynamical process, we get

$$\begin{cases} \Psi_\lambda(\mathcal{C} \text{ occurs for a.e. } t) & = & 1 & \text{for} & \lambda > \lambda_c \\ \Psi_\lambda(\neg\mathcal{C} \text{ occurs for a.e. } t) & = & 1 & \text{for} & \lambda < \lambda_c. \end{cases} \tag{10}$$

The following strengthening of (10), analogous to Proposition 1.1, is proved in [7].

Theorem 6.1: *For the above dynamical continuum percolation process on $\mathbf{R}^d$, $d \geq 2$, we have for $\lambda > \lambda_c(d)$ that*

$$\Psi_\lambda(\mathcal{C} \textit{ occurs for every } t) = 1$$

and for $\lambda < \lambda_c(d)$ that

$$\Psi_\lambda(\neg\mathcal{C} \textit{ occurs for every } t) = 1.$$

It should be stressed that while Proposition 1.1 can be proved in a few lines, a considerable effort is needed in order to prove Theorem 6.1. The critical case remains open.

Benjamini and Schramm [6] take a different approach to studying dynamical percolation-type phenomena for the Poisson Boolean model: they dispose of the time dynamics and instead let space play the role of time. For $d \geq 3$, let Φ_d denote the set of two-dimensional affine subsets (planes) of $\mathbf{R}^d$. For the Poisson Boolean model on $\mathbf{R}^d$ with intensity λ and a fixed plane $\phi \in \Phi_d$, the random set $U \cap \phi$ forms a two-dimensional continuum percolation process with Poisson balls of random (but bounded) radii. Let λ_d^* denote the intensity of the d-dimensional process which makes the induced two-dimensional process critical, and for $\phi \in \Phi_d$ let C_ϕ denote the event that $U \cap \phi$ contains an unbounded connected component. The following theorem from [6] is analogous to Proposition 1.1 and Theorem 6.1.

Theorem 6.2: *For the Poisson Boolean model on $\mathbf{R}^d$, $d \geq 3$, we have for $\lambda > \lambda_d^*$ that*

$$P_\lambda(C_\phi \text{ occurs for every } \phi \in \Phi_d) = 1$$

and for $\lambda < \lambda_d^*$ that

$$P_\lambda(\neg C_\phi \text{ occurs for every } \phi \in \Phi_d) = 1.$$

Interestingly enough, the situation at $\lambda = \lambda_d^*$ is more complicated. The result that $P_{\lambda_c}(C) = 0$ in two dimensions is known to extend to the case where the radii are random but bounded. Hence, for the d-dimensional process at intensity λ_d^* and any fixed plane $\phi \in \Phi_d$, there will a.s. be no unbounded connected component in $U \cap \phi$. On the other hand, Benjamini and Schramm show that for $d \geq 4$ there are exceptional planes:

Theorem 6.3: *For the Poisson Boolean model on $\mathbf{R}^d$, $d \geq 4$, we have*

$$P_{\lambda_d^*}(C_\phi \text{ occurs for some } \phi \in \Phi_d) = 1.$$

In fact, [6] contains a proof of the even stronger result that an unbounded connected component occurs in some plane ϕ containing the origin. The result is somewhat analogous to a result of Adelman *et al.* [1] about three-dimensional Brownian motion $\xi = \{\xi_t\}_{t \geq 0}$, stating that although the projection of ξ on any plane $\phi \in \Phi_3$ is a two-dimensional Brownian motion, there will nevertheless a.s. exist planes ϕ for which the projected path fails to be neighbourhood-recurrent.

The proof of Theorem 6.3 uses an elegant second moment argument, and exploits the fact that two generic elements of Φ_4 intersect only at a point.

This last observation fails for Φ_3, and for this reason the case $d = 3$ remains open:

Question 6.4: *Does Theorem 6.3 hold for $d = 3$?*

The final result from [6] that I would like to mention concerns the number of unbounded components in $U \cap \phi$. It is particularly striking when contrasted with the fact that the uniqueness of the infinite cluster result for static percolation on $\mathbf{Z}^d$ carries over to the static Poisson Boolean model with random radii (see [26]).

Theorem 6.5: *If d is sufficiently large, then we have for the Poisson Boolean model on $\mathbf{R}^d$ with $\lambda = \lambda_d^*$ that a.s. there is some $\phi \in \Phi_d$ for which $U \cap \phi$ contains more than one unbounded connected component.*

Acknowledgement. I thank Jeff Steif for helpful comments and corrections to an earlier version of the manuscript.

References

[1] Adelman, O., Burdzy, K. and Pemantle, R. (1997) Sets avoided by Brownian motion, *Ann. Probab.*, to appear.

[2] Aizenman, M., Kesten H. and Newman, C.M. (1987) Uniqueness of the infinite cluster and continuity of connectivity functions for short- and long-range percolation, *Commun. Math. Phys.* **111**, 505–532.

[3] Alexander, K. (1995) Simultaneous uniqueness of infinite clusters in stationary random labeled graphs, *Commun. Math. Phys.* **168**, 39–55.

[4] Benjamini, I., Kalai, G. and Schramm, O. (1997) Noise sensitivity of Boolean functions and application to percolation, in preparation.

[5] Benjamini, I. and Schramm, O. (1996) Percolation beyond $\mathbf{Z}^d$, many questions and a few answers, *Electr. Commun. Probab.* **1**, 71–82.

[6] Benjamini, I. and Schramm, O. (1997) Exceptional planes of percolation, preprint.

[7] van den Berg, J., Meester, R. and White, D. (1997) Dynamic Boolean models, *Stoch. Proc. Appl.* **69**, 247–257.

[8] Burton, R. and Keane, M. (1989) Density and uniqueness in percolation, *Commun. Math. Phys.* **121**, 501–505.

[9] Chayes, J.T., Chayes, L. and Durrett, R. (1987) Inhomogeneous percolation problems and incipient infinite clusters, *J. Phys. A* **20**, 1521–1530.

[10] Chayes, J.T., Chayes, L. and Newman, C.M. (1985) The stochastic geometry of invasion percolation, *Commun. Math. Phys,* **101**, 383–407.

[11] Grimmett, G. (1997) Percolation and disordered systems, *École d'été de probabilités de Saint-Flour, XXVI—1996*, Springer, to appear.

[12] Häggström, O. (1997) Infinite clusters in dependent automorphism invariant percolation on trees, *Ann. Probab.* **25**, 1423–1436.

[13] Häggström, O. and Pemantle, R. (1997) On near-critical and dynamical percolation in the tree case, preprint.

[14] Häggström, O. and Peres, Y. (1997) Monotonicity of uniqueness for percolation on Cayley graphs: all infinite clusters are born simultaneously, preprint.

[15] Häggström, O., Peres, Y. and Steif, J.E. (1997) Dynamical percolation, *Ann. Inst. H. Poincaré, Probab. Statist.* **33**, 497–528.

[16] Hara, T. and Slade, G. (1994) Mean field behavior and the lace expansion, in *Probability and Phase Transition* (ed. G. Grimmett), Proceedings of the NATO ASI meeting in Cambridge 1993, Kluwer.

[17] Harris, T.E. (1960) A lower bound for the critical probability in a certain percolation process, *Proc. Cambridge Phil. Soc.* **56**, 13–20.

[18] Kesten, H. (1980) The critical probability of bond percolation on the square lattice equals $\frac{1}{2}$, *Commun. Math. Phys.* **74**, 41–59.

[19] Kesten, H. (1986) Aspects of first passage percolation, *École d'été de probabilités de Saint-Flour, XIV—1984*, 125–264, *Lecture Nothes in Math.* **1180**, Springer, New York.

[20] Kesten, H. (1986) The incipient infinite cluster in two-dimensional percolation, *Probab. Th. Relat. Fields* **73**, 369–394.

[21] Kesten, H. and Zhang, Y. (1987) Strict inequalites for some critical exponents in 2D-percolation, *J. Statist. Phys.* **46**, 1031–1055.

[22] Lyons, R. (1989) The Ising model and percolation on trees and tree-like graphs, *Commun. Math. Phys.* **125**, 337–353.

[23] Lyons, R. (1990) Random walks and percolation on trees, *Ann. Probab.* **18**, 931–958.

[24] Lyons, R. (1992) Random walks, capacity, and percolation on trees, *Ann. Probab.* **20**, 2043–2088.

[25] Meester, R. (1994) Uniqueness in percolation theory, *Statist. Neerl.* **48**, 237–252.

[26] Meester, R. and Roy, R. (1996) *Continuum Percolation*, Cambridge University Press.

[27] Newman, C.M. and Piza, M. (1995) Divergence of shape fluctuations in two dimensions, *Ann. Probab.* **23**, 977–1005.

[28] Peres, Y. (1996) Intersection-equivalence of Brownian paths and certain branching processes, *Commun. Math. Phys.* **177**, 417–434.

[29] Peres, Y. and Steif, J.E. (1997) The number of infinite clusters in dynamical percolation, *Probab. Th. Relat. Fields*, to appear.

[30] Spitzer, F. (1975) Markov random fields on an infinite tree, *Ann. Probab.* **3**, 387–398.

[31] Stauffer, D. and Aharony, A. (1992) *Introduction to Percolation Theory*, Taylor & Francis, London.

[32] Tanemura, H. (1996) Critical behaviour in a continuum percolation model, *Proceedings of the Seventh Japan–Russia Symposium on Probability and Mathematical Statistics*, 485–495, World Scientific.

[33] Wierman, J.C. (1981) Bond percolation on honeycomb and triangular lattices, *Avd. Appl. Probab.* **13**, 298–313.

Olle Häggström
Mathematical Statistics
Chalmers University of Technology
412 96 Göteborg
Sweden

DIMACS Series in Discrete Mathematics
and Theoretical Computer Science
Volume 41, 1998

Distinguishing and reconstructing sceneries from observations along random walk paths

Harry Kesten

ABSTRACT. We discuss some results and some open problems of the following type: Let $\mathcal{G}$ be an infinite, connected, locally finite graph, and let ξ and η be maps from $\mathcal{V}$ to $\{0, 1, \dots, k-1\}$, where $\mathcal{V}$ is the vertex set of $\mathcal{G}$ (such maps are called k-colorings or sceneries). Let $\{S_n\}$ be a random walk on $\mathcal{G}$. Can we reconstruct ξ if we observe the sequence $\{\xi(S_n)\}_{n\geq 0}$? If ξ and η are known and we observe either $\{\xi(S_n)\}_{n\geq 0}$ or $\{\eta(S_n)\}_{n\geq 0}$, but we are not told which of these alternatives prevails, can we decide (with zero probability of error) which of the two sequences were observed ?

1. Introduction. The problems considered here were introduced independently by I. Benjamini and by F. den Hollander and M. Keane (private communications) approximately 10 years ago. They can be phrased as 'statistical' problems, but their interest comes more from their similarity to problems in ergodic theory. Roughly speaking, the question is how much can we tell about the coloring of the vertices of a graph $\mathcal{G}$, by observing the colors of a (partially unknown) random collection of vertices. For this collection of vertices we shall always take the path of a random walk $\{S_n\}_{n\geq 0}$ on $\mathcal{G}$ and the most interesting choice for the graph $\mathcal{G}$ is $\mathbb{Z}^d$. We need some definitions to state the problems more precisely.

Throughout $\mathcal{G}$ is taken to be infinite, connected and with every vertex of finite degree. A *coloring* ξ of $\mathcal{G}$ is a map ξ from the vertex set $\mathcal{V}$ of $\mathcal{G}$ to $\{0, 1, \dots, k-1\}$. Such a coloring will also be called a *scenery*. Usually we take the number of colors, k, finite, but Howard (1995), (1996), (1997) also allows $k = \infty$. The space of all colorings will be denoted by $\Xi = \Xi^{(k)} = \{0, 1, \dots, k-1\}^{\mathcal{V}}$. $\{S_n\}$ is a random walk on $\mathcal{G}$ which starts at a specified vertex, call it $\mathbf{0}$. The measure governing the random walk $\{S_n\}$ is denoted by P. Most work has been done in the case of a simple random walk $\{S_n\}$ which moves from a vertex v to one of its neighbors w with a probability equal to $1/($ number of neighbors of $v)$. A random walker moving in the scenery ξ will successively see the colors $\xi(S_0), \xi(S_1), \dots$ and the available observations will consist of such sequences. They are elements of the space of all

1991 *Mathematics Subject Classification.* Primary 28A99, 60J15.

Key words and phrases. Random walk, scenery, coloring, distinguishability, mutually singular measures.

Research supported in part by the NSF through a grant to Cornell University.

sequences of colors $\mathcal{X} := \{0, 1, \dots, k-1\}^{\mathbb{N}}$. A generic element of $\mathcal{X}$ will be denoted by $(x_0, x_1, \dots)$ and X_i denotes the i-th coordinate function given by $X_i(x) = x_i$. $\mathcal{F}_\ell$ will denote the σ-field generated by the coordinate functions $\{X_n : n \geq \ell\}$ and $\mathcal{F}_\infty = \cap_{\ell=0}^\infty \mathcal{F}_\ell$. Q_ξ^ℓ is the measure induced on $\mathcal{X}$ by $\{\xi(S_n)\}_{n\geq \ell}$, i.e., the distribution of $\xi(S_n), n = \ell, \ell+1, \dots$, when ξ is fixed and $\{S_n\}$ has the distribution P.

The following are the basic problems.

Recognition problem: This is the analogue of a hypothesis testing problem in statistics. For two *known* sceneries ξ and η, are Q_ℓ^ξ and Q_ℓ^η mutually singular for all ℓ ? This is the case if and only if there exists a set $B \in \mathcal{F}_\infty$ such that

$$Q_0^\xi(B) = 1 \text{ and } Q_0^\eta(B^c) = 0. \tag{1.1}$$

A pair of sceneries ξ and η for which this holds is called *distinguishable*. The interpretation of this is that ξ and η are two known potential sceneries. One is given one of two sequences of observations $X^\xi = \{X_n^\xi\}_{n\geq\ell} = \{\xi(S_n)\}_{n\geq\ell}$ or $X^\eta = \{X_n^\eta\}_{n\geq\ell} = \{\eta(S_n)\}_{n\geq\ell}$ but one is not told which of the two sequences is given. Can one tell from the observations from which of the two sceneries they come (with zero probability of error) ? (see Remarks (i) and (ii) below).

Reconstruction problem: This is the analogue of an estimation problem in statistics. Can one find ξ with probability 1 from a sequence of observations X^ξ ? Equivalently, does there exist $\mathcal{F}_\infty$-measurable map $\Phi : \mathcal{X} \mapsto \Xi$ so that $\Phi(\{X_n^\xi\}_{n\geq 0}) = \xi$ a.e $[P]$? Can one perhaps even reconstruct ξ from $\{X_n^\xi\}_{n\geq\ell}$ for all ℓ ? (As discussed in Remark (ii), this question needs to be rephrased to avoid a trivial negative answer.)

Remarks

(i) In the recognition problem the requirement that $Q_\ell^\xi \perp Q_\ell^\eta$ for *all* ℓ is reasonable since in many cases ξ and η can be distinguished on the basis of a few observations. E.g., if $\xi(\mathbf{0}) \neq \eta(\mathbf{0})$, then the single observation X_0 tells us uniquely whether the observations come from ξ or η. We do not know much about the combinatorial problem which pairs of sceneries can be distinguished by a finite number of observations $X_0, \dots, X_m$.

(ii) If $\mathcal{G} = \mathbb{Z}$ or $\mathbb{Z}^2$ and v is such that

$$P\{S'_\ell - S''_\ell = v\} > 0 \tag{1.2}$$

for some ℓ and two independent copies $\{S'_n\}, \{S''_n\}$ of $\{S_n\}$, then we cannot distinguish ξ from $\theta_v\xi$, defined by $(\theta_v\xi)(w) = \xi(w+v)$. Indeed, let u be such that $P\{S_\ell = u\} > 0, P\{S_\ell + v = u\} > 0$ and let B be such that $Q_\ell^\xi(B) = 1$. Then $P\{\{\xi(S_p)\}_{p\geq\ell} \in B\} = 1$, and therefore $P\{S_\ell = u, \{\xi(S_p)\}_{p\geq\ell} \in B\} = P\{S_\ell = u\} > 0$. There then exists a set $C \subset (\mathbb{Z}^d)^{\mathbb{N}}$ such that

$$\{S_\ell = u, \{\xi(S_p)\}_{p\geq\ell} \in B\} = \{S_\ell = u\} \cap \{\{S_p - S_\ell\}_{p\geq\ell} \in C\},$$

and

$$0 < P\{S_\ell = u, \{\xi(S_p)\}_{p\geq\ell} \in B\} = P\{S_\ell = u\} P\{\{S_p - S_\ell\}_{p\geq\ell} \in C\}.$$

But then also

$$\begin{aligned} Q_\ell^{\theta_v\xi}(B) &= P\{\{\xi(S_p + v)\}_{p\geq\ell} \in B\} \\ &\geq P\{S_\ell + v = u\} P\{\{S_p - S_\ell\}_{p\geq\ell} \in C\} > 0. \end{aligned}$$

Thus we cannot have $Q_\ell^\xi \perp Q_\ell^{\theta_v \xi}$ and hence also $Q_m^\xi \perp Q_m^{\theta_v \xi}$ does not hold for any $m \geq \ell$.

For the same reasons we cannot hope to find a $\Phi_\ell \in \mathcal{F}_\ell$ such that $\Phi_\ell(X^\xi) = \xi$ a.e. $[P]$, when (1.2) holds, for then necessarily $P\{\Phi_\ell(X^\xi) = \theta_v \xi\} > 0$ also. We can only hope to distinguish pairs ξ, η for which $\eta \neq \theta_v \xi$ for all v satisfying (1.2) for some ℓ. Similarly we can only hope to reconstruct ξ up to a shift.

It is even more obvious that if $\mathcal{G} = \mathbb{Z}$ or $\mathbb{Z}^2$ and $\{S_n\}$ is a symmetric random walk on $\mathcal{G}$, then we also cannot distinguish ξ from its reflection $\widetilde{\xi}$ given by $\widetilde{\xi}(v) = \xi(-v)$.

Benjamini, as well as den Hollander and Keane, conjectured that if $\mathcal{G} = \mathbb{Z}$ and $\{S_n\}$ is simple random walk, then ξ and η are indistinguishable if and only if

$$\xi = \theta_v \eta \text{ or } \xi = \theta_v \widetilde{\eta} \text{ for some even } v.$$

Recently, however, Elon Lindenstrauss (1997) gave an ingenious construction (for $k = 2$ or even for $k = \infty$) of an uncountable number of indistinguishable sceneries on $\mathbb{Z}$. Clearly this shows that there exist pairs ξ, η of indistinguishable sceneries in which η is not obtainable from ξ by translation and/or reflection.

When $\mathcal{G} = \mathbb{Z}^d$ we call ξ and η *equivalent* (notation: $\xi \sim \eta$) if

$$\eta = \theta_v \xi \text{ or } \eta = \theta_v \widetilde{\xi} \text{ for some } v \in \mathbb{Z}^d. \tag{1.3}$$

In the recognition problem we then only consider pairs ξ, η with $\xi \not\sim \eta$; in the reconstruction problem we only require that

$$\Phi_\ell(X^\xi) \sim \xi \text{ a.s. } [P].$$

Note that in the definition (1.3) we allowed any shift, not just shifts for which (1.2) holds. This version of the definition simplifies the statements of some of the results, even though it does not always give the sharpest form of the results.

(iii) Since results for fixed ξ and/or η are scarce, we usually consider 'typical' ξ and/or η. This means that for finite k we introduce the product measure

$$\rho = \rho^{(k)} = \prod_{v \in \mathcal{V}} \rho_v \text{ with } \rho_v = \text{ uniform measure on } \{0, 1, \dots, k-1\}$$

on Ξ, and that we make statements about distinguishability or possibility of reconstruction, valid for almost all ξ $[\rho]$.

(iv) As pointed out, we call two sceneries ξ and η distinguishable if there exists a set B in $\mathcal{F}_\infty$ for which (1.1) holds. For the case $\mathcal{G} = \mathbb{Z}$, Howard (1995), (1996) and (1997) uses the slightly stricter definition that ξ and η are distinguishable if one can satisfy (1.1) for some translationinvariant set B (that is with $\theta^{-1}B = B$). Even though this makes ξ and $\theta_1 \xi$ distinguishable in our sense for almost all $\xi\,[\rho]$, but indistinguishable in Howard's sense, this difference in definition has little influence on most results so far (see also the last page of Howard (1997)).

(v) Harris and Keane (1996) studied a closely related question. Let $\{S_n\}$ be a simple random walk on $\mathbb{Z}^d$ and define the probability measure Q_p on $\{0,1\}^{\mathbb{N}}$ by means of its conditional measure, given the realization of $\{S_n\}$. Given $\{S_n\}$, the coordinates X_ℓ are independent, with

$$1 - Q_p\{X_\ell = 0 | \{S_n\}\} = Q_p\{X_\ell = 1 | \{S_n\}\} = \begin{cases} \frac{1}{2} & \text{if } S_\ell \neq \mathbf{0} \\ p & \text{if } S_\ell = \mathbf{0}. \end{cases} \tag{1.4}$$

The question is whether $Q_p \perp Q_{1/2}$ when $p \neq 1/2$. Harris and Keane show that indeed $Q_p \perp Q_{1/2}$ if $d = 1$, but this fails if $d = 2$. (Of course this also fails if $d \geq 3$, for then $S_n = \mathbf{0}$ only finitely often, with probability 1.) Harris and Keane raise the following elegant more general problem. Let $\mathcal{E} = (E_1, E_2, \dots)$ be a recurrent event as defined in Feller (1968), Ch. XIII, with the probabilities of occurrence

$$u_n = P\{E_n\}.$$

Conditionally on the outcome of the sequence $(E_1, E_2, \dots)$ the coordinates X_ℓ are still independent under Q_p, but (1.4) is now replaced by

$$1 - Q_p\{X_\ell = 0 | \{E_n\}\} = Q_p\{X_\ell = 1 | \{E_n\}\} = \begin{cases} \frac{1}{2} & \text{if } E_\ell^c \text{ occurs} \\ p & \text{if } E_\ell \text{ occurs.} \end{cases}$$

Harris and Keane prove that $\sum u_n^2 = \infty$ is a sufficient condition for $Q_p \perp Q_{1/2}$ when $p \neq 1/2$. They conjecture that this condition is also necessary, but are only able to prove that $\sum_{n=0}^{\infty} u_n^2 < 1 + |2p - 1|^{-2}$ implies that Q_p and $Q_{1/2}$ are not mutually singular.

(vi) Burdzy (1993) considers a continuous analogue of the reconstruction problem. Let $\{B_i(t) = B_i(t, \omega_i)\}, i = 1, 2$, be independent one-dimensional Brownian motions with $B_i(0) = \mathbf{0}$. $B_1(t)$ is defined for all $t \in \mathbb{R}$, while $B_2(t)$ only needs to be defined on $\{t \geq 0\}$. *Iterated Brownian motion* is the process $\mathcal{B}(t) := \{B_1(B_2(t))\}$. This is the analogue of $\{\xi(S_n)\}$ when ξ is also random with distribution ρ. B_1 corresponds to our scenery, or more precisely to the sums of the scenery $\sum_{1 \leq j \leq n} \xi(j)$ if $n \geq 0$ and $-\sum_{n+1 \leq j \leq 0} \xi(j)$ if $n < 0$; B_2 corresponds to the random walk in our setting. Burdzy proves that one can almost surely reconstruct B_1 and B_2, up to reflections, from a realization of $\mathcal{B}$, that is, there exists a function Φ, such that for almost all pairs (ω_1, ω_2), $\Phi(B_1(B_2(\cdot, \omega_2), \omega_1)) = (B_1(\cdot, \omega_1), B_2(\cdot, \omega_2))$ or $= (B_1'(\cdot, \omega_1), B_2'(\cdot, \omega_2))$, where $B_1'(t) = B_1(-t), B_2'(t) = -B_2(t)$.

(vii) The problems considered here can also be viewed as dealing with a hidden Markov chain. The hidden Markov chain is the random walk $\{S_n\}$ with the countable state space $\mathcal{V}$. The observations take values in $\{0, 1, \dots, k-1\}$. What is unknown here is which function of the Markov chain is observed. This function is given by the unknown coloring of $\mathcal{V}$.

In the next sections we list some results which have been obtained so far.

2. The reconstruction problem on $\mathbb{Z}$ when $\{S_n\}$ is simple random walk.

In this section $\mathcal{G} = \mathbb{Z}$ and $\xi \sim \eta$ if (1.3) holds. Of course the best result is if one can reconstruct the scenery ξ from the observations without any prior knowledge of ξ. As discussed in Remark (iii) this is not possible, and the best we can hope for is to reconstruct the equivalence class of ξ. It is remarkable that this is actually possible for a typical ξ and realization $\{S_n\}$ of the random walk, as stated in the first theorem.

THEOREM 1. *(Matzinger (1997)) If $\mathcal{G} = \mathbb{Z}$, $2 \leq k < \infty$ and $\{S_n\}$ is simple random walk, then there exists an $\mathcal{F}_\infty$-measurable function $\Phi : \mathcal{X} \mapsto \Xi$ such that*

$$\Phi(\{\xi(S_n)\}) \sim \xi \textit{ for almost all pairs } (\xi, \{S_n\})\, [\rho^{(k)} \times P]. \tag{2.1}$$

Matzinger's method seems to depend strongly on properties of simple random walk on the integers, in particular that it is skipfree. It does seem possible, though, to allow S_n to stand still. That is, the theorem seems to remain valid for any random walk $\{S_n\}$ on $\mathbb{Z}$ with $P\{S_{n+1}-S_n = 0\} = r \in [0,1)$, $P\{S_{n+1}-S_n = \pm 1\} = \frac{1}{2}(1-r)$. Such special properties of the random walk allow an observer to conclude that certain intervals of observations come with high probability from the same interval of the scenery. Thus this method relies on the 'finestructure' of the observations; it looks at the appearance of certain patterns in the observations. In contrast, most methods for the recognition problem use 'global' aspects of the observations, namely frequencies of certain colors in long sequences of observations. There is of course no contradiction between Matzinger's theorem and the counterexample of Lindenstrauss mentioned in Remark (ii), because Matzinger's result only states that *almost all* sceneries can be reconstructed.

By Fubini's theorem, (2.1) says that for almost all $\xi\,[\rho]$, one can reconstruct ξ with P-probability 1. It is still of interest to give explicit classes of sceneries for which such almost sure reconstruction is possible. Howard (1996), Theorems 3.1, 5.1 and Section 6, shows that for $\mathcal{G} = \mathbb{Z}$ this is possible for all sceneries which are periodic, except for a finite defect, that is for the class of sceneries Ξ_{def}, defined by

$$\Xi_{\text{def}} = \{\xi \in \Xi : \exists c \geq 1, L < \infty, \text{ and a } \xi' \in \Xi \text{ with } \xi'(v) = \xi'(v+c),\ v \in \mathbb{Z}, \\ \text{and } \xi(v) = \xi'(v) \text{ for } |v| > L\}.$$

(Note that Ξ_{def} includes all strictly periodic sceneries.) Howard even allows the random walk to pause at a site (and in some cases even to be asymmetric). Specifically, if

$$p_i = P\{S_{n+1} - S_n = i\},$$

then Howard considers all distributions with

$$p_i = 0 \text{ for } |i| > 1,\ p_{-1} + p_1 > 0. \tag{2.2}$$

THEOREM 2. *(Howard (1996)) If $\mathcal{G} = \mathbb{Z}$, and the random walk $\{S_n\}$ satisfies (2.2) with $p_{-1} = p_1$, then there exists an $\mathcal{F}_\infty$-measurable function $\Phi : \mathcal{X} \mapsto \Xi$ such that for all $\xi \in \Xi_{\text{def}}$*

$$\Phi(\{\xi(S_n)\}) \sim \xi \textit{ for almost all realizations of } \{S_n\}\,[P].$$

When ξ is restricted to the strictly periodic sceneries, then one can even drop the symmetry condition $p_{-1} = p_1$ (see Howard (1996), Theorem 4.2).

3. The recognition problem on $\mathbb{Z}^d$.

Our definitions are such that one expects that there are setups for which many pairs are distinguishable, even though one cannot reconstruct a scenery. The next theorem shows that on $\mathbb{Z}$ and $\mathbb{Z}^2$ 'almost all' pairs are distinguishable.

THEOREM 3. *(Benjamini and Kesten (1996)) If $\mathcal{G} = \mathbb{Z}$ or $\mathbb{Z}^2$, and $\{S_n\}$ is random walk with mean zero and bounded steps, then for every fixed $\xi \in \Xi$, ξ is distinguishable from almost all $\eta\,[\rho^{(k)}]$.*

Benjamini and Kesten only proved this theorem for simple random walk and with $k = 2$. However, the same proof works in the more general case stated here (one can actually weaken the hypothesis of bounded steps still further). Basically this is so, because the sceneries are distinguished by means of frequencies of the colors in the observations, rather than by occurrence of special patterns.

Howard (1997) simplified the proof of Theorem 3 somewhat. By some general ergodic theory observations he showed that it suffices to show that almost all pairs $(\xi, \eta) \in \Xi^{(k)} \times \Xi^{(k)}$ $[\rho^{(k)} \times \rho^{(k)}]$ are distinguishable; the stronger statement of Theorem 3 then follows automatically.

One would expect that for fixed k, it becomes more difficult and perhaps impossible to distinguish two 'typical' sceneries when the dimension d is large enough, because the random walk $\{S_n\}$ becomes transient. It seems that no such result is known when $\{S_n\}$ is simple random walk. The next result goes in this direction for an 'extremely transient' random walk, namely one which can only take steps in the positive direction. However, the theorem after that shows that transience of the random walk is not the only determining factor. If one also raises the number of colors, then one can still distinguish sceneries even in high dimension.

THEOREM 4. *(Benjamini and Kesten (1996)) Let* $\mathcal{G} = \mathbb{Z}^d$, $k = 2$ *and let* $\{S_n\}$ *be the oriented random walk whose steps have the distribution*

$$P\{S_{n+1} - S_n = e_i\} = \frac{1}{d}, \ 1 \le i \le d$$

(e_i *denotes the vector* $(0, \dots, 0, 1, 0, \dots, 0)$ *with the 1 in the i-th place). Then for* $d \ge$ *some* d_0 *almost all pairs* (ξ, η) $[\rho^{(2)} \times \rho^{(2)}]$ *are indistinguishable.*

It seems that the result of Theorem 4 holds for any fixed $k < \infty$ (of course with d_0 dependent on k).

THEOREM 5. *(Benjamini and Kesten (1996)) If* $\mathcal{G} = \mathbb{Z}^d$ *for some fixed d and* $\{S_n\}$ *is simple random walk, then there exists a* $k_0 = k_0(d)$ *such that for* $k \ge k_0$ *and each fixed* $\xi \in \Xi^{(k)}$, ξ *is distinguishable from almost all* η $[\rho^{(k)}]$.

Howard (1995) and (1997) describes a number of specific pairs ξ, η which can be distinguished.

THEOREM 6. *(Howard (1995)) Let* $\mathcal{G} = \mathbb{Z}$ *or* $\mathbb{Z}^2$ *and assume that* $\{S_n\}$ *is a random walk with*

$$E\{S_{n+1} - S_n\} = \mathbf{0}, \quad E\{|S_{n+1} - S_n|^2\} < \infty.$$

Let

$$N(\xi, i) = \sum_v I[\xi(v) = i]$$

be the number of times the color i appears in the scenery ξ. *If there is some color* i_0 *with*

$$0 < N(\xi, i_0) < \infty, \tag{3.1}$$

then ξ *is distinguishable from every* η *with* $N(\eta, i_0) \neq N(\xi, i_0)$.

If $\mathcal{G} = \mathbb{Z}$ *and* $\{S_n\}$ *is simple random walk on* $\mathbb{Z}$ *and (3.1) holds, then* ξ *is distinguishable from every* $\eta \not\sim \xi$ *(even with* $N(\eta, i_0) = N(\xi, i_0)$*).*

In trying to determine which pairs are distinguishable, one is naturally led to test pairs which seem difficult to distinguish. In particular, one may consider pairs which differ only at finitely many sites, or even only on a single site. Theorem 2 above showed that if these are sceneries on $\mathbb{Z}$ which are periodic outside a finite interval, then they can be distinguished (in fact one can even reconstruct a single scenery of this type). The next theorem distinguishes a random scenery ξ from a scenery which differs from ξ only at the origin. When $\{S_n\}$ is simple random walk on $\mathbb{Z}$, then Theorem 1 does much better.

THEOREM 7. *(Kesten (1996)) Let $\mathcal{G} = \mathbb{Z}$ and let the random walk $\{S_n\}$ satisfy*

$$E\{S_{n+1} - S_n\} = \mathbf{0}, \quad E\{[S_{n+1} - S_n]^2\} < \infty. \tag{3.2}$$

Define

$$p = \max_{i \neq 0} p_i.$$

Then for

$$k > \frac{1}{p^2} \tag{3.3}$$

one can distinguish almost all $\xi \in \Xi^{(k)}\,[\rho^{(k)}]$ from a scenery ξ' with

$$\xi'(v) = \begin{cases} \xi(v) & \text{if } v \neq \mathbf{0} \\ \neq \xi(v) & \text{if } v = \mathbf{0}. \end{cases}$$

Actually Kesten (1996) only proves this result when $\{S_n\}$ is simple random walk on $\mathbb{Z}$, in which case (3.3) reduces to $k > 4$. However, the more general case when only (3.2) holds can be proven in the same way as in Kesten (1996). One merely has to choose m_0 such that $p_{m_0} = p$, and to replace the event $\{S_{\lambda_j + i} = i \text{ for } 1 \leq i \leq L_n\}$ in (2.12) of Kesten (1996) by $\{S_{\lambda_j + i} = im_0 \text{ for } 1 \leq i \leq L_n\}$. The rest then requires only minor technical changes.

4. A negative result and open problems.

All the results cited so far (with the exception of Theorem 4 and the example of Lindenstrauss in Remark (ii)) are positive results, in the sense that they show that certain sceneries can be reconstructed or certain pairs of sceneries can be distinguished. Such results are proven by demonstrating that an explicit statistic does the trick. Such statistics are found by more or less ad hoc methods. The most obvious problem now is to prove general negative results which say that in certain cases one can *not* distinguish a typical pair of sceneries. The method of Theorem 4 to prove such results is rather involved. This requires a rather explict comparison of $Q^{\xi}_{\ell}\{X_{\ell+j} = w_j, 0 \leq j \leq N\}$ and $Q^{\eta}_{\ell}\{X_{\ell+j} = w_j, 0 \leq j \leq N\}$ for many sequences $(w_0, \dots, w_N)$ with $w_i \in \{0, \dots, k-1\}$. It seems hard to make this into a general method, but it can be pushed a little further to give the following result when $\mathcal{G}$ is a tree.

THEOREM 8. *(Benjamini and Kesten (1996)) Let $\mathcal{G} = \mathcal{T}_b :=$ the regular b-ary tree, $k = 2$, and let $\{S_n\}$ be simple random walk on $\mathcal{T}_b$. Then for large enough b, almost all pairs $(\xi, \eta)\,[\rho^{(2)} \times \rho^{(2)}]$ are indistinguishable.*

In connection with Theorem 4 we already mentioned the following problem.

Problem 1. Prove that for fixed k, $\mathcal{G} = \mathbb{Z}^d$ and $\{S_n\}$ simple random walk on $\mathbb{Z}^d$, almost all pairs $(\xi, \eta)\,[\rho^{(k)} \times \rho^{(k)}]$ are indistinguishable when d is large.

For pairs of sceneries which differ only at one site we expect that they already cannot be distinguished when $d = 2$. We formulate this as the next problem.

Problem 2. Let $k = 2$ and $d = 2$. Let $\{S_n\}$ be simple random walk on $\mathbb{Z}^2$. For any scenery $\xi \in \Xi^{(2)}$ let $\widehat{\xi}$ be the scenery defined by

$$\widehat{\xi}(v) = \begin{cases} \xi(v) & \text{if } v \neq \mathbf{0} \\ 1 - \xi(\mathbf{0}) & \text{if } v = \mathbf{0}. \end{cases}$$

Prove or disprove that for almost all $\xi\ [\rho^{(2)}]$, ξ and $\widehat{\xi}$ are indistinguishable.

Of a much more combinatorial nature are the following two problems, due to I. Benjamini (private communication).

Problem 3. What is the 'right' definition of equivalence of sceneries ? In view of the example of E. Lindenstrauss this is of importance even on $\mathbb{Z}$, where we don't know exactly which pairs of sceneries are indistinguishable when $\{S_n\}$ is simple random walk. When $d = 2$, we have even less of an idea which pairs of sceneries in $\Xi^{(k)}$ on $\mathbb{Z}^2$ are clearly indistinguishable. Here is a pair which is rather different from the one-dimensional pairs discussed in Remark (ii). With $v = (v_1, v_2)$ let

$$\xi(v) = \begin{cases} 0 & \text{if } v_1 \text{ is even} \\ 1 & \text{if } v_1 \text{ is odd.} \end{cases}$$

(Thus ξ has alternating vertical lines of zeroes and ones.) Also let

$$\eta(v) = \begin{cases} 0 & \text{if } \lfloor v_1/2 \rfloor + \lfloor v_2/2 \rfloor \text{ is even} \\ 1 & \text{if } \lfloor v_1/2 \rfloor + \lfloor v_2/2 \rfloor \text{ is odd.} \end{cases}$$

(Thus η has a checkerboard of alternating 2×2 squares of zeroes and ones.) One easily checks that in both sceneries each vertex v has two neighbors with color 0 and two neighbors with color 1. Thus under all measures Q_ℓ^ξ and Q_ℓ^η, X_i has the distribution of a sequence of i.i.d. random variables taking the values 0 and 1 with probability 1/2. Are there other such pairs ?

Problem 4. If $\mathcal{G} = \mathbb{Z}$, but $\{S_n\}$ is not a simple random walk, can one still reconstruct a periodic scenery ? For a simple explicit example let

$$P\{S_{n+1} - S_n = 1\} = P\{S_{n+1} - S_n = 2\} = \frac{1}{2}.$$

Can one reconstruct periodic sceneries in this case ?

References

Benjamini, I. and Kesten, H., *Distinguishing sceneries by observing the scenery along a random walk path*, J. d'Anal. Math. **69** (1996), 97-135.

Burdzy, K., *Some path properties of iterated Brownian motion*, Seminar on Stochastic Processes (K. L. Chung, E. Çinlar and M. J. Sharpe, eds.), Birkhäuser, Boston, 1993, pp. 67–87.

Feller, W., *An Introduction to Probability Theory and its Applications*, Vol I, 3rd Ed., John Wiley & Sons, 1968.

Harris, M. and Keane, M., *Random coin tossing*, preprint (1996).

Howard, C. D., *The orthogonality of measures induced by random walk with scenery*, Ph.D. thesis, Courant Inst. Math. Sciences, New York, 1995.

Howard, C. D., *Detecting defects in periodic scenery by random walks on* $\mathbb{Z}$, Random Structures and Algorithms **8** (1996), 59–74.

Howard, C. D., *Distinguishing certain random sceneries on* $\mathbb{Z}$ *via random walks*, Statistics and Probability Letters (1997).

Kesten, H., *Detecting a single defect in a scenery by observing the scenery along a random walk path*, Itô's Stochastic Calculus and Probability Theory (N. Ikeda, S. Watanabe, M. Fukushima and H. Kunita, eds.), Springer-Verlag, 1996, pp. 171-183.

Lindenstrauss, E., *Indistinguishable sceneries*, in preparation (1997).

Matzinger, H., Ph.D. Thesis Cornell University, in preparation.

Department of Mathematics White Hall Cornell University Ithaca NY 14853
E-mail address: kesten@math.cornell.edu

DIMACS Series in Discrete Mathematics
and Theoretical Computer Science
Volume 41, 1998

Mixing Times

László Lovász and Peter Winkler

ABSTRACT. The critical issue in the complexity of Markov chain sampling techniques has been "mixing time", the number of steps of the chain needed to reach its stationary distribution. It turns out that there are many ways to define mixing time—more than a dozen are considered here—but they fall into a small number of classes. The parameters in each class lie within constant multiples of one another, independent of the chain. Furthermore, there are interesting connections between these classes related to time reversal.

This work is more in the nature of a long research paper than a survey, with many new results and proofs. Some of the results have appeared in recent articles or were known previously for the important special case of reversible chains.

CONTENTS

1. Introduction and preliminaries

In the past ten years there have been numerous applications (see, e.g., [**A2**], [**JS**], [**DFK**]) for sampling via finite Markov chains. A Markov chain is constructed from whose stationary distribution one wishes to sample, and then the chain is run for a fixed number of steps after which the distribution of the current state is "nearly" stationary. To determine this fixed number, the "mixing time", is usually the main difficulty in applying the method. Various methods have been developed to estimate the mixing time of a chain: eigenvalues, coupling, coupling from the past, conductance, strong stopping times, etc.

Various applications of this method require different interpretations of the above scheme. First of all, we do not always want to get close to the stationary distribution in the same sense; the measure of distance from the stationary

1991 *Mathematics Subject Classification.* Primary 60G40, 62L15; Secondary 60J10, 60J15.
The first author was supported in part by NSF Grant #CCR-9712403.

distribution can be chosen in many ways, or rather, the specific needs of the application can lead to different measures here: total variation distance, ℓ_2-distance, χ^2-distance, entropy distance, pointwise filling up etc.

There are other variations of the problem raised by different applications. Perhaps the most basic set-up is when we want to generate a single state from the stationary distribution, starting from some fixed state (determined by the rest of the algorithm). This leads to the definition of the mixing time from a given state. If we don't have more information about the starting state, we have to use the maximum of this over all starting states, getting the basic definition of mixing time.

However, in some applications the starting state is already random, and we only want to use the Markov chain to improve its quality. In this case, one might hope that one can take advantage of the randomness already present. We call this situation the "warm start", and our mixing times will have "warm start" versions.

Quite often, we need to generate several independent (or approximately independent) samples from the stationary distribution. In this case we might start the second run of the Markov chain where the first one stopped, and so the expected time needed for this will be the average, rather than the maximum, of mixing times from individual states. This leads us to the definition of the *reset time.* (Note that this is different from warm start: we cannot make use of the randomness present in the starting state if we want independence.)

As a further example, we may use the Markov chain to find an element from a given, but not directly accessible subset of the state space. The worst expected time needed for this (normalized by the measure of the subset) is the *set hitting time.*

There is no particular reason why a walk must be run for a fixed number of steps; in fact, more general stopping rules, some of which "look where they are going", are capable of achieving the stationary distribution exactly. It turns out to be useful to consider stopping rules that achieve any given distribution, when starting from some other given distribution. From a theoretical standpoint, our stopping rules result in a generalization of the notion of "hitting time" to state-distributions. This measure of distance between distributions behaves quite nicely.

(We put no restriction on the amount of computation needed to implement a stopping rule, making the use of these rules as sampling mechanisms unlikely; but such rules are useful in analysis and, as we shall see, often replaceable by simpler ones without too much loss in efficiency.)

By considering the least expected number of steps required by rules of a particular kind to reach the stationary distribution (or some approximation thereof) we obtain further measures of mixing time. To these we may add some additional parameters, related to eigenvalues and the reverse of the given chain. Altogether we obtain a substantial collection of parameters each of which has mixing time implications.

Our objective is to show that the situation is not so hopelessly complex. The main result is that we can place these numbers into a few equivalence classes, within which numbers differ only by constant factors independent of the chain (see Section 3). For the case of reversible chains, such results were obtained by Aldous [**A2**]; other results in this direction were published in [**ALW, LW4, LW5**]. To prove the main results, we develop a calculus of "exit frequencies", study the behavior of mixing properties under time reversal, and make use of some linear algebraic tools. We hope that these tools shed some light on the mechanism of mixing.

Acknowledgement. We are grateful to David Aldous for useful suggestions and to Fang Chen for her careful reading of this manuscript, which resulted in many corrections and clarifications.

1.1. Preliminaries. Throughout this paper we will assume a fixed irreducible Markov chain with transition matrix $M = \{p_{ij}\}$, with a finite state space V of finite cardinality n (see [**ALW**] for extensions to infinite state space). If we say "distribution" without specifying an underlying set, we mean distribution on V. If σ is a distribution on V and $A \subseteq V$, then $\sigma(A) = \sum_{i\in A} \sigma_i$ is the probability of A. If $\sigma(A) > 0$, then we can consider the distribution σ_A defined by

$$(\sigma_A)_i = \begin{cases} \frac{\sigma_i}{\sigma(A)} & \text{if } i \in A \\ 0 & \text{otherwise.} \end{cases}$$

We denote by π the stationary distribution of the chain. The *intensity* of a distribution σ is $\max_{i\in V} \sigma_i/\pi_i$.

Most of the time, we use the *total variation metric* to describe the distance of two distributions. This is defined by

$$d(\sigma, \tau) = \max_{A\subseteq V}(\sigma(A) - \tau(A)).$$

It is easy to see that $d(\sigma, \tau) = d(\tau, \sigma)$, and this value is just half of the ℓ_1-norm of $\|\sigma - \tau\|$.

For two distributions $\sigma \neq \tau$, we define the "difference" distribution $\sigma \setminus \tau$ by

$$(\sigma \setminus \tau)_i = \frac{\max\{0, \sigma_i - \tau_i\}}{\sum_j \max\{0, \sigma_j - \tau_j\}}.$$

Given a Markov chain with transition probabilities p_{ij}, we define the *reverse chain* as the Markov chain on the same set of states, with transition probabilities $\overleftarrow{p}_{ij} = \pi_j p_{ji}/\pi_i$. We will generally use the reverse arrow over a symbol to indicate that it refers to this reverse chain.

It is trivial but important that the reverse chain has the same stationary distribution as the original. This follows by straightforward substitution. It turns out that there is a very close relationship between mixing properties of a chain and its reverse.

1.2. Examples: stopping rules. In this section we discuss examples which show that "intelligent" stopping rules can sometimes achieve specified distributions in an elegant or surprising manner.

EXAMPLE 1.1 (Cycle). The following is an interesting fact from folklore. Let G be a cycle of length n and start a random walk on G from a node u. Then the probability that v is the last node visited (i.e., the a random walk visits every other node before hitting v) is the same for each $v \neq u$.

While this is not an efficient way to generate a uniform random points of the cycle, it indicates that there are entirely different ways to use random walks for sampling than walking a given number of steps. This particular method does not generalize; in fact, apart from the complete graph, the cycle is the only graph which enjoys this property (see [**LW1**]).

EXAMPLE 1.2 (Cube). Consider another quite simple graph, the cube, which we view as the graph of vertices and edges of $[0,1]^n$. Let us do a random walk on it as follows: at each vertex, we select a direction (that is, one of the n coordinate indices) at random, then flip a coin. If we get "heads" we walk along the incident edge corresponding to that direction; if "tails" we stay where we are. We stop when we have selected every direction at least once (whether or not we walked along the edge).

It is trivial that after each of the n directions has been selected, the corresponding coordinate will be 0 or 1 with equal probability, independently of the rest of the coordinates. So the vertex we stop at will be uniformly distributed over all vertices.

This method takes about $n \ln n$ coin flips on the average, thus about $n \ln n/2$ actual steps, so it is a quite efficient way to generate a random vertex of the cube (assuming we insist on using random walks; otherwise choosing the coordinates independently is simpler and faster). We will see that it is in fact optimal.

EXAMPLE 1.3 (Card Shuffling). A classic application of Markov chain mixing is shuffling a deck of playing cards; see e.g. [**BD**] where Bayer and Diaconis argue that seven riffle shuffles are necessary and sufficient to mix a deck of 52 cards. In Aldous and Diaconis [**AD**], the following simple shuffling algorithm is analyzed: a card is removed from the top of an n-card deck and replaced with equal probability in any of the n slots among the remaining $n-1$ cards.

It is not difficult to see that if we note when the card originally at the bottom of the deck has reached the top and perform just one more shuffle, then the deck will be precisely uniformly random. Again, this method takes about $n \ln n$ steps on the average, and also as in the cube case, constitutes a "strong" stopping rule in the sense of [**AD**] (the ending distribution is uniform even when conditioned on the stopping rule having taken some fixed number of steps). However, we will see later that this shuffling rule is *not* optimal.

Without going into details, let us remark that the algorithm of Aldous [**A3**] and Broder [**BR**] can be regarded as a stopping rule for a Markov chain on trees that generates the uniform distribution.

Finally, we note that to stop after a fixed number of steps in the "continuous" time model of Markov chains (where the time needed by a step is an exponentially distributed random variable) can be viewed as choosing a random number T from a Poisson distribution and stopping after T steps. So here a (randomized) stopping rule is considered which, in many respects, has better properties than the "stop after t steps" rule. A similar remark applies to stopping a "lazy" walk as defined, e.g., in [**LS**].

2. Access times and mixing times

2.1. General stopping rules. We denote by V^* be the space of finite "walks" on V, that is, the set of finite strings $w = w_0 w_1 \dots w_t$, $w_i \in S$. Relative to the Markov chain we then have

$$\mathsf{P}(w) = \sigma_{w_0} \prod_{i=0}^{t-1} p_{w_i, w_{i+1}} \cdot$$

The distribution of w_t will be denoted by σ^t, so that

$$\sigma^0 = \sigma \qquad \text{and} \qquad \sigma_i^t = \sum_{w_t = i} \mathsf{P}(w_0 \dots w_t).$$

A *stopping rule* Γ is a rule that observes the walk and tells us whether to stop or not, depending on the walk seen so far (but independently of the continuation of the walk). This decision may be reached using coin flips; so the stopping rule just has to specify, for each walk w, the probability of continuing the walk. Formally, Γ is a map from V^* to $[0,1]$ (it would be enough to define Γ for walks w with $\mathsf{P}(w) > 0$). We interpret $\Gamma(w)$ as the probability of continuing given that w is the walk so far observed, each such stop-or-go decision being made independently. We assume that with probability 1 the walk eventually stops.

We can also regard Γ as a *stopping time*, i.e., a random variable with values in $\{0, 1, \dots\}$, so that we stop at w_Γ.

The probability of stopping at state j, given starting distribution σ, is

$$\sigma_j^\Gamma := \sum_{w=(w_0,\dots,w_t=j)} \sigma_{w_0} \left(\prod_{i=0}^{t-1} \Gamma(w_0, \dots, w_i) p_{w_i, w_{i+1}} \right) (1 - \Gamma(w)) \,.$$

We often think of a stopping rule Γ as a means of moving from a starting distribution σ to a given target distribution $\tau = \sigma^\Gamma$; we say then that Γ is a stopping rule from σ to τ.

For any Markov chain and any two distributions σ and τ, there is at least one finite stopping rule Γ from σ to τ; namely, we select a random target state j in accordance with τ and walk until we reach j. We call this the "naive" stopping rule $\Omega_{\sigma,\tau}$.

Note that if $\mathbf{X}$ is a random variable whose values are stopping rules, then $\mathbf{X}$ is equivalent to a stopping rule Γ whose probability of continuing at w is a sum of $\mathbf{X}(w)$ conditioned on $\mathbf{X}$ not having stopped the chain so far. Thus our stopping rules are not less general for insisting on independent randomization at each step. For example, the stopping rule given above in Example 1.2 (Cube) does not at first appear to fit our model, but we may simply walk on the (loopless) cube and stop according to the probability that the "draw and flip" rule would stop, given that it reached the current time and state.

The *mean length* $\mathsf{E}_\sigma\Gamma$ of the stopping rule Γ is its expected duration, starting from the distribution σ. Most often we'll have the distribution σ tacitly included with the stopping rule, and we just write $\mathsf{E}\Gamma$.

If $\mathcal{H}(i,j)$ denotes the expected hitting time (mean number of steps to reach j from i), then the mean length of $\Omega_{\sigma,\tau}$ is given by

$$\mathcal{N}(\sigma, \tau) = \sum_{i,j} \sigma_i \tau_j \mathcal{H}(i,j)\,.$$

When the target is the stationary distribution π for the chain, $\mathsf{E}\Omega_{\sigma,\pi}$ is independent of the starting distribution σ, on account of what we call the "Random Target Identity" (also known as the right averaging principle, e.g. in [**AF**]) which says that there is a constant $\mathcal{N}$ for which

$$\mathcal{N}(\sigma, \pi) = \mathcal{N} \tag{2.1}$$

for every starting distribution σ. In particular, we have

$$\mathcal{N} = \sum_j \pi_j \mathcal{N}(\pi, j) = \sum_j \pi_j \mathcal{H}(i, j) \tag{2.2}$$

for every state i.

2.2. Access times. A stopping rule Γ is said to be *mean-optimal* or simply *optimal* (for σ and τ) if $\mathsf{E}\Gamma$ is minimal. The mean length of a mean-optimal stopping rule from σ to τ will be called the *access time* from σ to τ and denoted $\mathcal{H}(\sigma, \tau)$. As suggested by the notation, we think of $\mathcal{H}(\sigma, \tau)$ as a generalized hitting time. It will turn out that $\mathcal{H}(\sigma, \tau)$ leads to the most important mixing measures we are going to study.

The access time between distributions behaves in many respects as a "distance" between the two distributions. Trivially, $\mathcal{H}(\sigma, \tau) = 0$ if and only if $\sigma = \tau$. It is easy to see that the following *triangle inequality* is satisfied for any three distributions α, β and γ:

$$\mathcal{H}(\alpha, \gamma) \leq \mathcal{H}(\alpha, \beta) + \mathcal{H}(\beta, \gamma)\,. \tag{2.3}$$

(To generate γ from α, we can first use an optimal rule to generate β from α and then use the state obtained as a starting state for an optimal rule generating γ from β). We should warn the reader, however, that $\mathcal{H}(\sigma, \tau) \neq \mathcal{H}(\tau, \sigma)$ in general.

Thus the access time $\mathcal{H}(\sigma, \tau)$ has the properties of a metric on the space of distributions, except for symmetry; the latter is of course too much to expect since the ordinary hitting time, even for a reversible chain, is not generally symmetric.

If τ is concentrated at state j (for which we write, rather carelessly, "$\tau = j$") then

$$\mathcal{H}(\sigma, j) = \sum_i \sigma_i \mathcal{H}(i, j) \tag{2.4}$$

since clearly the only optimal stopping rule in this case is $\Omega_{i,j}$, "walk until state j is reached." This rule is therefore mean-optimal; it also minimizes the *maximum* number of steps, although in general $\max(\Omega_{i,j})$ will be infinite. By considering the naive rule $\Omega_{\sigma,\tau}$, we get the inequality

$$\mathcal{H}(\sigma, \tau) \leq \mathcal{N}(\sigma, \tau) = \sum_{i,j} \sigma_i \tau_j H(i, j)\,. \tag{2.5}$$

This may be quite far from equality; for example, $\mathcal{H}(\sigma, \sigma) = 0$ for any σ.

The following formula for access times was given in [**LW3**]:

THEOREM 2.1. *For all distributions σ and τ,*

$$\mathcal{H}(\sigma, \tau) = \max_j \left(\mathcal{H}(\sigma, j) - \mathcal{H}(\tau, j)\right).$$

By (2.4), we can write this as

$$\mathcal{H}(\sigma, \tau) = \max_j \sum_i (\sigma_i - \tau_i) \mathcal{H}(i, j)\,.$$

The inequality $\geq$ in the theorem is trivial by the triangle inequality. The reverse inequality depends on a simple characterization of optimal rules. For any rule $\Gamma: \sigma \to \tau$, a state j is called a *halting state*, if it is never exited provided the rule Γ is obeyed. Thus by definition we stop immediately if and when any halting state is entered. (Of course we may stop in other states too, just not all the time.) The following theorem from [**LW2**] and [**LW4**] is very handy for determining optimality of stopping rules.

THEOREM 2.2. *A stopping rule is mean-optimal if and only if it has a halting state.*

The maximum in Theorem 2.1 is achieved precisely when j is a halting state of some optimal rule which stops at τ when started at σ.

In the case of time-reversible chains, we can use the cycle-reversing identity from [**CTW**] to obtain the following expression:

$$\mathcal{H}(\sigma,\tau)=\sum_i(\tau_i-\sigma_i)\mathcal{H}(\pi,i)-\min_j\sum_i(\tau_i-\sigma_i)\mathcal{H}(j,i)\,.$$

We will be concerned mostly with the case when $\tau=\pi$ and σ is concentrated at a single state i. In this case this formula can be simplified using the Random Target Identity (2.1) to get that

$$\mathcal{H}(i,\pi)=\max_j\mathcal{H}(j,i)-\mathcal{H}(\pi,i)\,. \tag{2.6}$$

Thus the halting state of a reversible chain, in attaining the stationary distribution from a fixed state i, is the state j *from which* the mean time to hit i is greatest.

The quantity $\mathcal{H}(\sigma,j)-\mathcal{H}(\tau,j)$ in Theorem 2.1 turns out to be a very useful measure of the "distance" between σ and τ. Instead of the maximum in j, we can also introduce the average

$$\mathcal{Z}(\sigma,\tau)=\sum_j\pi_j|\mathcal{H}(\sigma,j)-\mathcal{H}(\tau,j)|$$

Note that $\sum_j\pi_j(H(\sigma,j)-H(\tau,j))=0$ by the Random Target Identity, and hence we could replace the absolute values by the "surpluses" $\max\{0,H(\sigma,j)-H(\tau,j)\}$, at the cost of a factor of exactly 2. It is obvious from this definition that $\mathcal{Z}(\sigma,\tau)\le\mathcal{H}(\sigma,\tau)$.

2.3. Different mixing measures: definitions. We can use the notion of the access time $\mathcal{H}(\sigma,\tau)$ between distributions to define a parameter-free notion of mixing time. In fact, as described in the introduction, we have several different, but reasonably well-motived ways to do so, depending on assumptions about the starting distribution, approximation of the target distribution, and possible additional requirements like independence of the starting and ending state.

Notation. Before defining some of these notions of mixing time, it will be useful to establish a reasonably consistent notation scheme. We will use calligraphic letters (like $\mathcal{H}$ or $\mathcal{N}$) to denote mixing measures with different letters often used to distinguish classes of stopping rules. The dependence of a mixing measure on the (fixed) transition matrix of a Markov chain will always be understood.

As before we permit initial and target distributions as parameters, e.g. $\mathcal{H}(\sigma,\tau)$ for the access time from σ to τ, but we will extend the scope of the arguments even further, to sets of distributions. If A and B are two nonempty sets of distributions, we define

$$\mathcal{H}(A,B):=\max_{\alpha\in A}\min_{\beta\in B}\mathcal{H}(\alpha,\beta)$$

(and analogously for other mixing measures).

Thus $\mathcal{H}(A,B)\le t$ means that for every starting distribution in A there is a stopping rule with mean length at most t that yields a distribution in B. For example, $\mathcal{H}(V, \leq 2\pi)$ would be the minimum expected time needed to reach a distribution $\tau \leq 2\pi$ from a worst single state (equivalently, in this case, from a worst distribution of states). Note that in this notation, the triangle inequality extends to access times between sets:

$$\mathcal{H}(A,C) \leq \mathcal{H}(A,B) + \mathcal{H}(B,C). \tag{2.7}$$

Often mixing times will be computed from the worst starting distribution among *all* distributions—as in the above case this is usually equivalent to worst *state*—and that will be the assumption if there is only one argument to $\mathcal{H}$. Thus

$$\mathcal{H}(B) := \mathcal{H}(V,B) = \max_{s\in V} \min_{\tau\in B} \mathcal{H}(s,\tau).$$

If there are *no* arguments to $\mathcal{H}$ then the starting distribution is worst case and the target distribution π, as it is in most applications. In symbols:

$$\mathcal{H} := \mathcal{H}(\pi) = \max_{\sigma} \mathcal{H}(\sigma,\pi) = \max_{s\in V} \mathcal{H}(s,\pi) = \mathcal{H}(V,\pi)\,.$$

We bestow upon $\mathcal{H}$ the honor of calling it *the* mixing time of the chain. The state $\dot{s}$ attaining the maximum in the definition of $\mathcal{H}$ will be called a *pessimal* state.

We have already seen the notation

$$\mathcal{N} := \mathcal{N}(\pi) := \mathcal{N}(\sigma,\pi)$$

for the naive rule and the constant in the Random Target Identity, where σ is any distribution.

Approximate mixing. Often we want to reach only a distribution close to π. Let $d_\varepsilon(\tau)$ denote the ball of radius ε and center τ in the total variation metric. Most often we consider the case $\tau = \pi$, and we set $d_\varepsilon = d_\varepsilon(\pi)$.

We indicate pointwise approximation from above and from below by a bar over or under the subscript, and define

$$D_{\overline{\varepsilon}} = \{\sigma \in D:\ \sigma \leq (1+\varepsilon)\pi\}\,,$$
$$D_{\underline{\varepsilon}} = \{\sigma \in D:\ \sigma \geq (1-\varepsilon)\pi\}\,,$$

and

$$D_\varepsilon = D_{\overline{\varepsilon}} \cap D_{\underline{\varepsilon}} = \{\sigma \in D:\ (1-\varepsilon) \leq \sigma \leq (1+\varepsilon)\pi\}\,.$$

The set $D_1 = \{\sigma : \sigma \leq 2\pi\}$ will play a particularly important role.

We define

$$\mathcal{H}_\varepsilon := \mathcal{H}(D,d_\varepsilon) = \max_{s} \min_{\tau:\ d(\tau,\pi)<\varepsilon} \mathcal{H}(s,\tau)\,.$$

In terminology, $\mathcal{H}_\varepsilon$ becomes the "approximate mixing time". We also define

$$\mathcal{H}_{\overline{\varepsilon}} := \mathcal{H}(D, \leq (1+\varepsilon)\pi) = \max_{s\in V} \min_{\tau\leq(1+\varepsilon)\pi} \mathcal{H}(s,\tau)$$

and analogously

$$\mathcal{H}_{\underline{\varepsilon}} := \mathcal{H}(D, \geq (1-\varepsilon)\pi).$$

The former will be called the "dispersion time" and the latter the "filling time". At this point, the distinction between mixing in the "filling", "disperse" and "total variation" sense may seem pedantic, but in fact these three mixing measures behave quite differently.

Warm start. We also consider the effect of a "warm start", where the starting distribution σ has bounded intensity (most often intensity at most 2, which is an arbitrary but convenient choice). We will indicate a warm start in our mixing time notation by placing a tilde over the calligraphic letter.

We define

$$\tilde{\mathcal{H}} := \mathcal{H}(D_1, \pi) = \max_{\sigma \le 2\pi} \mathcal{H}(\sigma, \pi)$$

and

$$\tilde{\mathcal{H}}_\varepsilon := \mathcal{H}(D_1, d_\varepsilon) = \max_{\sigma \le 2\pi} \min_{\tau:\ d(\tau,\pi)<c} \mathcal{H}(\sigma, \tau)\,.$$

Another, somewhat different way of handling warm start is to consider a non-empty set $S \subseteq V$ of states, and start from the stationary distribution, but condition on starting in S. We are then interested in, say, $\mathcal{H}(\pi_S, \pi)$. Of course, this quantity increases if S gets smaller, so it is natural to scale and consider

$$\hat{\mathcal{H}} = \max_S \pi(S) H(\pi_S, \pi).$$

We call $\mathcal{H}$ the *mixing time with restricted start*, or briefly *restricted mixing time*. Again, we can define the restricted version of $\mathcal{H}_\varepsilon$ etc.

Blind rules. A general stopping rule, which may make use of a complete knowledge of the chain, and may do an unlimited amount of computation to decide when to stop, is useless from a practical point of view. Therefore we also introduce a type of stopping rule which, while not capable of achieving arbitrary target distributions, and almost never optimal, is easily implementable and will often provide a good approximation of more sophisticated stopping rules. We call a stopping rule Γ *blind* if $\Gamma(w)$ depends only on the length $|w|$ of the walk.

The simplest blind stopping rule is the stopping rule "stop after t steps". We call this the *t-step-rule.* Several other practical methods to generate elements from the stationary distribution (approximately) can also be viewed as blind rules. For example, the *uniform averaging rule* (to be discussed below) of walking u steps and then choosing one of the exited points uniformly can be viewed as a blind rule: after t steps, we stop with probability $1/(u-t)$. *Lazy* random walks have been considered (see Lovász and Simonovits [**LS**]) because they have better convergence to the stationary distribution; the lazy version of a Markov chain is obtained by flipping a coin before each move and staying where we are if we see "heads". Stopping the lazy version of a Markov chain after u steps is equivalent to following the original walk $(w_0, w_1, \dots, w_u)$ for u steps and then choosing a w_t according to the binomial distribution, i.e. with probability $\binom{u}{t}2^{-u}$. This is again equivalent to a blind rule, where we stop after t steps with probability

$$\binom{u}{t} \Big/ \left(\binom{u}{t} + \binom{u}{t+1} + \cdots + \binom{u}{u}\right)\,.$$

It is often more convenient to describe a blind stopping rule by the probabilities a_t that state v_t is selected (not conditioning on not having stopped before). Trivially $a_t \ge 0$ and the finite termination of the rule is equivalent to $\sum_t a_t = 1$. The rule is bounded if and only if the sequence a_t has a finite number of non-0 terms. Thus a blind rule can always be thought of as an averaging, using the distribution (a_t).

One cannot generate any distribution by a blind stopping rule; for example, starting from the stationary distribution, every blind rule generates the stationary distribution itself. Blind stopping rules will be used to generate the stationary distribution, or at least approximations of it.

Unfortunately, blind rules to achieve the stationary distribution exist only for a rather restricted class of chains, as the following result shows, stated here and in [**LW4**] without proof.

THEOREM 2.3. *There is a finite blind stopping rule that generates π from any starting state if and only if no eigenvalue of the transition matrix is a positive real number.*

Interestingly, the condition formulated in the theorem is most restrictive for time-reversible chains; then all the eigenvalues are real, and typically many of them are positive. Only very special Markov chains admit a blind rule for generating π which is mean-optimal.

We introduce the letter $\mathcal{B}$ to denote mixing time using blind rules, so that

$$\mathcal{B}(\sigma,\tau) := \min_{\Gamma \text{ blind}, \ \sigma^\Gamma=\tau} \mathsf{E}\Gamma$$

and for distribution-sets A and B,

$$\mathcal{B}(A,B) := \max_{\alpha\in A} \min_{\beta\in B} \mathcal{B}(\alpha,\beta)\,.$$

This quantity may well be infinite (for example, when $A = \{\pi\}$ and B consists of a single state different from π). But blind rules will be important in approximate mixing. In particular, we define the "blind approximate mixing time"

$$\mathcal{B}_\varepsilon := \mathcal{B}(d_\varepsilon(\pi)) = \max_\sigma \mathcal{B}(\sigma, d_\varepsilon(\pi)).$$

We can further restrict the rules used in the definition of these mixing times, to get closer to practically implementable algorithms. To the letter $\mathcal{B}$, reserved for blind rules, we can add the even more restrictive $\mathcal{U}$ for uniform averaging rules. The exact forms $\mathcal{B}(\sigma,\tau)$ and $\mathcal{U}(\sigma,\tau)$ may be infinite, but the approximate forms, e.g. $\mathcal{U}_\varepsilon$, make sense.

We define the warm *blind* and *uniform* mixing times $\tilde{\mathcal{B}}_\varepsilon$ and $\tilde{\mathcal{U}}_\varepsilon$ in a manner analogous to $\tilde{\mathcal{H}}_\varepsilon$ but for blind and uniform stopping rules.

The maximum number of steps. The *maximum length* $\max(\Gamma)$ of a stopping rule Γ is the maximum length of a walk that has positive probability. Γ is said to be *bounded* if this quantity is finite. The naive rule is in general not bounded, but bounded stopping rules are available when the target distribution τ has sufficiently large support, thus in particular when $\tau = \pi$ (see [**LW3**]).

A stopping rule is *max-optimal* (for given σ and τ) if $\max(\Gamma)$ is minimal. The maximum length of a max-optimal stopping rule from σ to τ will be denoted by $\mathcal{M}(\sigma,\tau)$. Thus

$$\mathcal{M}(\sigma,\tau) := \min_{\Gamma:\ \sigma^\Gamma=\tau} \max(\Gamma)\,.$$

Then

$$\mathcal{M} := \mathcal{M}(V,\pi) := \max_\sigma \mathcal{M}(\sigma,\pi)$$

is the related mixing measure, which we cannot resist calling the "maxing time" of the chain.

Independent rules. In many applications repeated independent samples are required from the stationary distribution of a chain. If we use a stopping rule to generate these states, we need to make the following assumption about the rules. A stopping rule is said to be *independent* if the state at which it stops the chain is independent of the one in which the chain was started. If we write

$$\mathcal{I}(\sigma,\tau) := \min\{\mathsf{E}\Gamma : \ \Gamma \text{ independent}\}\ ,$$

then we have

$$\mathcal{I}(\sigma,\tau) = \sum_{i\in V} \sigma_i \mathcal{H}(i,\tau)$$

since an independent rule must produce the target distribution "separately" from each starting state.

To get the time needed to generate an independent sample state, we let both the default initial and target distributions for an independent rule be π:

$$\mathcal{I} := \mathcal{I}(\pi,\pi) = \sum_i \pi_i \mathcal{H}(i,\pi)$$

obtaining what we call the *reset time* of the chain (see [**LW5**]). This is only formally similar to Formula (2.2) for $\mathcal{N}$; in general, $\mathcal{N}$ is much larger. (Instead of π, one might want to generate independent samples from some other distribution. This leads to the notion of the *regeneration time* $\max_\alpha \mathcal{I}(\alpha,\alpha)$, studied in [**BL**]; however, this turns out to be closely related to commute times rather than mixing times, and will not be discussed here.)

We also introduce an approximate version of the independence time. We say that two random variables X and Y (with values from a set V) are *ε-independent*, if for every two sets $A \subseteq V$ and $B \subseteq V$, we have

$$\Big|\mathsf{P}(X\in A, Y\in B) - \mathsf{P}(X\in A)\mathsf{P}(Y\in B)\Big| \le \varepsilon\,.$$

This is a very weak notion of independence, but in some applications of Markov chain techniques to sampling [**KLS**], this is exactly what is needed.

Let the *warm reset time* $\tilde{\mathcal{I}}_\varepsilon$ be the smallest t such that for every starting distribution σ with $\sigma \le (1+\varepsilon)\pi$, there exists a stopping rule Γ with $\mathsf{E}\Gamma \le t$ such that if v^0 is from σ then v^Γ is ε-independent of the starting state, and $\sigma^\Gamma \le (1+\varepsilon)\pi$.

Hitting sets. Another very natural extension of hitting time is1 set-hitting time: for $U \subseteq V$, $U \ne \emptyset$, let $\mathcal{H}(s,U)$ be the expected number of steps before hitting U, when starting from s. More generally, if σ is any starting distribution, then we define $\mathcal{H}(\sigma,U)$ as the expected number of steps before hitting U, when starting from a random node drawn from σ; clearly, $H(\sigma,U) = \sum_i \sigma_i H(i,U)$. For two distributions σ and τ, we define

$$\mathcal{S}(\sigma,\tau) = \max_{A\subseteq V} \tau(A)\mathcal{H}(\sigma,A)\,.$$

It is easy to see that if $\sigma' \le A\sigma$ and $\tau' \le B\tau$, then

$$\mathcal{S}(\sigma',\tau') \le AB\mathcal{S}(\sigma,\tau). \tag{2.8}$$

The *set access time* $\mathcal{S}$ is the mean number of steps required to hit the "toughest" set of states (adjusted by the stationary probability of the set) from the worst start:

$$\mathcal{S} := \max_s \mathcal{S}(s,\pi) = \max_{s\in V:\ U\subseteq V} \pi_U \mathcal{H}(s,U)\,.$$

We can define the warm start version of the set access time simply as $\tilde{\mathcal{S}} = \mathcal{S}(\pi,\pi)$. It is immediate that if $\sigma \le 2\pi$, then $\mathcal{S}(\sigma,\pi) \le 2\tilde{\mathcal{S}}$.

Forgetting where we started. The next mixing measure which we introduce at this stage does not overtly involve the stationary distribution; instead, we look for the *best* target distribution, in the sense of expected access time from a worst starting state for that distribution. We call this mixing measure the *forget time* because it measures, in a sense, the least time it can take to "forget" what state the chain was started in. We denote the forget time by $\mathcal{F}$ and this time no arguments are needed:

$$\mathcal{F} := \min_{\tau} \max_{s \in V} \mathcal{H}(s, \tau) = \min_{\tau} \max_{\sigma} \mathcal{H}(\sigma, \tau).$$

The target distribution τ which minimizes $\max_s \mathcal{H}(s, \tau)$ will be called the "forget distribution" and denoted by φ. It was proved in [**LW5**] that this distribution is uniquely determined; see [**LW5**] for an explicit formula for μ.

We can define approximate versions of the forget time, by $\mathcal{F}_{\underline{\varepsilon}} := \min_\tau \mathcal{H}(\geq (1-\varepsilon)\tau)$ etc.

Discrepancy. This quantity is quite closely related to mixing measures, although its definition seems somewhat artificial. We define

$$\mathcal{Z} = \max_i \mathcal{Z}(i, \pi) = \max_i \sum_j \pi_j |H(i,j) - H(\pi, j)| \tag{2.9}$$

and

$$\tilde{\mathcal{Z}} = \mathcal{Z}(D_1, \pi) = \max_{\sigma \leq 2\pi} \mathcal{Z}(\sigma, \pi).$$

We call $\mathcal{Z}$ the *discrepancy* of the chain, because it could be considered as a measure of "unevenness" of these hitting times.

Recall that the mixing time can be defined as follows:

$$\mathcal{H} = \max_i \max_j (H(i,j) - H(\pi, j)).$$

So $\mathcal{H}$ can also be considered as the maximum "surplus" of hitting times to a fixed state, over the average hitting time to that state, while $\mathcal{Z}$ is the average "surplus".

We can also average in the other variable, and get a mixing measure introduced when discussing warm start:

$$\hat{\mathcal{H}} = \max_j \sum_i \pi_i |H(i,j) - H(\pi, j)|. \tag{2.10}$$

Finally, we mention one the most used parameters, the "eigenvalue gap". For a time-reversible chain, this is defined as follows. Let $1 = \lambda_1 > \lambda_2 \geq \cdots \geq \lambda_n$ denote the eigenvalues of the transition matrix M. The *eigenvalue gap* is the value $1 - \lambda_2$. The *relaxation time* is defined by

$$\mathcal{L} = \frac{1}{1 - \lambda_2}$$

(we use $\mathcal{L}$ to remind us of λ).

Unfortunately, for chains that are not time-reversible, there are several ways to define this number, and these play different roles in mixing. We take the easy route and do not define it at all.

Strong rules. We mention strong stopping rules, introduced by Aldous and Diaconis [**AD**], which have been used very successfully to estimate mixing times. A stopping rule Γ is said to be *strong* if the final state v^Γ is independent of the number of steps Γ. In other words, for any time t with $\mathsf{P}(\Gamma = t) > 0$, the conditional distribution $\{\sigma^t | \Gamma = t\}$ is the same as σ^Γ.

2.4. Different mixing measures: examples. We give some examples of Markov chains and various mixing measures. We state the values without proof (just with some hints), and without any attempts of providing complete analysis. The computation of some of these mixing times is based on the results in Chapter 4.

EXAMPLE 2.4 (Two States). It turns out that already the simplest nontrivial chain shows some interesting phenomena. Suppose that there are just two states 1 and 2, with transition matrix

$$\begin{pmatrix} 1-a & a \\ b & 1-b \end{pmatrix}, \qquad 0 < a \le b < 1 .$$

It will be convenient to set $d = a + b$ and $c = a/(a+b)$, so that $a = cd$ and $b = (1-c)d$.

It is easy to see that $\pi_1 = 1 - c$, $\pi_2 = c$. Hence the chain is time-reversible. It is also clear that $\mathcal{H}(1,2) = 1/a = 1/(cd)$ and $\mathcal{H}(2,1) = 1/b = 1/(1-c)d$.

The naive rule is optimal from any starting state, since the other one will be a halting state. This gives that

$$\mathcal{H} = \mathcal{H}(1,\pi) = \mathcal{H}(2,\pi) = \frac{1}{d}.$$

It follows that $\mathcal{I} = 1/d$ and it is not difficult to compute that $\mathcal{F} = 1/d$. The eigenvalues of the transition matrix are 1 and $1-d$, and hence $\mathcal{L} = 1/d$ as well.

Approximate mixing times are similar. We only discuss the quantity $\mathcal{H}_\varepsilon$, the time needed to get closer to π than ε in the total variation distance, where $\varepsilon < 1/4$ is given. If we start at 2, then clearly the best target distribution τ, with $d(\tau,\pi) \le \varepsilon$, is $\tau' = (1-c-\varepsilon, c+\varepsilon)$, and it is easy to compute that

$$\mathcal{H}(2, d_\varepsilon) = \mathcal{H}(2,\tau') = \frac{1-c-\varepsilon}{(1-c)d} .$$

The situation is somewhat more complicated if we start at 1. If $\varepsilon \le c$, then a similar computation yields

$$\mathcal{H}(1, d_\varepsilon) = \frac{c-\varepsilon}{cd} .$$

But if $\varepsilon > c$, then already the starting distribution is closer to π than ε, and thus $\mathcal{H}(1, d_\varepsilon) = 0$. It is easy to check that 2 is always the worse starting state, and hence

$$\mathcal{H}_\varepsilon = \frac{1-c-\varepsilon}{(1-c)d} > \frac{1}{2d}.$$

So we gain at most a factor of two by allowing only approximate mixing, independently of ε.

Another quantity worth computing is $\tilde{\mathcal{H}}_\varepsilon$, the approximate mixing time from warm start. There are basically two starting distributions to worry about, namely $(1,0)$ (which we denote by '1') and $\sigma = (1-2c, 2c)$ (all others are convex combinations of these, and it is easy to argue that they are not worse than these two). If $\varepsilon \ge c$ then these distributions are already in d_ε, so in this case $\tilde{\mathcal{H}}_\varepsilon = 0$. If $\varepsilon \le c$, then we already computed $\mathcal{H}(1, d_\varepsilon)$. A similar argument gives that

$$\mathcal{H}(\sigma, d_\varepsilon) = \frac{c-\varepsilon}{(1-c)d},$$

and so 1 is the worse starting state. Thus we get

$$\tilde{\mathcal{H}}_\varepsilon = \begin{cases} 0 & \text{if } \varepsilon \ge c \\ (c-\varepsilon)/(cd) & \text{if } \varepsilon \le c. \end{cases}$$

Note the strange dependence on ε.

The maximum time of any rule from 2 to π is quite different. If we happen to stay at 2 then we cannot stop until the probability of still being at 2 is reduced to its stationary probability. This gives

$$\mathcal{M} \geq \left\lceil \frac{\log c}{\log(1-(1-c)d)} \right\rceil$$

(one can show that equality holds here). If, say, $d = 1/2$, then $\mathcal{H} = 2$ while $\mathcal{M} > \log(1/c)$.

Blind rules don't fare well either. Just to give a rough argument, suppose that a blind rule, starting from 2, yields a distribution with intensity at most 2. If the rule stops after $\mathcal{M} - 2$ or fewer steps, then it stops at 2 with at least four times the stationary probability. Hence at least half of the time it stops after more than $\mathcal{M} - 2$ steps. This shows that

$$\mathcal{B}_{\overline{1}} > \frac{\mathcal{M}-2}{2}.$$

EXAMPLE 2.5 (Path). Consider the classic case of a random walk on the path with nodes labeled $1, 2, \dots, n$. The hitting times from endpoints are $\mathcal{H}(1, j+1) = \mathcal{H}(n, n-j) = j^2$ and the stationary distribution is

$$\frac{1}{2n-2}, \frac{1}{n-1}, \frac{1}{n-1}, \dots, \frac{1}{n-1}, \frac{1}{2n-2},$$

since for any random walk on a graph π_i is proportional to the degree of the node i.

It is clear that halting states from any state to π can only be the endnodes; the two endnodes are also the pessimal starting states (cf. Corollary 4.12(c)). Suppose that we start at 1. Owing to the special topology of the path, the naive rule has a halting state (namely the state n), and hence it is optimal by Theorem 2.2. Thus we can compute the mixing time easily:

$$\mathcal{H}(1, \pi) = \sum_{k=0}^{n-2} \frac{1}{n-1} k^2 + \frac{1}{2n-2}(n-1)^2 = \frac{2n^2 - 4n + 3}{6}.$$

The naive rule is not bounded, so to get any idea of the maxing time, one has to consider more complicated rules. It can be shown that $\mathcal{M} = O(n^2)$, but the exact value does not seem to be known.

To forget, one can walk to the midpoint of the path and stop there (assume for simplicity that n is odd). It turns out that this is optimal, and hence

$$\mathcal{F} = \frac{(n-1)^2}{4}.$$

Thus the forget time is smaller than the mixing time, but has the same order of magnitude. The relaxation time behaves similarly: its value is

$$\mathcal{L} = \frac{1}{1 - \cos(\pi/(n+1))} \sim \frac{2n^2}{\pi^2}.$$

EXAMPLE 2.6 (Fat Path). Let us modify the path to give the walk a large drift. We connect nodes i and $i+1$ by 2^{n-i-1} parallel edges. We also add a (two-way) loop at n and 2^{n-2} loops at 1, just to make the numbers come out nicer (see Fig. 1).

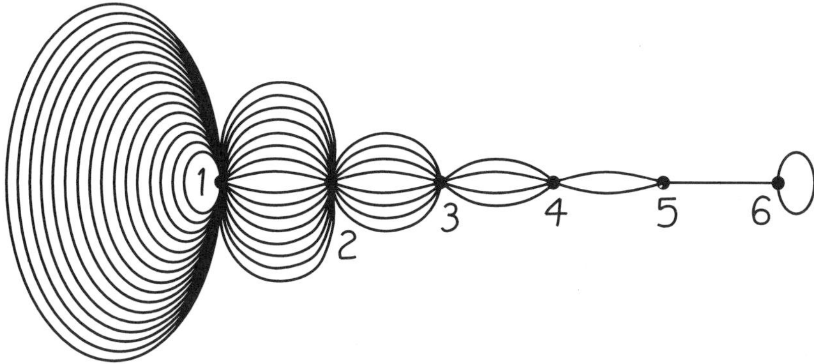

FIGURE 1. The fat path, for $n = 6$

The random walk on the resulting "fat path" is time-reversible. The transition probabilities are $p_{i,i+1} = 1/3$, $p_{i,i-1} = 2/3$ for $2 \le i \le n-1$, $p_{1,2} = 1/3$, $p_{1,1} = 2/3$, $p_{n,n-1} = 1/3$, $p_{n,n} = 2/3$. The stationary probabilities are

$$\frac{1}{2}, \frac{1}{4}, \quad \dots, \quad \frac{1}{2^{n-1}}, \quad \frac{1}{2^{n-1}}.$$

It can be shown by induction that

$$\mathcal{H}(i,j) = \begin{cases} 3(i-j) & \text{if } j \le i \\ 3(i-j+2^j-2^i) & \text{if } j \ge i. \end{cases}$$

Again, it follows e.g. from Corollary 4.12(c) that the endnodes are pessimal, and the naive rule is optimal when starting at an endnode, whence elementary computation gives

$$\mathcal{H}(1,\pi) = \mathcal{H}(n,\pi) = 3n - 6 + \frac{3}{2^{n-1}} \sim 3n.$$

Note that it takes exactly the same time to reach π from both endnodes, but through a very different arithmetic!

If we want to get only close to π (in total variation distance), then the two endnodes will behave quite differently. From n, we need to get to the other "most of the time", and hence $\mathcal{H}_\varepsilon \sim 3n$ for every fixed $\varepsilon < 1$. But from 1, we only need to get to the last $\log(1/\varepsilon)$ nodes (the rest has stationary measure less than ε), and hence the time needed for this is only about $\log_2(1/\varepsilon)$.

A similar argument would work to show that the total variation mixing time, with warm start, is only $O(\log(1/\varepsilon))$.

EXAMPLE 2.7 (Cycle). Consider an (undirected) cycle of length n, where (say) n is even. Then of course each node is pessimal, and it is not difficult to compute that

$$\mathcal{H} = \frac{n^2}{12} + \frac{1}{6}.$$

(Compare this with expected time

$$\frac{n-1}{n}\frac{n(n-1)}{2} = \frac{(n-1)^2}{2}$$

for staying at 0 with probability $1/n$ else walking until the last new vertex is hit as in Example 1.1.)

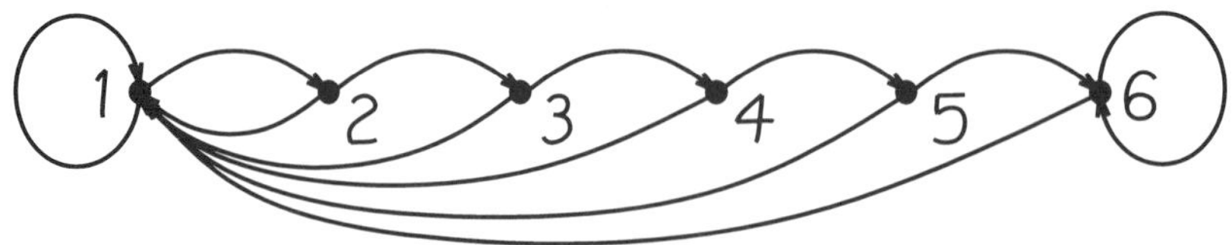

FIGURE 2. The winning streak, for $n = 6$

The fact that the cycle has a node-transitive automorphism group makes the determination of the reset and forget times easy. Clearly π is the uniform distribution, $H(i,\pi)$ is the same for all i, and hence $\mathcal{I} = \mathcal{H}$. The uniqueness of the forget distribution implies that it must be just π, and hence $\mathcal{F} = \mathcal{I}$.

The eigenvalues of the cycle are well known, and hence it is easy to compute the relaxation time:

$$\mathcal{L} = \frac{1}{1-\cos(\pi/n)} \sim \frac{2n^2}{\pi^2}.$$

It is also worth taking a look at the directed cycle, which is perhaps the simplest case that is not time-reversible. The walk on this is deterministic, but the mixing times are defined nevertheless. The naive rule is optimal (it has a halting state), and hence $\mathcal{H} = (n-1)/2$. It is also clear that $\mathcal{M} = n-1$. As before, rotational symmetry implies that $\mathcal{F} = \mathcal{I} = \mathcal{H}$.

EXAMPLE 2.8 (Winning Streak). Let $V = \{1,\dots,n\}$, and define the transition probabilities by

$$p_{ij} = \begin{cases} 1/2 & \text{if } j = i+1 \\ 1/2 & \text{if } j = 1 \text{ and } 1 \le i \le n \\ 1/2 & \text{if } j = i = n \\ 0 & \text{otherwise.} \end{cases}$$

See Fig. 2. It is easy to check that the stationary distribution is

$$\pi_i = \begin{cases} 2^{-i} & \text{if } 1 \le i \le n-1 \\ 2^{-(n-1)} & \text{if } i = n. \end{cases}$$

A very special property of the winning streak chain is that the $(n-1)$-step-rule generates the stationary distribution from every starting state. To see this, note that every walk can be described by a sequence $HTTH\dots$ of independent coin flips, where "head" means to go right (or stay in state n), while "tail" means to fall back to 1. This sequence leads to state i if and only if the sequence ends with exactly i–1 heads; the probability that this happens is just 2^{-i} if $i < n$ and $2^{-(n-1)}$, if $i = n$.

The $(n-1)$-step rule is optimal if we start at 1, since n is then a halting state; it is also optimal if we start at n, since $n-1$ is then a halting state; but it is not optimal when starting at any other state, since it is easy to check that it has no halting state. Hence

$$\mathcal{H} = n-1, \qquad \mathcal{I} < n-1.$$

For the reset time, a more tedious calculation shows that

$$\mathcal{I} = n - k + 1 - \frac{n-k}{2^k} - \frac{1}{2^{n-1}},$$

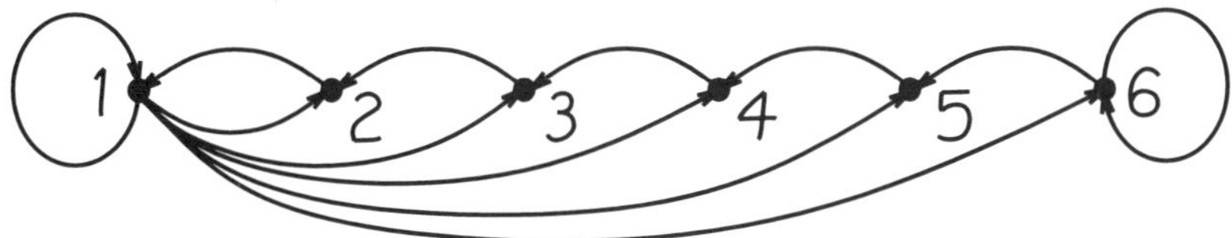

FIGURE 3. The reverse winning streak, for $n = 6$

where k is the largest integer with $2^k \le n - k + 1$ (so $k \approx \log_2 n$; $\mathcal{H}(i, \pi)$ behaves differently for $i < k$ and $i \ge k$). This value is strictly smaller than the mixing time $\mathcal{H}$, but still asymptotically n.

If we want to forget where we started, a very simple strategy is to stop at 1. This happens in an expected number of at most 2 steps, no matter where we start, and hence

$$\mathcal{F} \le 2.$$

It turns out that it is slightly better if we allow a tiny probability of stopping elsewhere; the exact value of the forget time is

$$\mathcal{F} = 2 - 2^{-(n-2)},$$

achieved by the rule "walk until you hit 1 or make $n-1$ steps, whichever comes first".

EXAMPLE 2.9 (Reverse Winning Streak). Now look at the reverse chain of the previous example (see Fig. 3). Here the transition probabilities are $p_{i+1,i} = 1$ for $1 \le i \le n-1$, $p_{n,n-1} = p_{n,n} = 1/2$, $p_{1,i} = 2^{-i}$ for $1 \le i \le n-1$ and $p_{1,n} = 2^{-(n-1)}$ (other transition probabilities are 0). So the transition from any state is deterministic except for the transition from state 1, which leads directly to the stationary distribution. From any other state $i < n$, we can walk to 1 and then make one more move, which takes i moves. This rule is optimal (n is a halting state). We could use the same rule if we start at the last state, but it would not be optimal; instead, the following strategy is optimal: walk for $n-1$ steps.

It follows that the mixing time is $\mathcal{H} = n-1$, while the reset time is

$$\mathcal{I} = \frac{1}{2} \cdot 1 + \frac{1}{4} \cdot 2 + \cdots + \frac{1}{2^{n-1}}(n-1) + \frac{1}{2^{n-1}}(n-1) = 2 - \frac{1}{2^{n-2}}.$$

To forget, we can again walk to node 1; in this chain, this will take n steps from the worst starting node. It turns out, somewhat surprisingly, that this rule is almost optimal; we have $\mathcal{F} \sim n$.

3. Main results

This section states the main results of this paper. We discuss five groups of mixing times. The relationship between these groups is shown in Fig. 4. Each group contains several mixing measures, which are all within absolute constant factors of each other (for the relaxation group, some these equivalences are only conjectured, and the statements of the results are more complicated for some of the parametric measures; see Theorems 3.5 and 3.6. Mixing measures in a group higher up in the figure are larger (again, up to an absolute constant factor). We will see that parameters in the same group are often quite differently defined and their equivalence is unexpected.

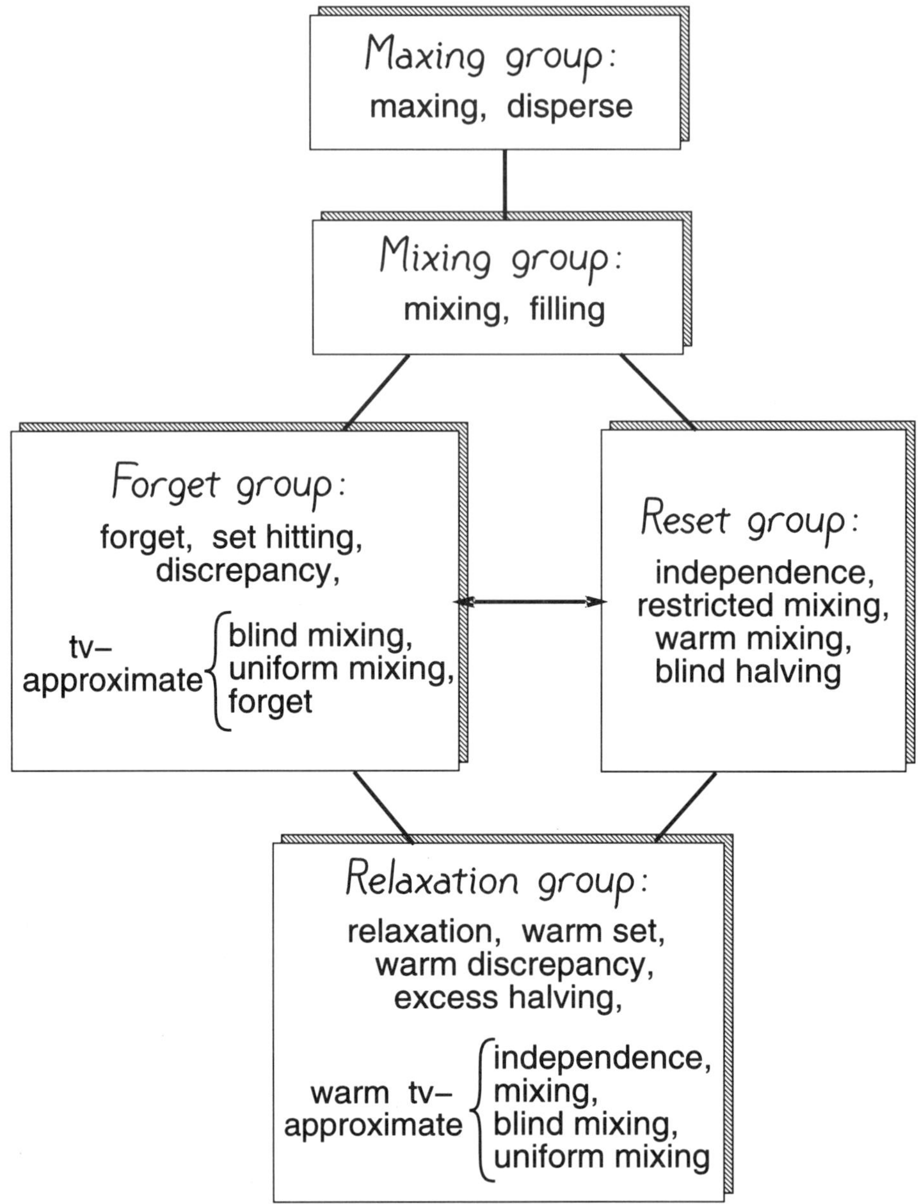

FIGURE 4. Main groups of mixing times and their relationship

We'll also see that reflecting the figure in the vertical axis corresponds to time reversal. Thus, the Relaxation Group, Mixing Group and Maxing Group are invariant under time reversal (again, up to absolute constants), while the Forget Group and Reset Group are interchanged.

For every chain, either the Forget Group or the Reset Group is within an absolute constant factor of the Mixing group, but generally not both. Thus in every finite Markov chain, either "forgetting is difficult" or "resetting is difficult".

We don't know if this dichotomy is reflected in other properties of finite Markov chains. In time-reversible chains, all three groups coincide.

Throughout, we consider the error parameter $0 < \varepsilon < 1/4$ fixed. Below, if we write $\mathcal{A} \approx \mathcal{B}$ for two mixing measures, it means that there are constants $c_1, c_2 > 0$ such that $c_1\mathcal{A} \le \mathcal{B} \le c_2\mathcal{A}$. Similarly, $A \ll B$ means that there is a constant $c > 0$ such that $\mathcal{A} \le c\mathcal{B}$. These constants may depend on the error parameter ε, but not on the chain. It should also be remarked that the constants are moderate (see Section 5 for a detailed statement of these relations).

3.1. Groups of mixing times. The *Mixing Group* is a small group containing the most important mixing measure, the mixing time $\mathcal{H}$. The following theorem shows that the approximate filling time is also in this group (see [**AF**] for this result and many others in the setting of reversible Markov chains). While either one of the forget and reset times may be much smaller than the mixing time, their maximum (or, equivalently, sum) is also in this group.

THEOREM 3.1.
$$H \approx \mathcal{H}_{\underline{\varepsilon}} \approx \max\{\mathcal{F}, \mathcal{I}\}.$$

The *Forget Group* contains a number of interesting invariants: the forget time $\mathcal{F}$; the set access time $\mathcal{S}$; for any fixed $\varepsilon > 0$, the approximate mixing times $\mathcal{H}_\varepsilon$ and $\mathcal{H}_{\overline{\varepsilon}}$; and both approximate forget times $\mathcal{F}_\varepsilon$ and $\mathcal{F}_{\underline{\varepsilon}}$. Note that the approximate mixing time $\mathcal{H}_{\underline{\varepsilon}}$ is not in this group.

THEOREM 3.2.
$$\mathcal{F} \approx \mathcal{F}_\varepsilon \approx \mathcal{F}_{\underline{\varepsilon}} \approx \mathcal{H}_\varepsilon \approx \mathcal{H}_{\overline{\varepsilon}} \approx \mathcal{S}.$$

The *Reset Group* contains the reset time $\mathcal{I}$ and the warm mixing time $\tilde{\mathcal{H}}$, as well as the restricted mixing time $\hat{\mathcal{H}}$ and the *blind halving time*
$$\tilde{\mathcal{B}}_{\overline{1/2}} = \mathcal{B}(\le 2\pi, \le (3/2)\pi)\,.$$
It will then follow by general results that $\tilde{\mathcal{B}}_{\overline{\varepsilon}}$ is in this group for every fixed ε.

THEOREM 3.3.
$$\mathcal{I} \approx \tilde{\mathcal{H}} \approx \hat{\mathcal{H}} \approx \tilde{\mathcal{B}}_{\overline{\varepsilon}}\,.$$

The *Maxing Group* contains two, seemingly unrelated quantities: the maxing time $\mathcal{M}$ and the blind disperse time $\mathcal{B}_{\overline{\varepsilon}}$.

THEOREM 3.4.
$$\mathcal{M} \approx \mathcal{B}_{\overline{\varepsilon}}.$$

The *Relaxation Group* contains several important parameters: the warm set access time $\tilde{\mathcal{S}}$, the warm discrepancy $\tilde{\mathcal{Z}}$, the average discrepancy $\mathcal{G}$, the relaxation time $\mathcal{L}$. Among the approximate mixing times, we consider
$$\mathcal{H}^* = \max_\varepsilon \mathcal{H}(\le (1+2\varepsilon)\pi, \le (1+\varepsilon)\pi),$$
which we call the *excess halving time.*

THEOREM 3.5.
$$\tilde{\mathcal{S}} \approx \mathcal{H}^* \ll \tilde{\mathcal{Z}}\,.$$
For a reversible chain,
$$\tilde{\mathcal{Z}} \ll \mathcal{L}, \qquad \tilde{\mathcal{S}} \approx \mathcal{L}\,.$$

We list two further interesting mixing measures in this group: the warm independence time $\tilde{\mathcal{I}}_\varepsilon$, and the warm approximate mixing time $\tilde{\mathcal{H}}_\varepsilon$. The following theorem states that these are "essentially equivalent"; but we have to explicitly state the dependence on ε. Recall that in Example 2.4, $\tilde{\mathcal{H}}_\varepsilon$ was 0 for $\varepsilon \geq c$ but positive for $\varepsilon < c$, where $0 < c \leq 1/2$ was an arbitrary parameter of the chain; so no equivalence can be established, say, between $\tilde{\mathcal{H}}_\varepsilon$ and $\tilde{\mathcal{H}}_{2\varepsilon}$.

THEOREM 3.6. *For every finite Markov chain and every $\varepsilon > 0$, we have*

$$\varepsilon\tilde{\mathcal{U}}_{2\varepsilon} \leq \tilde{\mathcal{H}}_\varepsilon \leq \tilde{\mathcal{B}}_\varepsilon \leq \tilde{\mathcal{U}}_\varepsilon$$

and

$$\tilde{\mathcal{I}}_{6\varepsilon} \ll \tilde{\mathcal{H}}_\varepsilon \ll \tilde{\mathcal{I}}_{\varepsilon/3}\,.$$

3.2. Relations between groups. The next result shows that the ratio of mixing measures in different groups may be unbounded, but even the ratio quantities in different groups remains relatively small: the reset group and the maxing group are not more than a factor $\ln\hat{\pi}$ apart, where $\hat{\pi} := \min_i\{\pi_i\}$. Considering the reverse chain, we get a similar inequality between the forget and maxing groups. We conjecture that a similar inequality must hold between the relaxation and maxing groups.

THEOREM 3.7.

$$\mathcal{M} \leq \left(\ln\frac{1}{\hat{\pi}}\right)\tilde{\mathcal{B}}_{\overline{1/2}}.$$

Note that in a typical application in sampling algorithms the state space V is exponentially large (in terms of the natural "size" of the problem), and in the setting of approximate counting π is most often uniform. In such cases $\ln(1/\hat{\pi})$ is polynomial, and it follows that all mixing measures are are within polynomial factors of one another.

The winning streak chain (Example 2.8 shows that the logarithmic factor in the upper bound in the theorem can occur already as the quotient $\mathcal{H}/\mathcal{F}$; in this case of course $\mathcal{M} \approx \mathcal{I} \approx \mathcal{H}$ and $\mathcal{L} \approx \mathcal{F}$. The reverse winning steak chain of course shows a similar situation with $\mathcal{F}$ and $\mathcal{I}$ interchanged. The chain with two states in Example 2.4 shows that this logarithmic factor can also occur as the ratio $\mathcal{M}/\mathcal{H}$, even in the time-reversible case (and then of course $\mathcal{H} \approx \mathcal{F} = \mathcal{I} \approx \mathcal{L}$). Finally, Example 2.6 shows that this logarithmic factor can also occur as $\mathcal{I}/\mathcal{L}$ in a time-reversible chain (in this case of course $\mathcal{M} \approx \mathcal{H} \approx \mathcal{F} = \mathcal{I}$).

We conclude by stating a theorem that implies that the mixing, maxing and relaxation groups are invariant under time reversal, while the forget and reset groups are interchanged. Parts (a) and (c) appear, with proofs, in [**LW5**].

THEOREM 3.8. *For every finite Markov chain,*

(a) $\overleftarrow{\mathcal{H}} = \mathcal{H}$;

(b) $\overleftarrow{\mathcal{B}_{\overline{\varepsilon}}} = \mathcal{B}_{\overline{\varepsilon}}$.

(c) $\overleftarrow{\mathcal{I}} = \mathcal{F}$ *and* $\overleftarrow{\mathcal{F}} = \mathcal{I}$.

(d) $\overleftarrow{\hat{\mathcal{H}}} = \mathcal{Z}$ *and* $\overleftarrow{\mathcal{Z}} = \hat{\mathcal{H}}$.

If the chain is time-reversible, then $\mathcal{I} = \mathcal{F}$, and hence by Theorem 3.1, $\mathcal{F} \approx \mathcal{H}$. Thus for time-reversible chains the mixing, forget and reset groups form one group of equivalent mixing measures.

4. Tools

4.1. Exit frequencies. Let us now fix a transition matrix M, an initial distribution σ and a finite stopping rule Γ.

Definition. The *exit frequencies* $x_i = x_i(\Gamma)$ $(i \in V)$ are defined by setting x_i equal to the expected number of times the walk leaves state i before stopping. The *scaled exit frequencies* are defined by $y_i = \frac{1}{\pi_i}x_i$. In the finite case, the use of x_i or y_i is only a matter of taste; in the case of Markov chains with infinite state space, the difference between x_i and y_i is quite significant.

Exit frequencies are a special case of what Pitman [**P**] calls "pre-T occupation measures" for stopping times T. Indeed, versions of Lemma 4.2 and Theorem 4.4 below are proved in [**P**] using Pitman's "occupation measure identity".

Exit frequencies for the naive rule are easy to find; see [**AF**]. Several related formulas could be derived using relations to electrical networks, as in [**CRRST**], [**DS**], or [**T**].

LEMMA 4.1. *The exit frequencies $\tilde{x}$ for the naive stopping rule $\Omega_{i,j}$ from state i to j are given by*

$$\tilde{x}_k = \pi_k \left(\mathcal{H}(i,j) + \mathcal{H}(j,k) - \mathcal{H}(i,k)\right) .$$

More generally, the exit frequencies for the naive stopping rule $\Omega_{\sigma,\tau}$ are given by

$$\begin{aligned} \dot{x}_k &= \pi_k \sum_{i,j} \sigma_i \tau_j (\mathcal{H}(i,j) + \mathcal{H}(j,k) - \mathcal{H}(i,k)) \\ &= \pi_k \left(\sum_{i,j} \sigma_i \tau_j \mathcal{H}(i,j) + \mathcal{H}(\tau,k) - \mathcal{H}(\sigma,k) \right) . \end{aligned}$$

Exit frequencies are related to the starting and ending distributions by a simple formula, found in Pitman [**P**], which can be obtained by counting entrances and exits at a given state:

LEMMA 4.2. *The exit frequencies of any stopping rule Γ that reaches distribution τ from distribution σ satisfy the equation*

$$\sum_i p_{i,j} x_i(\Gamma) - x_j(\Gamma) = \tau_j - \sigma_j .$$

One can rewrite this identity as follows.

$$\sum_i p_{i,j} x_i(\Gamma) - \sum_i p_{ji} x_j(\Gamma) = \tau_j - \sigma_j .$$

In other words, the values $p_{ij}x_i(\Gamma)$ can be viewed as values of a flow through the underlying graph, from supply σ to demand τ. This is also easily seen by observing that $p_{ij}x_i(\Gamma)$ is the expected number of passes from i to j while following Γ.

The next theorem, found also in [**LW4**], implies that the exit frequencies of a stopping rule are "almost" determined by the starting and target distributions.

THEOREM 4.3. *Fix two distributions σ and τ, and let Γ and Γ' be two finite stopping rules from σ to τ. Then $x(\Gamma') - x(\Gamma) = (\mathsf{E}\Gamma' - \mathsf{E}\Gamma)\pi$.*

Conversely if Γ and Γ' are two stopping rules such that when starting from σ, $x(\Gamma)$ and $x(\Gamma')$ differ by a multiple of π, then $\sigma^\Gamma = \sigma^{\Gamma'}$.

It follows from Theorem 4.3 that the exit frequencies of any mean-optimal stopping rule from σ to τ are the same. We denote them by $x_i(\sigma,\tau)$.

Another way of formulating Theorem 4.3 is to say that the values $x_i - \pi_i \mathsf{E}\Gamma$ are the same for *every* stopping rule Γ (optimal or not) from a given σ to a given τ. We call these values the *exit discrepancies* and denote them by $z_i(\sigma,\tau)$. They will be very convenient quantities to use.

In particular, we can compute $z_k(\sigma,\tau)$ by considering the naive rule:

$$
\begin{aligned}
z_k(\sigma,\tau) &= \pi_k\left(\sum_{i,j}\sigma_i\tau_j\mathcal{H}(i,j)+\mathcal{H}(\tau,k)-\mathcal{H}(\sigma,k)\right)-\pi_k\sum_{i,j}\sigma_i\tau_j\mathcal{H}(i,j)\\
&= \pi_k\big(\mathcal{H}(\tau,k)-\mathcal{H}(\sigma,k)\big)\,. \qquad (4.1)
\end{aligned}
$$

Although derived from the naive rule, this formula holds for the exit discrepancies of any stopping rule from σ to τ, a fact we shall exploit throughout. Applying it to any optimal rule, we get the following result:

THEOREM 4.4. *The scaled exit frequencies of a mean-optimal stopping rule from σ to τ are given by*

$$y_k(\sigma,\tau)=\mathcal{H}(\tau,k)-\mathcal{H}(\sigma,k)+\mathcal{H}(\sigma,\tau).$$

Formula (4.1) yields immediately that $z_k(\sigma,\tau)$ is a linear function in each of σ and τ, and that it has a number of useful properties:

$$z_k(\sigma,\tau)+z_k(\tau,\rho)=z_k(\sigma,\rho)$$

$$z_k(\sigma,\tau)=-z_k(\tau,\sigma).$$

We obtain another interesting formula for z by considering a simple stopping rule: walk one step! Its exit frequencies are σ_i, and hence

$$z_k(\sigma,\sigma^1)=\sigma_k-\pi_k\ .$$

It follows that

$$z_k(\sigma,\sigma^t)=\sum_{m=0}^{t-1}(\sigma_k^m-\pi_k)\ . \qquad (4.2)$$

If the chain is ergodic, then we can let $t\to\infty$ and obtain

$$z_k(\sigma,\pi)=\sum_{m=0}^{\infty}(\sigma_k^m-\pi_k)$$

(it is easy to see that the series on the right is absolutely convergent). Hence we get that for any two distributions σ and τ,

$$z_k(\sigma,\tau)=\sum_{m=0}^{\infty}(\sigma_k^m-\tau_k^m)\ .$$

Time-reversibility is also reflected by the exit discrepancies in a very natural way:

LEMMA 4.5. *The Markov chain is reversible if and only if*

$$\pi_k z_i(k,\pi)=\pi_i z_k(i,\pi)$$

for all $i,k\in S$.

Exit discrepancies are related to access times by some simple but useful inequalities. Trivially

$$z_i(\sigma,\tau) \geq -\pi_i \mathcal{H}(\sigma,\tau) \tag{4.3}$$

and hence

$$z_i(\sigma,\tau) \leq \pi_i \mathcal{H}(\tau,\sigma)\,. \tag{4.4}$$

In terms of $\mathcal{H}(\sigma,\tau)$, we only have the much weaker inequality

$$z_i(\sigma,\tau) \leq (1-\pi_i)\mathcal{H}(\sigma,\tau)\,. \tag{4.5}$$

To get a stronger inequality, note that since $\sum_i z_i(\sigma,\tau)=0$, we have

$$\max_A \sum_{i\in A} z_i(\sigma,\tau) = \sum_i \max\{0, z_i(\sigma,\tau)\} = \frac{1}{2}\|z(\sigma,\tau)\|\,.$$

Then for the sum of z_i over any subset:

$$\sum_{i\in A} z_i(\sigma,\tau) = \sum_{i\in A} x_i(\sigma,\tau) - \pi(A)\mathcal{H}(\sigma,\tau) \leq \mathcal{H}(\sigma,\tau) - \pi(A)\mathcal{H}(\sigma,\tau)\,. \tag{4.6}$$

Hence

$$\frac{1}{2}\|z(\sigma,\tau)\| \leq \mathcal{H}(\sigma,\tau)\,. \tag{4.7}$$

Exit discrepancies are closely related to some other measures introduced before. From (4.1) we get immediately that

$$\mathcal{Z} = \max_i \|z(i,\pi)\|\,, \qquad \mathcal{G} = \sum_i \pi_i \|z(i,\pi)\|.$$

We can also define the warm start version of $\mathcal{Z}$,

$$\tilde{\mathcal{Z}} := \max_{\sigma\leq 2\pi} \|z(\sigma,\pi)\|\,.$$

4.2. Optimal stopping rules. We describe two optimal stopping rules whose analysis will be needed in proving our main results. A more detailed analysis of these and other optimal stopping rules will be deferred to another publication.

Chain rule. Let ρ be a probability distribution on the *subsets* of the state space V, and observe that it provides a stopping rule: "choose a subset U from ρ, and walk until some state in U is hit". The naive rule is of course a special case, with ρ concentrated on singletons.

THEOREM 4.6. *For every starting distribution σ and target distribution τ, there exists a unique distribution ρ which is concentrated on a chain of subsets and gives an optimal stopping rule for generating τ.*

PROOF. Starting the chain from distribution σ, let α^U be the distribution of the first node in U, for every subset U. For example, $\alpha^V=\sigma$. Let $U_0=V$, $\mu^0=\tau$. We define distinct nodes $i_1,\dots,i_n$, non-negative numbers $\rho_1,\dots,\rho_n$, and non-negative vectors $\mu^1,\dots,\mu^n\in\mathbf{R}^V$, by induction. Assume that we have defined i_j, ρ_j and μ_j for $j\leq k$. Let $U_{k+1}=S\setminus\{i_1,\dots,i_k\}$. Select a node i_{k+1} for which $i=i_{k+1}$ minimizes $\mu_i^k/\alpha_i^{U_{k+1}}$, and let

$$\rho_{k+1} = \mu^k_{i_{k+1}}/\alpha^{U_{k+1}}_{i_{k+1}}$$

and

$$\mu_i^{k+1} = \mu_i^k - \rho_{k+1}\alpha_i^{U_{k+1}}\,.$$

Note that $\rho_1 = \tau_{i_1}/\sigma_{i_1}$.

It follows by induction and by the choice of i_{k+1} that $\rho_{k+1} \geq 0$, $\mu^{k+1} \geq 0$, and $\mu_i^{k+1} = 0$ for $i \notin U_{k+2}$. Moreover, we have

$$\sum_{k=1}^{n} \rho_k = 1\,.$$

□

We call this rule the "chain rule" and denote it by Ξ. A rather neat way to think of this rule is to assign a "price", a real value $r(i) = 1 - \sum\{\rho_U :\ i \in U\}$ to each state i. The rule is then implemented by choosing a random real "budget" b uniformly from $[0,1]$ and walking until a state j with $r(j) \leq b$ (an item that we can buy) is reached.

As an application of the chain rule, we prove a general inequality concerning set hitting times.

LEMMA 4.7. *Let σ, τ be two arbitrary distributions and $\varepsilon > 0$. Then*

$$\mathcal{H}(\sigma, \leq (1+\varepsilon)\tau) \leq (1+\varepsilon)\ln(1+1/\varepsilon))\mathcal{S}(\sigma,\tau)\,.$$

If $\tau_i \geq d > 0$ for all i, then

$$\mathcal{H}(\sigma,\tau) \leq \ln(1/d)\mathcal{S}(\sigma,\tau)\,.$$

PROOF. We use the chain rule. By Theorem 4.6, there exists a labelling $\{0,\ldots,n-1\}$ of the nodes and a probability distribution ρ such that setting $S_j = \{j, j+1, \ldots, n-1\}$, selecting a j, $0 \leq j \leq n-1$ according to ρ and then walking until S_j is hit, generates τ. Note that

$$\rho(S_j) \leq \tau(S_j), \tag{4.8}$$

since $\rho(S_j) = \rho_j + \cdots + \rho_{n-1}$ is the probability of choosing a subset of S_j as our target, and in this case we certainly end up in S_j.

To prove the first inequality, fix the subscript m such that $\rho(S_{m+1}) \leq 1/(1+\varepsilon) < \rho(S_m)$. Now modify the chain rule by selecting S_j with probability

$$\rho'_j = \begin{cases} (1+\varepsilon)\rho_j & \text{if } j \leq m-1 \\ 1-(1+\varepsilon)(\rho_1+\cdots+\rho_{m-1}) & \text{if } j = m \\ 0 & \text{otherwise.} \end{cases}$$

Note that $\rho'_m \leq (1+\varepsilon)\rho_m$ by the choice of m.

Let this new rule generate distribution τ'. Then obviously

$$\tau'_i \leq (1+\varepsilon)\tau_i.$$

The mean length of this new rule is

$$\begin{aligned}\mathcal{H}(\sigma,\tau') &= \sum_{j=1}^{m} \rho'_j \mathcal{H}(\sigma,S_j)\\ &\le \left(\sum_{j=1}^{m} \frac{\rho'_j}{\tau(S_j)}\right) \mathcal{S}(\sigma,\tau)\\ &\le \left(\sum_{j=1}^{m} \frac{\rho'_j}{\rho(S_j)}\right) \mathcal{S}(\sigma,\tau)\\ &= (1+\varepsilon)\left(\sum_{j=1}^{m-1} \frac{\rho_j}{\rho(S_j)} + \frac{\rho(S_m)-\varepsilon/(1+\varepsilon)}{\rho(S_m)}\right) \mathcal{S}(\sigma,\tau).\end{aligned}$$

Using the inequality $x \le \ln 1/(1-x)$, we get

$$\begin{aligned}\mathcal{H}(\sigma,\tau') &\le (1+\varepsilon)\left(\sum_{j=1}^{m-1} \ln\frac{\rho(S_j)}{\rho(S_{j+1})} + \ln\frac{\rho(S_m)}{\varepsilon/(1+\varepsilon)}\right) \mathcal{S}(\sigma,\tau)\\ &= (1+\varepsilon)\ln\left(1+\frac{1}{\varepsilon}\right)\mathcal{S}(\sigma,\tau).\end{aligned}$$

This proves the first inequality. The second follows by a similar computation. □

If we know a bound on the ratio σ_i/τ_i, then a sharper inequality can be proved:

COROLLARY 4.8. *Let σ,τ be two distributions, let $\varepsilon > 0$, and assume that $\sigma \le (1+2\varepsilon)\tau$. Then*

$$\mathcal{H}(\sigma, \le (1+\varepsilon)\tau) \le 2\mathcal{S}(\sigma,\tau)$$

PROOF. Let $W = \{i : \sigma_i > (1+\varepsilon)\tau_i\}$ and $U = V \setminus W$. If $W = \emptyset$, then we have nothing to prove, so suppose that $W \ne \emptyset$.

Let $a_i = \max\bigl(0, \sigma_i - (1+\varepsilon)\tau_i\bigr)$ and $b_i = \max\bigl(0, (1+\varepsilon)\tau_i - \sigma_i\bigr)$. Let $a = \sum_i a_i$. Then simple computation yields that $\sum_i b_i = a + \varepsilon$. Obviously, $a \le \varepsilon$. Setting $\alpha_i = a_i/a$, $\beta_i = b_i/(a+\varepsilon)$, we get two distributions α and β.

Now fix a stopping rule Δ from α to some distribution $\beta' \le 2\beta$, such that $\mathsf{E}\Delta = \mathcal{H}(\alpha, \le 2\beta)$. Consider the following rule Γ: if you start at i, then halt right away with probability $(\sigma_i - a_i)/\sigma_i$; else follow Δ.

First, we show that $\sigma^\Gamma \le (1+\varepsilon)\tau$. Indeed, if $i \in W$ then $\beta_i = 0$ and hence $\beta'_i = 0$, which implies that

$$\sigma_i^\Gamma = \sigma_i - a_i = (1+\varepsilon)\tau_i.$$

If $i \in U$ then

$$\sigma_i^\Gamma = \sigma_i + a\beta'_i \le \sigma_i + 2a\beta_i = \sigma_i + \frac{2a}{a+\varepsilon}\bigl((1+\varepsilon)\tau_i - \sigma_i\bigr) \le (1+\varepsilon)\tau_i.$$

Second, we show that $\mathsf{E}\Gamma \le 2\mathcal{S}(\sigma,\tau)$:

$$\mathsf{E}\Gamma = a\mathsf{E}\Delta = a\mathcal{H}(\alpha, \le 2\beta).$$

By Lemma 4.7, we have $\mathcal{H}(\alpha, \le 2\beta) \le 2\mathcal{S}(\alpha,\beta)$. But it is easy to see that

$$\alpha \le \frac{\varepsilon}{a(1+2\varepsilon)}\sigma,$$

and

$$\beta \leq \frac{1+\varepsilon}{\varepsilon}\tau,$$

and hence by (2.8)

$$\mathcal{S}(\alpha,\beta) \leq \frac{\varepsilon}{a(1+2\varepsilon)}\frac{1+\varepsilon}{\varepsilon}\mathcal{S}(\sigma,\tau) < \frac{1}{a}\mathcal{S}(\sigma,\tau).$$

Thus

$$\mathsf{E}\Gamma \leq 2\mathcal{S}(\sigma,\tau)$$

as claimed. □

Threshold rule. A natural way of generalizing the simplest stopping rule ("walk for k steps") is to have a "threshold" or "release time" $0 \leq h_i \leq \infty$ associated with each state i. The stopping rule is defined by

$$\Gamma(w_0,\dots,w_k) = \begin{cases} 0 & \text{if } k \geq h_{w_k} \\ 1 & \text{if } k < h_{w_k} - 1 \\ h_{w_k} - k & \text{otherwise.} \end{cases}$$

Thus if we hit a state past the release time we stop; otherwise we continue, except that if we're within 1, we use the fractional part to randomize. Theorems 4.9 and 4.10, from [**LW2**] and [**LW4**], establish the universality and max-optimality of threshold rules.

THEOREM 4.9. *For every starting distribution σ and target distribution τ, there exists a threshold function h which gives an optimal stopping rule for generating τ.*

PROOF. We define this rule recursively as follows. Let p_i^t be the probability of being at state i after t steps (and thus not having stopped at a prior step); let q_i^t be the probability of stopping at state i in *fewer* than t steps. Let x_i^t be the expected number of exits from state i, again in fewer than t steps. Let x_i be the exit frequency of state i in any optimal rule from σ to τ. We'll maintain that $x_i^t \leq x_i$. Then *if we are at state i after step t, we continue with probability* $\min\{1,(x_i - x_i^t)/p_i^t\}$.

It is clear that this rule gives the recurrence

$$x_i^{t+1} = \min(x_i, x_i^t + p_i^t),$$

and hence $x_i^{t+1} \leq x_i$ remains valid. Let us also observe $q_j^t \leq \tau_j$ remains valid. Indeed, if we stop at j by time t at all, then we have $x_j^{t+1} = x_j$; but then the expected number of times j is entered by time $t+1$ is

$$\sigma_j + \sum_i p_{ij}x_i^t \leq \sigma_j + \sum_i p_{ij}x_i = \tau_j + x_j = \tau_j + x_j^{t+1},$$

and hence the expected number of times we stop there is at most τ_j.

It is also clear that we stop with probability 1 (since the sum of its exit frequencies is bounded) and does achieve an ending distribution. Thus this distribution must be τ. The rule is optimal, since the state with $x_i = 0$ is a halting state.

To show that this rule is a threshold rule, we define the thresholds h_i as follows. For a given state i, consider the largest t_i for which $x_i^{t_i} < x_i$, and let

$$h_i = t_i + \frac{x_i - x_i^{t_i}}{p_i^{t_i}}.$$

If $x_i^t < x_i$ for all T, then $h_i = \infty$. If t_i is finite, then $x_i^{t_i+1} = x_i \leq x_i^{t_i} + p_i^{t_i}$, and hence $p_i^{t_i} > 0$ and $h_i \leq t_i + 1$). □

The threshold vector may not be uniquely determined by a threshold rule Θ (e.g. all possible thresholds h_i smaller than the time before any possible walk reaches i are equivalent), but by convention we always associate with Θ the vector each of whose coordinates is minimal. Then h is finite if and only if Θ is bounded. In fact more is true.

THEOREM 4.10. *For any two distributions σ and τ, the threshold rule is max-optimal. In other words, if h_i $(i \in V)$ are the thresholds for an optimal threshold rule from σ to τ, then*

$$\mathcal{M}(\sigma, \tau) = \max_i \lceil h_i \rceil .$$

PROOF. Let, as before, p_i^t be the probability that following the threshold rule Θ, we are at state i after t steps, and let x_i^t be the expected number of exits from state i in fewer than t steps. Let Γ be any stopping rule from σ to τ, and let $p_i^t(\Gamma)$ and $x_i^t(\Gamma)$ be the corresponding quantities for Γ.

We claim that $x_i^t(\Gamma) \leq x_i^t$ for all i and t. This clearly implies that the $\max(\Gamma) \geq \max(\Theta)$ as claimed. Using induction on t, we have

$$x_i^{t+1}(\Gamma) \leq \min(x_i, x_i^t(\Gamma) + p_i^t(\Gamma)) \leq \min(x_i, x_i^t + p_i^t) = x_i^{t+1}.$$

□

4.3. Reverse chains. We defined reverse chains in the introduction and mentioned that they have the same stationary distribution as the "forward" chain. Here we study exit frequencies and access times in reverse chains. The key to the proof of many properties of the reverse chain is the following general "duality formula", proved in [**LW5**].

LEMMA 4.11. *Let $\alpha, \beta, \gamma, \delta$ be four distributions on the states. Then*

$$\sum_i (\beta_i - \alpha_i) \overleftarrow{y}_i(\gamma, \delta) = \sum_i (\delta_i - \gamma_i) y_i(\alpha, \beta) .$$

We recall some consequences of this lemma.

COROLLARY 4.12. (a) *For every state j, we have*

$$\mathcal{H}(\pi, j) = \overleftarrow{\mathcal{H}}(\pi, j) .$$

For every two states i and j, we have

$$\overleftarrow{\mathcal{H}}(i, j) = \mathcal{H}(j, i) - \mathcal{H}(\pi, i) + \mathcal{H}(\pi, j) .$$

(b) $\overleftarrow{\mathcal{H}} = \mathcal{H}$; $\overleftarrow{\mathcal{N}} = \mathcal{N}$.

(c) *If i is a pessimal state and j is a halting state from i to π, then j is a pessimal state for the reverse chain and i is a halting state from j to π in the reverse chain.*

In particular, every time-reversible chain has (at least) two pessimal starting states, which are halting states for getting from the other to π.

4.4. Matrix formulas. We describe some useful formulas connecting the transition matrix M with mixing parameters. All matrices below are $n \times n$ (where $n = |V|$), and their rows and columns are indexed by states. H is the matrix of hitting times: $H_{ij} = \mathcal{H}(i,j)$. We denote by R the diagonal matrix with the return time $1/\pi_i$ in the i-th position of the diagonal. The identity matrix is denoted by I, the all-1 matrix, by J.

It is easy to see that these matrices satisfy the equation

$$(I - M)H = J - R. \tag{4.9}$$

Unfortunately, the matrix $I - M$ is singular, and so (4.9) does not uniquely determine H. But the only left eigenvector of $I - M$ with eigenvalue 0 is π^{T}, and the only right eigenvector with eigenvalue 0 is $\mathbf{1}$, hence $I - M + \mathbf{1}\pi^{\mathsf{T}} = I - M + JR^{-1}$ is non-singular. Moreover, we have

$$(I - M + JR^{-1})(I - JR^{-1}) = I - M$$

(using that $MJ = J$ and $JR^{-1}J = J$). Hence

$$(I - M + JR^{-1})(I - JR^{-1})H = (I - M)H = J - R,$$

and thus

$$H = (I - M + JR^{-1})^{-1}(J - R) + JR^{-1}H\,. \tag{4.10}$$

For convenience, let us set

$$G := (I - M + JR^{-1})^{-1}(J - R) = J - (I - M + JR^{-1})^{-1}R\,.$$

Equation (4.10) is still not the right formula for H, since H also occurs on the right hand side. But we can determine $\pi^{\mathsf{T}}H$ by looking at the diagonal: $0 = G_{ii} + (\pi^{\mathsf{T}}H)_i$ and hence

$$-G_{ii} = (\pi^{\mathsf{T}}H)_i = \mathcal{H}(\pi, i)\,.$$

This implies that

$$G_{ij} = \mathcal{H}(i,j) - \mathcal{H}(\pi,j) = \frac{1}{1\pi_j} z_j(i,\pi).$$

Some important properties of the matrix G are summarized below:

LEMMA 4.13. *(a) The G-matrix of the reverse chain is the transpose of G. In particular, if the chain is time-reversible then G is symmetric.*

The eigenvalues of $R^{-1}G$ are 1 and $1/(1-\lambda_2), \ldots, 1/(1-\lambda_n)$, where $1 = \lambda_1, \lambda_2, \ldots, \lambda_n$ are the eigenvalues of M.

The first assertion is immediate from Corollary 4.12, the second, from the definition of G.

By Theorem 2.1, we get

$$\mathcal{H}(i,\pi) = \max_j(\mathcal{H}(i,j) - \mathcal{H}(\pi,j)) = \max_j G_{ij}.$$

Hence the mixing time is the largest entry of G.

More generally, the matrix G can be used to express several mixing measures. Obviously,

$$\mathcal{H}(\sigma,\pi) = |G^T\sigma|_\infty.$$

Since

$$z_j(\sigma,\pi) = \pi_j(\mathcal{H}(\sigma,j) - \mathcal{H}(\pi,j)) = (R^{-1}G^T\sigma)_j\,,$$

we get

$$\mathcal{Z} = \max_i \|z(i,\pi)\| = \max_i \|R^{-1}G^T e_i\|.$$

Furthermore,

$$\hat{H} = \max_A \pi(A)\mathcal{H}(\pi_A,\pi) = \max_A \max_j \pi_A^T G e_j = \frac{1}{2}\max_j \|R^{-1}Ge_j\|.$$

Let's make a rather trivial but useful remark. Let $w \in \mathbb{R}^V$ be a vector with $\sum_i w_i = 0$. We have considered the ℓ_1-norm

$$\|w\| = \sum_i |w_i| = \max\{w^T x : \ |x|_\infty \le 1\},$$

and used the fact that because of the condition that $\sum_i w_i = 0$, we could as well require that $0 \le x_i \le 2$, which leads to the formula

$$\|w\| = 2\max_{A\subseteq V} \sum_{i\in A} w_i.$$

Another quantity that comes up often is the norm

$$\|w\|' = \max\{w^T x : \ |x|_\infty \le 1, \ \sum_i \pi_i x_i = 0\}.$$

Using that $\sum_i w_i$, we can also write this as

$$\|w\|' = \max\{w^T x : \ 0 \le x_i \le 2, \ \sum_i \pi_i x_i = 1\}.$$

Here $\sigma_i = \pi_i x_i$ may be viewed as a distribution with $\sigma \le 2\pi$.

Now the two norms are not far from each other:

$$\|w\|' \le \|w\| \le 2\|w\|'. \tag{4.11}$$

The first inequality is trivial. To see the second, let $A = \{i : \ w_i \ge 0\}$. We may assume without loss of generality that $\pi(A) \ge 1/2$ (else, consider $-w$). Then

$$\|w\| = 2\sum_{i\in A} w_i \le 2\pi(A)\sum_{i\in A}\frac{1}{\pi(A)}w_i \le 2\pi(A)\|w\|' \le 2\|w\|'.$$

In particular, we get for any distribution σ

$$\begin{aligned}\mathcal{Z}(\sigma,\pi) &= \|z(\sigma,\pi)\| \ge \|z(\sigma,\pi)\|' = \max_{\tau\le 2\pi}\sum_j \frac{\tau_j}{\pi_j} z_j(\sigma,\pi) \\ &= \max_{\tau\le 2\pi} \tau^T G^T \sigma \ge \frac{1}{2}\|z(\sigma,\pi)\| = \frac{1}{2}\mathcal{Z}(\sigma,\tau).\end{aligned}$$

Thus we get that

$$\tilde{\mathcal{Z}} = \max_{\sigma\le 2\pi}\|z(\sigma,\pi)\| \ge \max_{\sigma\le 2\pi}\max_{\tau\le 2\pi}\sigma^T G\tau \ge \frac{1}{2}\tilde{\mathcal{Z}}. \tag{4.12}$$

4.5. The filling method. This section describes a folklore procedure to access a distribution, if we can access a positive fraction of it.

LEMMA 4.14. *For every two distributions τ and σ, and every $0 \le \varepsilon < 1$,*

$$\mathcal{H}(\sigma,\tau) \le \mathcal{H}(\sigma, \ge (1-\varepsilon)\tau) + \frac{\varepsilon}{1-\varepsilon}\mathcal{H}(\ge (1-\varepsilon)\tau).$$

PROOF. Let σ be any starting distribution, and consider the following stopping rule: follow an optimal stopping rule Γ rule from σ to σ^Γ such that $\sigma^\Gamma \ge (1-\varepsilon)\tau$. We get a random state w from distribution σ^Γ. We flip a biased coin and stop with probability $(1-\varepsilon)\tau_w/\sigma_w^\Gamma$ (by our assumption on Γ, this is at most 1). The probability that we stop at state i is

$$\sigma_i^\Gamma \frac{(1-\varepsilon)\tau_i}{\sigma_i^\Gamma} = (1-\varepsilon)\tau_i,$$

and hence the probability that we stop at all is $1-\varepsilon$.

If we do not stop, the continuation of the walk can be considered as a walk starting from the distribution $\sigma' = \frac{1}{\varepsilon}\sigma^\Gamma - \frac{1-\varepsilon}{\varepsilon}\tau$. We follow an optimal stopping rule Γ' from σ' to $(\sigma')^{\Gamma'}$, such that $(\sigma')^{\Gamma'} > (1-\varepsilon)\tau$, to get a random state j; there we stop with probability $(1-\varepsilon)\tau_j/(\sigma')_j^{\Gamma'}$, else follow an optimal stopping rule Γ'' from $\sigma'' = \frac{1}{\varepsilon}(\sigma')^{\Gamma'} - \frac{1-\varepsilon}{\varepsilon}\tau$ to get close to τ etc. The probability that we stop (eventually) at state i is

$$(1-\varepsilon)\tau_i + \varepsilon(1-\varepsilon)\tau_i + \varepsilon^2(1-\varepsilon)\tau_i + \cdots = \tau_i.$$

These numbers add up to 1, hence we stop with probability 1. The expected number of steps is

$$\mathsf{E}\Gamma + \varepsilon\mathsf{E}\Gamma' + \varepsilon^2\mathsf{E}\Gamma'' + \cdots \le \frac{1}{1-\varepsilon}\mathcal{H}(\ge (1-\varepsilon)\tau).$$

□

COROLLARY 4.15. *For any target distribution τ and $0 \le \varepsilon < 1$,*

$$\mathcal{H}(\tau) \le \frac{1}{1-\varepsilon}\mathcal{H}(\ge (1-\varepsilon)\tau).$$

In particular,

$$\mathcal{H} \le \frac{1}{1-\varepsilon}\mathcal{H}_{\underline{\varepsilon}}.$$

4.6. Elementary properties of access times. As a preliminary remark, we note that making a move independently of where we are does not hurt; more exactly, recalling that d is total variation distance and σ^1 is the state distribution after one step of the chain, we have

LEMMA 4.16.

$$d(\sigma^1,\pi) \le d(\sigma,\pi); \qquad \mathcal{H}(\sigma^1,\pi) \le \mathcal{H}(\sigma,\pi).$$

PROOF. The first inequality is easily checked. To prove the second, consider the following rule from σ to π: make one step, then follow an optimal rule from σ^1 to π. Comparing this with an optimal rule from σ to π using Theorem 4.3, we get

$$\pi_i\big(\mathcal{H}(\sigma,\pi) - \mathcal{H}(\sigma^1,\pi)\big) = x_i(\sigma,\pi) - x_i(\sigma^1,\pi) + \pi_i - \sigma_i \ge -x_i(\sigma^1,\pi)$$

by Lemma 4.2. Since there is a state i for which the right hand side is 0, the second inequality also follows. □

As an immediate corollary of Theorem 2.1 we obtain that $\mathcal{H}(\sigma,\tau)$ depends on the difference $\sigma-\tau$ only. In fact a little more follows:

COROLLARY 4.17. *Let $\alpha,\beta,\gamma,\delta$ be four distributions such that $\alpha-\beta=c(\gamma-\delta)$. Then*

$$\mathcal{H}(\alpha,\beta)=c\mathcal{H}(\gamma,\delta)\,.$$

It is easy to see that the access time $\mathcal{H}(\sigma,\tau)$ is convex in both its arguments.

Next we study how access times behave if the source, or the target, distribution is changed "a little". Theorem 4.14 implies the following bound on the "sensitivity" of the target distribution:

LEMMA 4.18. *Let τ and τ' be distributions and $0<\varepsilon<1$ such that $\tau'\geq(1-\varepsilon)\tau$. Then for any starting distribution σ,*

$$\mathcal{H}(\sigma,\tau)\leq\mathcal{H}(\sigma,\tau')+\frac{\varepsilon}{1-\varepsilon}\mathcal{H}(\tau')\,.$$

In particular,

$$\mathcal{H}(\tau)\leq\frac{1}{1-\varepsilon}H(\tau')\,.$$

There is an analogous property of starting distributions:

LEMMA 4.19. *Assume that $\sigma'\geq(1-\varepsilon)\sigma$. Then for every target distribution τ,*

$$\mathcal{H}(\sigma',\tau)\leq(1-\varepsilon)\mathcal{H}(\sigma,\tau)+\varepsilon\mathcal{H}(\tau)\,.$$

In particular,

$$\mathcal{H}(\sigma',\sigma)\leq\varepsilon\mathcal{H}(\sigma)\,.$$

PROOF. This follows from the decomposition

$$\sigma=(1-\varepsilon)\rho+\varepsilon\left(\frac{1}{\varepsilon}\sigma-\left(\frac{1}{\varepsilon}-1\right)\rho\right)$$

and the convexity of $\mathcal{H}(\sigma,\tau)$ as a function of σ. □

We also have

$$\begin{aligned}\|z(\sigma,\rho)\| &= d(\sigma,\rho)\|z(\sigma\backslash\rho,\ \rho\backslash\sigma)\|\\ &\leq d(\sigma,\rho)\Big(\|z(\sigma\backslash\rho,\ \pi)\|+\|z(\rho\backslash\sigma,\ \pi)\|\Big)\\ &\leq 4d(\sigma,\rho)\mathcal{Z}\,.\end{aligned}\tag{4.13}$$

We conclude with another simple lemma:

LEMMA 4.20. *For any three distributions σ and $\overline{\sigma}$ and τ there exists a distribution $\overline{\tau}$ such that*

$$d(\tau,\overline{\tau})\leq d(\sigma,\overline{\sigma})$$

and

$$\mathcal{H}(\overline{\sigma},\overline{\tau})\leq\mathcal{H}(\sigma,\tau)\,.$$

Another way of saying this is that if $d(\sigma,\overline{\sigma})=\varepsilon$ then for every τ,

$$\mathcal{H}_\varepsilon(\overline{\sigma},\tau)\leq\mathcal{H}(\sigma,\tau)\,.$$

PROOF. Let Γ be an optimal stopping rule from σ to τ. Let v^0 be a state from distribution σ and $\overline{v}^0$, a state from distribution $\overline{\sigma}$, coupled so that

$$\mathsf{P}(v^0 = \overline{v}^0) \geq 1 - \varepsilon\,.$$

Define a new stopping rule $\overline{\Gamma}$ by

$$\overline{\Gamma} = \begin{cases} \Gamma & \text{if } v^0 = \overline{v}^0 \\ 0 & \text{otherwise.} \end{cases}$$

Then $v^{\overline{\Gamma}} = v^{\Gamma}$ with probability at least $1 - \varepsilon$, and so

$$d(\overline{\sigma}^{\overline{\Gamma}}, \sigma^{\Gamma}) = d(\overline{\sigma}^{\overline{\Gamma}}, \tau) \leq \varepsilon\,.$$

Since trivially $\mathsf{E}\overline{\Gamma} \leq \mathsf{E}\Gamma$, this proves the lemma. □

4.7. Approximation by averaging. If we consider only stopping at some fixed time, even approximating π may take far more time than a mean-optimal rule (for example when the chain is almost periodic). However, there are very simple blind *approximate* stopping rules that are easily implementable and give a good approximation of the stationary distribution, in expected time only a constant factor more than the mixing time.

The *uniform averaging rule* $\Upsilon = \Upsilon_t$ $(t \geq 0)$ is defined as follows: choose a random integer Y uniformly from the interval $\{0, \ldots, t-1$, and stop after Y steps. (To describe this as a stopping rule: stop after the u-th step with probability $1/(t-u)$ $(u = 0, \ldots, t-1)$.)

We shall give estimates on how close the distribution σ^{Υ} of the state generated by the averaging rule is to the stationary. To this end, we derive an explicit formula for the distribution of the state produced by the averaging rule.

LEMMA 4.21.

$$\sigma^{\Upsilon} = \pi + \frac{1}{t} z(\sigma, \sigma^t)\,.$$

PROOF. Let x_i denote the expected number of times state i is exited during the first t steps. Note that $\sigma_i^{\Upsilon} = x_i/t$. Then the x_i are the exit frequencies of a rule from σ to σ^t, and hence by Theorem 4.3, we have

$$x_i - \pi_i t = z_i(\sigma, \sigma^t)\,,$$

whence the lemma follows. □

With a little more care we can obtain a result that estimates the error of the uniform averaging rule in terms of the approximate mixing time *starting from the same distribution.*

THEOREM 4.22. *For any starting distribution σ and $0 \leq \varepsilon \leq 1$,*

$$d(\sigma^{\Upsilon}, \pi) \leq \varepsilon + \frac{1}{t}\mathcal{H}(\sigma, d_{\varepsilon})\,.$$

In particular,

$$d(\sigma^{\Upsilon}, \pi) \leq \frac{1}{t}\mathcal{H}(\sigma, \pi)\,.$$

PROOF. Let τ be a distribution such that $d(\tau, \pi) \leq \varepsilon$ and $\mathcal{H}(\sigma, \tau) = \mathcal{H}(\sigma, d_{\varepsilon})$. Then

$$z(\sigma, \sigma^t) = z(\sigma, \tau) + z(\tau, \tau^t) + z(\tau^t, \sigma^t)$$

and hence for every $U \subseteq V$,

$$\begin{aligned}\sum_{k\in U}(\sigma_k^Y - \pi_k) &= \frac{1}{t}\sum_{k\in U} z_k(\sigma,\sigma^t)\\ &= \frac{1}{t}\sum_{k\in U} z_k(\sigma,\tau) + \frac{1}{t}\sum_{k\in U} z_k(\tau,\tau^t) + \frac{1}{t}\sum_{k\in U} z_k(\tau^t,\sigma^t).\end{aligned}$$

Putting $\pi_U := \sum_{i\in U}\pi_i$, the first term above is at most $(1-\pi_U)\mathcal{H}(\sigma,\tau)$ and the last term is at most

$$\pi_U\mathcal{H}(\sigma^t,\tau^t) \le \pi_U\mathcal{H}(\sigma,\tau)$$

by (4.4) and (4.6). For the middle term, we use 4.2:

$$\sum_{k\in U} z_k(\tau,\tau^t) = \sum_{k\in U}\sum_{m=0}^{t-1}(\tau_k^m - \pi_k) \le \sum_{m=0}^{t-1} d(\tau^m,\pi) \le td(\tau,\pi) \le t\varepsilon.$$

Substituting these bounds, the theorem follows. (See [**LW4**] for a simple direct proof.) □

Theorem 4.22 asserts closeness in the total variation distance; approaching π pointwise (as noted, for the reversible case, in [**AF**]) turns out to be somewhat harder in general.

Below we establish results about the pointwise distance of distributions obtained by various averaging rules. However, we shall have to use the worst-case bound on the mixing time.

Thus we cannot generally control the quotient σ_i^Y/π_i by averaging over times of order $\mathcal{H}$. In fact, we have see in Section 3.1 that the time needed to make a two-sided approach to π using averaging is at least the order of the *maximum* length of a max-optimal stopping rule that achieves π exactly.

We can show, however, that in time $O(\mathcal{H})$, the excess of σ^Υ over π can be decreased to an arbitrarily small fraction of its original value.

THEOREM 4.23. *Assume that $\sigma \le K\pi$. Then*

$$\sigma^\Upsilon \le \left(1 + \frac{1}{t}K\mathcal{H}\right)\pi.$$

PROOF. By Lemma 4.21, the triangle inequality, and Lemma 4.16,

$$\begin{aligned}\sigma_i^Y &= \pi_i + \frac{1}{t}z_i(\sigma,\sigma^t) = \pi_i - \frac{1}{t}z_i(\sigma^t,\sigma)\\ &\le \pi_i + \frac{1}{t}\pi_i\mathcal{H}(\sigma^t,\sigma) \le \pi_i + \frac{1}{t}\pi_i\big(\mathcal{H}(\sigma^t,\pi) + \mathcal{H}(\pi,\sigma)\big)\\ &\le \pi_i + \frac{1}{t}\pi_i\big(\mathcal{H}(\sigma,\pi) + \mathcal{H}(\pi,\sigma)\big).\end{aligned}$$

Here we know that $\pi \ge \frac{1}{K}\sigma$, and so by Lemma 4.19 we get

$$H(\pi,\sigma) \le (K-1)\mathcal{H},$$

and thus

$$\sigma_i^Y \le \left(1 + \frac{1}{t}K\mathcal{H}\right)\pi_i.$$

□

Our next goal is to describe a rule giving a point that has a probability of at least $(1-\varepsilon)\pi_i$ of being at state i. We assume that the starting distribution is already close to π in the total variation distance (this can be achieved, by the above, using the averaging rule), and do another averaging. As before, let Y be chosen uniformly from $\{0,\ldots,t-1\}$. The main result is the following.

THEOREM 4.24. *Let $0 \le \delta \le \varepsilon \le 1$ and assume that $d(\sigma,\pi) = \varepsilon$ and $\sigma \ge (1-\delta)\pi$. Then*

$$\sigma^Y \ge \left(1-\varepsilon-\frac{\delta}{t}\mathcal{H}\right)\pi .$$

PROOF. Let $\alpha = \sigma\backslash\pi$ and $\beta = \pi\backslash\sigma$, so that we can write

$$\sigma = \pi + \varepsilon\alpha - \delta\beta .$$

Clearly, the supports of α and β are disjoint, and hence $\sigma \ge (1-\delta)\pi$ implies that $\beta \le (\delta/\varepsilon)\pi$. Clearly

$$\sigma^Y = \pi^Y + \varepsilon\alpha^Y - \varepsilon\beta^Y = \pi + \varepsilon\alpha^Y - \varepsilon\beta^Y \ge \pi - \varepsilon\beta^Y .$$

Now by Theorem 4.23, we have

$$\beta^Y \le \left(1+\frac{\delta}{\varepsilon t}\mathcal{H}\right)\pi ,$$

and hence

$$\sigma^Y \ge \left(1-\varepsilon-\frac{\delta}{t}\mathcal{H}\right)\pi$$

as claimed. □

It follows that in order to fill π up to a factor $1-\varepsilon$, it suffices to choose two integers Y_1 and Y_2 uniformly and independently from $\{0,\ldots,\mathcal{H}/\varepsilon\}$, and then do Y_1+Y_2 steps; symbolically, $\mathcal{H}_{\underline{\varepsilon}} \le \mathcal{B}_{\underline{\varepsilon}} \le \mathcal{H}/\varepsilon$.

This result is not entirely satisfactory, however; one would like to see that the error diminishes exponentially with t or, in other words, that the time needed is proportional to $\log(1/\varepsilon)$ rather than to $1/\varepsilon$. Next we describe another simple averaging rule that achieves this. The result below also follows by adaptation of the "multiplicativity property" in Aldous [**AF**].

Let $M > 0$, $t = 8\lceil\mathcal{H}\rceil$, and let X be the sum of M independent random variables $Y_1,\ldots,Y_M$, distributed uniformly in $\{0,\ldots,t-1\}$. Stop at v^X.

To analyze this rule, let $X_k = Y_1 + \ldots Y_k$.

LEMMA 4.25. *For all $k \ge 1$, we have*

$$d(\sigma^{X_k},\pi) \le 2^{-(k+1)}$$

and

$$\sigma^{X_k} \ge \left(1-2^{-(k-1)}\right)\pi .$$

We prove this inequality by induction on k. For $k=1$ the inequality follows by Theorem 4.22. Assume that $k>1$. Then σ^{X_k} can be obtained from $\sigma^{X_{k-1}}$ by the uniform averaging rule, and hence we get by Theorem 4.24 that

$$\sigma^{X_k} \ge \left(1-2^{-k}-\frac{1}{8}2^{-(k-2)}\right)\pi > (1-2^{-(k-1)})\pi . \tag{4.14}$$

On the other hand, we have by the induction hypothesis and Lemma 4.19 that

$$\mathcal{H}(\sigma^{X_{k-1}}, \pi) \le 2^{-(k-2)}\mathcal{H},$$

and hence by Theorem 4.22,

$$d(\sigma^{X_k}, \pi) \le \frac{1}{8}2^{-(k-2)} = 2^{-(k+1)}.$$

This completes the induction. □

Choose $M = \lceil \log(1/\varepsilon) \rceil$, and denote the resulting rule by Γ_ε. Then we have, as an immediate consequence of Lemma 4.25, the following theorem.

THEOREM 4.26. *For any starting distribution σ, the rule Γ_ε produces a distribution τ satisfying*

$$\tau \ge (1-\varepsilon)\pi,$$

and has mean length $O(\mathcal{H}\log(1/\varepsilon))$.

The next lemma allows us in some cases to consider only bounded averaging rules.

LEMMA 4.27. *Let σ be any starting distribution. Suppose that there exists an averaging rule Y and a $0 < \delta < 1$ such that*

$$\sigma^{\mathrm{Y}} \le (1+\delta)\pi.$$

Then there exists an averaging rule W such that $\mathsf{E}\mathrm{W} \le \mathsf{E}\mathrm{Y}$, $\max \mathrm{W} \le (2/\delta)\mathsf{E}Y$ and

$$\sigma^{\mathrm{W}} \le (1+3\delta)\pi.$$

PROOF. Let t be the smallest integer such that $\mathsf{P}(Y > t) \le \delta/2$. Define a non-negative integer valued random variable W by

$$\mathsf{P}(\mathrm{W} = k) = \mathsf{P}(Y = k \mid Y \le t).$$

Then clearly

$$\sigma_i^{\mathrm{W}} = \mathsf{P}(v^{\mathrm{W}} = i) \le \frac{\sigma_i^Y}{\mathsf{P}(Y \le t)} \le \left(\frac{1+\delta}{1-\delta/2}\right)\pi \le (1+3\delta)\pi,$$

and by Markov's inequality,

$$\max(\mathrm{W}) \le t \le \frac{2}{\delta}\mathsf{E}\mathrm{Y}.$$

□

4.8. Parameter dependence. Some of our mixing measures involve one or two somewhat arbitrary parameters: approximate mixing times depend on the error in the approximation (usually denoted by ε), while quantities defined for warm start depend on the bound on the intensity of the starting distribution (often set to 2). In this section we give a brief discussion of this dependence.

Of course, when $\varepsilon = 0$ all approximate mixing times are the same:

$$\mathcal{H}_{\underline{0}} = \mathcal{H}_{\overline{0}} = \mathcal{H}_0 = \mathcal{H}$$

and

$$\mathcal{F}_{\underline{0}} = \mathcal{F}_{\overline{0}} = \mathcal{F}_0 = \mathcal{F}$$

and in general

$$\mathcal{H}_\varepsilon \le \mathcal{H}_{\underline{\varepsilon}} \le \mathcal{H}$$

and

$$\mathcal{F}_\varepsilon \le \mathcal{F}_{\underline{\varepsilon}} \le \mathcal{F}\,.$$

Similar inequalities hold for the disperse times and also for the blind rules.

The conventional wisdom is that the error drops exponentially with time, or put it in another way, the mixing time depends logarithmically on the error. This is indeed true in several situations, but not always. Example 2.4 illustrates that the dependence on the error parameter can be quite tricky.

A nice clean result is due to Aldous and Diaconis [**AD**]. Let $e(t) = \max d(\sigma^t, \rho^t)$, where the maximum is taken over all starting distributions σ and ρ. Clearly the maximum is attained when σ and ρ are concentrated on singletons. Note that obviously $d(\sigma^t, \pi) = d(\sigma^t, \pi^t) \le e(t)$ for every σ, and $e(t) \le 2\max_\sigma d(\sigma^t, \pi)$.

LEMMA 4.28. *For every $t_1, t_2 \ge 0$,*

$$e(t_1 + t_2) \le e(t_1)e(t_2).$$

Thus if we know, say, that $d(\sigma^t, \pi) \le 1/4$ for every starting distribution σ, then it follows that $e(t) \le 1/2$, and hence $d(\sigma^{kt}, \pi) \le e(kt) \le 1/2^k$.

One can carry over this argument to blind rules, and prove

$$\mathcal{B}_{2\varepsilon_1\varepsilon_2} \le \mathcal{B}_{\varepsilon_1}\mathcal{B}_{\varepsilon_2}.$$

Note that we do not get a similar result for the uniform averaging rule, because there the error in σ decreases only as $1/t$ as the parameter t increases.

For the approximate mixing (filling) time $\mathcal{H}_{\overline{\varepsilon}}$, Theorem 5.9 implies that it is essentially independent of ε. Similar assertion holds for the approximate mixing times $\mathcal{H}_\varepsilon$ and $\mathcal{H}_{\underline{\varepsilon}}$ in the time-reversible case. In the case of general Markov chains, we know from Theorem 5.1 below that

$$\mathcal{H}_{\underline{\varepsilon}} \le H_\varepsilon \le 16(1 + \ln(1/\varepsilon))\mathcal{H}_{1/8},$$

which implies a similar exponential convergence.

In what follows, we study dependence of parameters when we also put restrictions on the starting distribution. Our main interest is in the case of warm start, but we need results about more general restrictions.

LEMMA 4.29. *For every $0 < a, b \le 1$,*

$$\mathcal{B}(D_1, D_{ab}) \le \mathcal{B}(D_1, D_a) + \mathcal{B}(D_1, D_b)\,.$$

PROOF. Let $\sigma \le 2\pi$ and let Γ be a blind stopping rule such that

$$(1-a)\pi \le \sigma^\Gamma \le (1+a)\pi \qquad \text{and} \qquad \mathsf{E}\Gamma \le \mathcal{B}(D_1, D_a)\,.$$

Let $\rho = (1/a)\sigma^\Gamma + (1 - (1/a))\pi$. Then ρ is a distribution with $\rho \le 2\pi$, and hence there exists a blind stopping rule Δ, independent of Γ, such that

$$(1-b)\pi \le \rho^\Delta \le (1+b)\pi \qquad \text{and} \qquad \mathsf{E}\Delta \le \mathcal{B}(D_1, D_b)\,.$$

Substituting the definition of ρ, this implies that

$$(1-ab)\pi \le \sigma^{\Gamma+\Delta} \le (1+ab)\pi.$$

Since

$$\mathsf{E}(\Gamma + \Delta) \le \mathcal{B}(D_1, D_a) + \mathcal{B}(D_1, D_b)\,,$$

this proves the lemma. □

Corollary 4.17 implies the first half of the following lemma; the second half is trivial.

LEMMA 4.30. *Let $A, B \subset D_1$. Then*

$$\mathcal{H}(A,B) = \mathcal{H}(2\pi - A, 2\pi - B),$$

and

$$\mathcal{B}(A,B) = \mathcal{B}(2\pi - A, 2\pi - B).$$

We say that a set A of distributions is *closed*, if it is convex and for every $\sigma \in A$, also $\sigma^1 \in A$.

LEMMA 4.31. *Let A, B, C be sets of distributions, and assume that B is closed. Then*

$$\mathcal{B}(A, B \cap C) \leq \mathcal{B}(A,B) + \mathcal{B}(B,C)$$

and

$$\mathcal{B}(B \cap A, B \cap C) \leq \mathcal{B}(A,C).$$

PROOF. Let $\sigma \in A$. Apply an optimal blind rule Γ from σ to some $\tau \in B$. Then apply an optimal blind rule Δ from τ to some $\rho \in C$. Since B is closed, it follows that $\rho \in B$. Thus $\mathcal{B}(\sigma, B \cap C) \leq \mathsf{E}\Gamma + \mathsf{E}\Delta$, which proves the first inequality. The second inequality is trivial, if we observe that a blind stopping rule from a $\sigma \in A$ to some $\tau \in C$ yields a distribution in $B \cap C$ is σ happens to be in $B \cap A$. □

COROLLARY 4.32.

$$\mathcal{B}(D_1, D_{3/2}) \leq 2\mathcal{B}(\leq 2\pi, \leq (3/2)\pi).$$

For $0 \leq c \leq 1$, let $W_c : D \to D$ denote the linear map

$$W_c\sigma = c\sigma + (1-c)\pi.$$

LEMMA 4.33. *Let A and B be two sets of distributions. Then*

$$\mathcal{H}(W_cA, W_cB) = c\mathcal{H}(A,B),$$

and

$$\mathcal{B}(W_cA, W_cB) = \mathcal{B}(A,B).$$

Note that $\mathcal{H}$ and $\mathcal{B}$ show surprisingly different behavior in terms of c: to get, say, from pointwise relative error ε to relative error $\varepsilon/2$ takes the same time for a blind rule independently of ε, but time proportional to ε for a general rule.

PROOF. By Corollary 4.17, we have for every σ and τ,

$$\mathcal{H}(W_c\sigma, W_c\tau) = c\mathcal{H}(\sigma,\tau).$$

This implies the first equation.

For blind rules, it suffices to prove that

$$\mathcal{B}(W_c\sigma, W_c\tau) = \mathcal{B}(\sigma,\tau)$$

for every σ and τ (note that both sides may be infinite). Indeed, let Γ be a blind rule such that $\sigma^\Gamma = \tau$ and $\mathsf{E}\Gamma = \mathcal{B}(\sigma,\tau)$. Then

$$(W_c\sigma)^\Gamma = c\sigma^\Gamma + (1-c)\pi = W_c\tau$$

(here we use that $\pi^\Gamma = \pi$ for a blind rule), and hence

$$\mathcal{B}(W_c\sigma, W_c\tau) \leq \mathsf{E}\Gamma = \mathcal{B}(\sigma,\tau).$$

The reverse inequality follows similarly. □

LEMMA 4.34. *For every set A of distributions and every $0 < \varepsilon < 1$,*

$$\mathcal{H}(A,\pi) \le \frac{1}{1-\varepsilon}\mathcal{H}(A, W_\varepsilon A).$$

(Note that if $\pi \in A$ then trivially $\mathcal{H}(A,\pi) \ge \mathcal{H}(A, W_{1/2}A$.)

PROOF. Using (2.7) and Lemma 4.33, we have

$$\mathcal{H}(A,\pi) \le \sum_{k=0}^{\infty} \mathcal{H}(W_{\varepsilon^k}A, W_{\varepsilon^{k+1}}A) = \left(\sum_{k=0}^{\infty}\varepsilon^k\right)\mathcal{H}(A, W_\varepsilon A) = \frac{1}{1-\varepsilon}\mathcal{H}(A, W_\varepsilon A)\,.$$

□

The last lemma in this section bounds $\mathcal{H}_\varepsilon(\rho,\pi)$ if we only know that $\rho \le c\pi$ for some $c > 2$.

LEMMA 4.35. *Let ρ be a distribution such that $\rho \le r\pi$ for some $c \ge 2$. Then*

$$\mathcal{H}_\varepsilon(\rho,\pi) \le (c-1)\tilde{\mathcal{H}}_{\varepsilon/(2c-3)}\,.$$

PROOF. Let

$$h = \frac{\varepsilon}{2c-3} \qquad \text{and} \qquad \alpha = \frac{\rho + (c-2)\pi}{c-1}\,.$$

Then α is a distribution with $\alpha \le 2\pi$, and hence there exists a distribution τ such that $d(\tau,\pi) = h$ and $\mathcal{H}(\alpha,\tau) \le \tilde{\mathcal{H}}_h$. Then we can write $\tau = \pi - h\beta + h\gamma$ for some distributions β, γ.

Let

$$u = \frac{1}{1+h(c-2)}\,,$$

and consider the distributions $\rho' = u\rho + (1-u)\beta$ and $\tau' = u\tau + (1-u)\gamma$. Then

$$\rho' - \tau' = u(\rho-\tau) + (1-u)(\beta-\gamma) = \frac{c-1}{1+h(c-2)}(\alpha-\tau)\,.$$

Hence

$$\mathcal{H}(\rho',\tau') = \frac{c-1}{1+h(c-2)}H(\alpha,\tau) < (c-1)\tilde{\mathcal{H}}_h\,.$$

But $d(\rho,\rho') \le 1-u < h(c-2)$, and thus by Lemma 4.20, there exists a distribution τ'' such that $d(\tau'',\tau') < h(c-2)$ and $H(\rho,\tau'') \le H(\rho',\tau')$. Now we have

$$d(\tau'',\pi) \le d(\tau'',\tau') + d(\tau',\tau) + d(\tau,\pi) \le h(c-2) + h(c-2) + h = (2c-3)h = \varepsilon\,.$$

Thus

$$\mathcal{H}_\varepsilon(\rho,\pi) \le \mathcal{H}(\rho,\tau'') \le \mathcal{H}(\rho',\tau') < (c-1)\tilde{\mathcal{H}}_h\,.$$

□

5. Proofs of the main results

We prove the theorems stated in Chapter 3.1. In fact, we start each proof by stating the theorems with explicit constants; in some cases, we introduce new mixing parameters that facilitate the proof and are perhaps interesting on their own. We treat the groups in a slightly different order, since some partial results from one proof are needed in another.

5.1. The forget group: proof of Theorem 3.2. We start with a more complete statement of the theorem.

THEOREM 5.1. *For every finite Markov chain and* $0 < \varepsilon < 1/2$, *we have the following inequalities.*

$$(\frac{1}{2} - \varepsilon)\mathcal{Z} \le \mathcal{F}_\varepsilon \le \mathcal{F}_{\underline{\varepsilon}} \le 2\mathcal{H}_{\varepsilon/2} \le 2\mathcal{H}_{\overline{\varepsilon/2}} \le 4(1+\ln(2/\varepsilon))\mathcal{S} \le 4(1+\ln(2/\varepsilon))\mathcal{Z}\,.$$

In addition,

$$\mathcal{Z} \le \mathcal{F} \le \frac{1}{1-\varepsilon}\mathcal{F}_{\underline{\varepsilon}}\,.$$

COROLLARY 5.2.

$$\mathcal{S} \le \mathcal{Z} \le \mathcal{F} \le 8\mathcal{S}\,.$$

We break down the proof into several steps, one for each inequality claimed. The first step is (up to the constant) a stronger version of (4.7):

LEMMA 5.3. *Let* $0 < \varepsilon < 1/2$, *then*

$$\mathcal{Z} \le \frac{2}{1-2\varepsilon}\mathcal{F}_\varepsilon\,.$$

PROOF. Let τ be the optimal distribution in the definition of the approximate mixing time $\mathcal{F}_\varepsilon$, and let σ be a distribution maximizing $\|z(\sigma,\tau)\|$. By definition, there exists a distribution μ such that $d(\mu,\tau) \le \varepsilon$ and $\mathcal{H}(\sigma,\mu) \le \mathcal{F}_\varepsilon$. Then for an appropriate set $A \subseteq V$, we have

$$\mathcal{Z}(\sigma,\tau) = \sum_{i\in A} z_i(\sigma,\tau) = \sum_{i\in A} z_i(\sigma,\mu) + \sum_{i\in A} z_i(\mu,\tau)\,.$$

Here by (4.7),

$$\sum_{i\in A} z_i(\sigma,\mu) = \sum_{i\in V\setminus A} z_i(\mu,\sigma) \le (1-\pi(A))\mathcal{H}(\sigma,\mu) \le (1-\pi(A))\mathcal{F}_\varepsilon.$$

To estimate the other term, we have by (4.13)

$$\sum_{i\in A} z_i(\mu,\rho) \le \frac{1}{2}\|z(\mu,\rho)\| \le 2d(\mu,\rho)\mathcal{Z}(\sigma,\tau) \le 2\varepsilon\mathcal{Z}(\sigma,\tau)\,.$$

Hence

$$\mathcal{Z}(\sigma,\tau) \le \mathcal{F}_\varepsilon + 2\varepsilon\mathcal{Z}(\sigma,\tau)\,,$$

and hence

$$\mathcal{Z}(\sigma,\tau) \le \frac{1}{1-2\varepsilon}\mathcal{F}_\varepsilon.$$

Now for any distribution α,

$$\mathcal{Z}(\alpha,\pi) \le \mathcal{Z}(\alpha,\tau) + \mathcal{Z}(\pi,\tau) \le 2\mathcal{Z}(\sigma,\tau) \le \frac{2}{1-2\varepsilon}\mathcal{F}_\varepsilon,$$

and the lemma follows. □

An immediate consequence of Lemma 4.21 is the following inequality. Choosing $t = (2/\varepsilon)\mathcal{Z}$, the uniform averaging rule yields a distribution which is within variation distance ε of π. Since the averaging rule Υ is a blind rule, we have shown

COROLLARY 5.4. *For every* $0 \le \varepsilon \le 1$,

$$\mathcal{H}_\varepsilon \le \mathcal{B}_\varepsilon \le \mathcal{U}_\varepsilon \le \frac{2}{\varepsilon}\mathcal{Z}\,.$$

The inequality

$$\mathcal{F}_\varepsilon \le \mathcal{F}_{\underline{\varepsilon}}$$

is trivial.

LEMMA 5.5. *For* $\varepsilon < 1/2$,

$$\mathcal{F}_{\underline{2\varepsilon}} \le 2\mathcal{H}_\varepsilon \,.$$

PROOF. For each state k, let $\tau(k)$ be a distribution such that $d(\tau(k),\pi) \le \varepsilon$ and $\mathcal{H}(k,\tau(k)) \le \mathcal{H}_\varepsilon$. Let

$$f_i = \min_\mu \sum_k \tau(k)_i \mu_k \,,$$

where μ ranges over all distributions with $d(\mu,\pi) \le \varepsilon$. Let $\mu(i)$ be the distribution achieving this minimum. Write $\mu(i)_k = \pi_k + \alpha(i)_k$ and $\tau(k)_i = \pi_i + \beta(k)_i$. Let $a(i)_k = \min(0,\alpha(i)_k)$ and $b(k)_i = \min(0,\beta(k)_i)$. Then

$$\tau(k)_i\mu(i)_k \ge (\pi_i + b(k)_i)(\pi_k + a(i)_k) \ge \pi_i\pi_k + \pi_i a(i)_k + \pi_k b(k)_i$$

and hence

$$\sum_i f_i = \sum_{i,k} \tau(k)_i \mu(i)_k \ge 1 + \sum_i \pi_i \sum_k a(i)_k + \sum_k \pi_k \sum_i b(k)_i \,.$$

Now here, for any fixed i,

$$\sum_k a(i)_k = -d(\pi,\mu(i)) \ge -\varepsilon \,,$$

and similarly

$$\sum_i b(k)_i = -d(\pi,\tau(k)) \ge -\varepsilon \,.$$

Hence

$$\sum_i f_i \ge 1 - \sum_i \pi_i \varepsilon - \sum_k \pi_k \varepsilon = 1 - 2\varepsilon \,.$$

Consider the distribution

$$\rho_i = f_i \Big/ \sum_j f_j \,,$$

and the following stopping rule: starting at state k, follow an optimal rule from k to $\tau(k)$; if you end at j, then follow an optimal rule from j to $\tau(j)$. Let $\theta(k)$ be the distribution produced. Then

$$\theta(k)_i = \sum_j \tau(k)_j \tau(j)_i \ge f_i \ge (1-2\varepsilon)\rho_i \,.$$

Thus $\mathcal{H}_{\underline{2\varepsilon}}(\rho) \le 2\mathcal{H}_\varepsilon$. □

Lemma 4.7 implies that

COROLLARY 5.6. *For every* $0 < \varepsilon < 1$,

$$\mathcal{H}_{\overline{\varepsilon}} \le (1+\varepsilon)\ln(1+1/\varepsilon))\mathcal{S} \,.$$

LEMMA 5.7.

$$\mathcal{S} \le \mathcal{Z}/2.$$

PROOF. Let σ be any distribution and let $A \subseteq V$. Let ρ be the distribution of the first element in A, starting the chain from σ. Note that $\mathcal{H}(\sigma, \rho) = \mathcal{H}(\sigma, A)$ is just the expected number of steps before hitting A, starting from σ. Also note that trivially $x_i(\sigma, \rho) = 0$ for $i \in A$, and hence

$$\pi_i \mathcal{H}(\sigma, \rho) = -z_i(\sigma, \rho) = z_i(\rho, \sigma) .$$

Hence

$$\pi(A)\mathcal{H}(\sigma, A) = \sum_{i \in A} z_i(\rho, \sigma) \leq \frac{1}{2}\mathcal{Z}.$$

□

COROLLARY 5.8.

$$\mathcal{S} \leq \mathcal{Z} .$$

To complete the proof of Theorem 5.1, it suffices to notice that letting $\varepsilon \to 0$ in Lemma 5.3 we get

$$\mathcal{Z} \leq \mathcal{F} ,$$

while Corollary 4.15 implies that

$$\mathcal{F} \leq \mathcal{H}(\rho) \leq \frac{1}{1-\varepsilon}\mathcal{H}_{\underline{\varepsilon}}(\rho) = \frac{1}{1-\varepsilon}\mathcal{F}_{\underline{\varepsilon}} .$$

5.2. The mixing group: proof of Theorem 3.1. We state and prove two theorems separately, which together imply Theorem 3.1.

THEOREM 5.9.

$$\mathcal{H}_{\underline{\varepsilon}} \leq \mathcal{H} \leq \frac{1}{1-\varepsilon}\mathcal{H}_{\underline{\varepsilon}} .$$

PROOF. The first inequality is trivial; the second follows immediately from Corollary 4.15. □

THEOREM 5.10.

$$\mathcal{H} \leq 2\Big(\mathcal{S} + \mathcal{I}\Big) .$$

PROOF. Let $S := \{j : \mathcal{H}(j, \pi) \leq 2\mathcal{I}\}$. Then

$$\mathcal{I} = \sum_i \pi_i \mathcal{H}(i, \pi) \geq \sum_{i \in V \setminus S} \pi_i \mathcal{H}(i, \pi) \geq \pi(V \setminus S)(2\mathcal{I}),$$

and hence $\pi(S) \geq 1/2$. Now for any starting state s, we can walk from s until we hit some state $j \in S$, and then follow an optimal rule from j to π. The expected length of this walk is at most

$$\mathcal{H}(s, S) + 2\mathcal{I} \leq \frac{1}{\pi(S)}\mathcal{S} + 2\mathcal{I} \leq 2\mathcal{S} + 2\mathcal{I}.$$

□

Using the inequality $\mathcal{S} \leq \mathcal{F}$ (see Corollary 5.2), we get

$$\mathcal{H} \leq 2(\mathcal{F} + \mathcal{I}) \leq 4 \max\{\mathcal{F}, \mathcal{I}\}.$$

Since trivially $\max\{\mathcal{F}, \mathcal{I}\} \leq \mathcal{H}$, this completes the proof of Theorem 3.1.

5.3. The reset group: proof of Theorem 5.11. Besides the reset time $\mathcal{I}$ and the warm mixing time $\tilde{\mathcal{H}}$, this group also contains the mixing measures in the forget group, defined for the reverse chain. Of these, our proof uses the discrepancy of the reverse chain,

$$\overleftarrow{\mathcal{Z}} = \max_{v\in V,\ S\subseteq V} \sum_{i\in S} \overleftarrow{z}_i(v,\pi)$$

THEOREM 5.11.

$$\frac{1}{16}\mathcal{I} \le \overleftarrow{\mathcal{Z}} \le \tilde{\mathcal{H}} \le 2\mathcal{I}\,.$$

The first inequality is immediate by Corollary 5.2 and Theorem 3.8. To prove the second, let $u, v \in V$ and $A \subset V$ attain the maximum in the definition of $\overleftarrow{\mathcal{Z}}$. Since

$$\overleftarrow{z}_i(v,\pi) = \frac{\pi_i}{\pi_v} z_v(i,\pi) = \pi_i(\mathcal{H}(i,v) - \mathcal{H}(\pi,v)),$$

this implies that

$$A = \{i \in V :\ \mathcal{H}(i,v) \ge \mathcal{H}(\pi,v)\}.$$

Case 1. Assume that $\pi(A) \ge 1/2$. Then π_A is a distribution with $\pi_A \le 2\pi$. Moreover,

$$\begin{aligned}
\mathcal{H}(\pi_A,\pi) &\ge \mathcal{H}(\pi_A,v) - \mathcal{H}(\pi,v) = \sum_{i\in A} \frac{\pi_i}{\pi(A)} \big(\mathcal{H}(i,v) - \mathcal{H}(\pi,v)\big) \\
&= \frac{1}{\pi(A)} \sum_{i\in A} \frac{\pi_i}{\pi_v} z_v(i,\pi) = \frac{1}{\pi(A)} \sum_{i\in A} \overleftarrow{z}_i(v,\pi) = \frac{1}{\pi(A)} \overleftarrow{\mathcal{Z}} > \overleftarrow{\mathcal{Z}}\,.
\end{aligned}$$

Case 2. Assume that $\pi(A) < 1/2$. Set $t = (1 - 2\pi(A))/(1 - \pi(A))$ and define

$$\rho_i = \begin{cases} 2\pi_i & \text{if } i \in A \\ t\pi_i & \text{otherwise.} \end{cases}$$

Then again ρ is a distribution with $\rho \le 2\pi$. Moreover,

$$\begin{aligned}
\mathcal{H}(\rho,\pi) &\ge \mathcal{H}(\rho,v) - \mathcal{H}(\pi,v) \\
&= \sum_{i\in A} 2\pi_i\big(\mathcal{H}(i,v) - \mathcal{H}(\pi,v)\big) + \sum_{i\in V\setminus A} t\pi_i\big(\mathcal{H}(i,v) - \mathcal{H}(\pi,v)\big) \\
&= (2-t)\sum_{i\in A} \pi_i\big(\mathcal{H}(i,v) - \mathcal{H}(\pi,v)\big) = \frac{1}{1-\pi(A)} \sum_{i\in A} \frac{\pi_i}{\pi_v} z_v(i,\pi) \\
&= \frac{1}{1-\pi(A)} \sum_{i\in A} \overleftarrow{z}_i(v,\pi) = \frac{1}{1-\pi(A)} \overleftarrow{\mathcal{Z}} > \overleftarrow{\mathcal{Z}}\,.
\end{aligned}$$

This proves the second inequality.

Finally, the third inequality is easy. Assume that $\sigma \le 2\pi$, and consider the following rule from σ to π: look at the starting state and follow an optimal rule from that state. Hence

$$\mathcal{H}(\sigma,\pi) \le \sum_i \sigma_i \mathcal{H}(i,\pi) \le 2\sum_i \pi_i \mathcal{H}(i,\pi) = \mathcal{I}\,.$$

5.4. The maxing group: proof of Theorem 3.4.

LEMMA 5.12. *Let σ and τ be any two distributions. If $x_i(\sigma,\tau) \le \sigma_i$ for every state i, then $\mathcal{M}(\sigma,\tau) \le 1$.*

PROOF. Consider the rule "if you are at i, make one step with probability $x_i(\sigma,\tau)/\sigma_i$ and stop, else stop right away". This has the right exit frequencies, and hence it yields τ. □

LEMMA 5.13. *Suppose that there exists a blind rule Δ such that for any initial distribution σ,*

$$4\pi/5 \le \sigma^\Delta \le 5\pi/4\,.$$

Then for every σ there is a (non-blind) stopping rule Γ with $\sigma^\Gamma = \pi$ and $\max(\Gamma) \le 2\max(\Delta)$.

PROOF. Let $\max(\Delta) = t$ and $a_j = \mathsf{P}(\Delta = j)$ for $j = 0, 1, \dots, t$. Replacing the transition matrix M by $\sum_{j=0}^t a_j M^j$ preserves the stationary distribution, while the condition of the theorem becomes

$$4\pi/5 \le \sigma^1 \le 5\pi/4$$

for all σ. It now suffices to show that $\mathcal{M}(\sigma^1,\pi) \le 1$.

To do this we bound the exit frequencies $x_i(\sigma^1,\pi)$. Note that for our new fast-mixing Markov chain, we have $\mathcal{H}_{1/5} < 1$, and hence by Corollary 4.15, we have $\mathcal{H} \le 5/4$. Thus Corollary 4.18 gives

$$\mathcal{H}(\sigma^1) \le \frac{5}{4}\mathcal{H} \le \frac{25}{16}\,.$$

Now we can apply Lemma 4.19 twice to get

$$\mathcal{H}(\sigma^1,\pi) \le \frac{1}{5}\mathcal{H} \le \frac{1}{4}$$

and

$$\mathcal{H}(\pi,\sigma^1) \le \frac{1}{5}\mathcal{H}(\sigma^1) \le \frac{5}{16}\,.$$

It follows from (4.4) that for each state i,

$$x_i(\sigma^1,\pi) = z_i(\sigma^1,\pi) + \pi_i\mathcal{H}(\sigma^1,\pi) \le \pi_i\mathcal{H}(\pi,\sigma^1) + \pi_i\mathcal{H}(\sigma^1,\pi) \le \frac{9}{16}\pi_i \le \sigma^1.$$

By Lemma 5.12, this implies that $\mathcal{M}(\sigma^1,\pi) \le 1$ and hence $\mathcal{M}(\sigma,\pi) \le 2$ as claimed. □

LEMMA 5.14. *Let $\Phi : \sigma \to \tau$ and $\Psi : \sigma \to \rho$ be two stopping rules such that $\Phi \le \Psi$ for any walk. Then*

$$H(\tau,\rho) \le \mathsf{E}\Psi - \mathsf{E}\Phi\,.$$

PROOF. Define a stopping rule from τ to ρ as follows: the starting state from distribution τ may as well be generated by starting from σ and following Φ. But then we can just consider our walk as a continuation and stop it following the rule Ψ. This way we get a rule from τ to ρ with mean length $\mathsf{E}\Psi - \mathsf{E}\Phi$. □

LEMMA 5.15. *Assume that $t = \mathcal{M}(\sigma,\tau)$ is finite. Then*

$$\mathcal{H}(\sigma,\tau) + \mathcal{H}(\tau,\sigma^t) \le t\,.$$

PROOF. Since $\mathcal{H}(\sigma,\tau)$ is the mean time of the threshold rule $\Theta_{\sigma,\tau}$, t is the mean time of the rule "walk t steps", and $\Theta_{\sigma,\tau} \le t$, Lemma 5.14 implies that $\mathcal{H}(\tau,\sigma^t) \le t - \mathcal{H}(\tau,\sigma)$. □

LEMMA 5.16. *Let $t = \mathcal{M}(\sigma,\pi)$. Choose a random starting state from σ. Choose a random integer Y uniformly from the interval $[t, 2t-1]$, and walk for Y steps. Then the probability of being at state i is at most $2\pi_i$.*

PROOF. The probability that we stop at state i is

$$\frac{1}{t}\sum_{k=t}^{2t-1}\sigma_i^k = \pi_i + \frac{1}{t}z_i(\sigma^t,\sigma^{2t}) = \pi_i - \frac{1}{t}z_i(\pi,\sigma^t) - \frac{1}{t}z_i(\sigma^{2t},\pi).$$

Here we use that

$$z_i(\pi,\sigma^t) = x_i(\pi,\sigma^t) - \pi_i\mathcal{H}(\pi,\sigma^t) \ge -\pi_i\mathcal{H}(\pi,\sigma^t) \ge \pi_i(\mathcal{H}(\sigma,\pi) - t)$$

(by Lemma 5.15), and

$$z_i(\sigma^{2t},\pi) = x_i(\sigma^{2t},\pi) - \pi_i\mathcal{H}(\sigma^{2t},\pi) \ge -\pi_i\mathcal{H}(\sigma^{2t},\pi) \ge -\pi_i\mathcal{H}(\sigma,\pi)$$

(by Lemma 4.16). Hence we get that

$$\frac{1}{t}\sum_{k=t}^{2t-1}\sigma_i^k \le \pi_i - \frac{1}{t}\pi_i(\mathcal{H}(\sigma,\pi) - t) + \frac{1}{t}\pi_i\mathcal{H}(\sigma,\pi) = 2\pi_i.$$

This proves that $\mathcal{B}_1 \le \frac{3}{2}t \le \frac{3}{2}\mathcal{M}$. □

5.5. The relaxation group: proofs.

LEMMA 5.17.

$$\frac{1}{4}H^* \le \tilde{\mathcal{S}} \le 2H^*.$$

PROOF. The first inequality is a special case of Corollary 4.8. To prove the second, let A be a set such that $\pi(A)\mathcal{H}(\pi,A) = \tilde{\mathcal{S}}$. Let $\varepsilon = \pi(A)/2(1-\pi(A))$ and let $B = V \setminus A$. Then $\pi_B \le (1+2\varepsilon)\pi$. Let $\tau \le (1+\varepsilon)\pi$ be a distribution such that

$$\mathcal{H}(\pi_B,\tau) \le \mathcal{H}(\le (1+2\varepsilon)\pi, \le (1+\varepsilon)\pi).$$

Let

$$\sigma_i = \begin{cases} \frac{1}{\tau(A)}((1+2\varepsilon)\pi_i - \tau_i) & \text{if } i \in B \\ 0 & \text{otherwise.} \end{cases}$$

Then it is easy to check that σ is a distribution, and that $\sigma_i \ge (\varepsilon/\tau(A))\pi_i$ for $i \in B$. Hence by Corollary 4.17 and Inequality 2.8,

$$\begin{aligned} \mathcal{H}(\pi_B,\tau) &= \tau(A)\mathcal{H}(\sigma,\tau_A) \ge \tau(A)\mathcal{H}(\sigma,A) \\ &\ge \varepsilon\mathcal{H}(\pi,A) = \frac{1}{2(1-\pi(A))}\pi(A)\mathcal{H}(\pi,A) \ge \frac{1}{2}\tilde{\mathcal{S}}. \end{aligned}$$

This shows that $\mathcal{H}(\le (1+2\varepsilon)\pi, \le (1+\varepsilon)\pi) \ge (1/2)\tilde{\mathcal{S}}$. □

Next we prove inequalities for reversible chains. The next lemma was proved by Aldous and Fill [**AF**]:

LEMMA 5.18. *For every time-reversible Markov chain,*

$$\tilde{\mathcal{S}} \le \mathcal{L}.$$

We have a similar inequality for $\tilde{\mathcal{Z}}$:

LEMMA 5.19. *For every time-reversible Markov chain,*

$$\tilde{\mathcal{Z}} \le 8\mathcal{L}.$$

PROOF. By (4.12), we have

$$\tilde{\mathcal{Z}} \le 2 \max_{\sigma,\tau \le 2\pi} \sigma^T G\tau.$$

Here we have $\sigma^T G\tau = (R^{1/2}\sigma)^T(R^{-1/2}GR^{-1/2})(R^{1/2}\tau)$. From $\sigma \le 2\pi$ we get that

$$|R^{1/2}\sigma| = \left(\sum_i \frac{\sigma_i^2}{\pi_i}\right)^{1/2} \le \left(\sum_i 4\pi_i\right)^{1/2} = 2,$$

and hence

$$\sigma^T G\tau \le 4\|R^{-1/2}GR^{-1/2}\|,$$

where $\|.\|$ denote spectral norm of a matrix. Now the matrix $R^{-1/2}GR^{-1/2}$ has the same eigenvalues as $R^{-1}G$, and hence its largest eigenvalue is $1/(1-\lambda_2) = \mathcal{L}$. □

LEMMA 5.20.

$$\tilde{\mathcal{Z}} \le 2\mathcal{H}(D_1, D_{1/2}).$$

PROOF. Let $\sigma \le 2\pi$ such that $\tilde{\mathcal{Z}} = \frac{1}{2}\|z(\sigma,\pi)\|$. Let τ be a distribution such that $(1/2)\pi \le \tau \le (3/2)\pi$ and $\mathcal{H}(\sigma,\tau) \le \mathcal{H}(D_1, W_{1/2}D_1)$. Then

$$\tilde{\mathcal{Z}} = \frac{1}{2}\|z(\sigma,\pi)\| \le \frac{1}{2}\|z(\sigma,\tau)\| + \frac{1}{2}\|z(\tau,\pi)\|.$$

Now here

$$\frac{1}{2}\|z(\sigma,\tau)\| \le H(\sigma,\tau) \le \mathcal{H}(D_1, W_{1/2}D_1),$$

and

$$\frac{1}{2}\|z(\tau,\pi)\| = \frac{1}{4}\|z(2\tau-\pi,\pi)\| \le \frac{1}{2}\tilde{\mathcal{Z}},$$

and thus we get

$$\mathcal{Z} \le \mathcal{H}(D_1, D_{1/2}),$$

which implies the lemma. □

Next, let us give a more complete statement of Theorem 3.6.

THEOREM 5.21.

$$2\varepsilon\tilde{\mathcal{I}}_{4\varepsilon} \le 2\varepsilon\tilde{\mathcal{U}}_{2\varepsilon} \le \tilde{\mathcal{H}}_\varepsilon \le \tilde{\mathcal{B}}_\varepsilon \le \tilde{\mathcal{U}}_\varepsilon \le \frac{1}{\varepsilon}\tilde{\mathcal{Z}} \tag{5.1}$$

and

$$\tilde{\mathcal{H}}_\varepsilon \le 2\tilde{\mathcal{I}}_{\varepsilon/3}. \tag{5.2}$$

If the chain is time-reversible, then

$$\tilde{\mathcal{U}}_\varepsilon \le (\ln\varepsilon)\mathcal{L}. \tag{5.3}$$

The proof takes two lemmas.

LEMMA 5.22. *Let $0 < \varepsilon < 1$ and $t \ge 2\tilde{\mathcal{U}}_\varepsilon$. Let v^0 be chosen from a starting distribution $\sigma \le 2\pi$ and let Y be chosen uniformly from $\{0,\ldots,t-1\}$. Then v^0 and v^Y are (2ε)-independent.*

Note that Theorem 4.22 implies that the distribution of v^Y is closer than 2ε to π in total variation distance. Also, if $\sigma \le (1+\varepsilon)\pi$ then clearly $\sigma^Y \le (1+\varepsilon)\pi$. This lemma is proved (essentially) in [**KLS**].

PROOF. We have to show that

$$\left|\mathsf{P}(v^0 \in A, v^Y \in B) - \mathsf{P}(v^0 \in A)\mathsf{P}(v^Y \in B)\right| < \varepsilon$$

for every two sets A and B of states. We have

$$\begin{aligned} &\left|\mathsf{P}(v^0 \in A, v^Y \in B) - \mathsf{P}(v^0 \in A)\mathsf{P}(v^Y \in B)\right| \\ &= \sigma(A)\left|\mathsf{P}(v^Y \in B \mid v^0 \in A) - \mathsf{P}(v^Y \in B)\right| \\ &= \sigma(A)\left|\sigma_A^Y(B) - \sigma^Y(B)\right| \\ &\le \sigma(A) d(\sigma_A^Y, \sigma^Y). \end{aligned}$$

Notice that $\sigma_A \le (2/\sigma(A))\pi$. By the choice of t, we have

$$d(\sigma^Y, \pi) \le \varepsilon.$$

Furthermore, consider the distribution

$$\alpha = \frac{\sigma(A)}{2-\sigma(A)}\sigma_A + \frac{2-2\sigma(A)}{2-\sigma(A)}\pi.$$

Then $\alpha \le 2\pi$, and hence by the choice of t, $d(\alpha^Y, \pi) \le \varepsilon$. But

$$\alpha^Y - \pi = \frac{\sigma(A)}{2-\sigma(A)}(\sigma_A^Y - \pi),$$

and hence

$$d(\sigma_A^Y, \pi) \le \frac{2-\sigma(A)}{\sigma(A)} d(\alpha^Y, \pi) \le \frac{2-\sigma(A)}{\sigma(A)}\varepsilon.$$

Thus

$$d(\sigma_A^Y, \sigma^Y) \le \varepsilon + \frac{2-\sigma(A)}{\sigma(A)}\varepsilon = \frac{2}{\sigma(A)}\varepsilon,$$

and thus

$$\left|\mathsf{P}(v^0 \in A, v^Y \in B) - \mathsf{P}(v^0 \in A)\mathsf{P}(v^Y \in B)\right| \le \sigma(A) d(\sigma_A^Y, \sigma^Y) \le \sigma(A)\frac{2}{\sigma(A)}\varepsilon = 2\varepsilon.$$

□

LEMMA 5.23. *Let Γ be a stopping rule such that $d(\pi^\Gamma, \pi) < \varepsilon$ and if v^0 is drawn according to π, then v^0 and v^Γ are ε-independent. Then $\tilde{\mathcal{H}}_{3\varepsilon} \le 2\mathsf{E}_\pi\Gamma$.*

PROOF. Consider any starting distribution $\sigma < 2\pi$, and the result of stopping rule Γ starting from σ. Trivially

$$\mathsf{E}_\sigma\Gamma \le 2\mathsf{E}_\pi\Gamma.$$

Thus it suffices to prove that $d(\sigma^\Gamma, \pi) < 3\varepsilon$. Indeed, we have

$$d(\sigma^\Gamma, \pi) \le d(\sigma^\Gamma, \pi^\Gamma) + d(\pi^\Gamma, \pi) \le d(\sigma^\Gamma, \pi^\Gamma) + \varepsilon.$$

To estimate the first term, we write

$$\sigma = \sum_{k=1}^m a_k \pi_{A_k},$$

where $a_k > 0$ and $\emptyset \subset A_1 \subset \cdots \subset A_m \subseteq V$. It is easy to see that such a decomposition exists. Looking at an element $i \in A_1$, we see that $\sum_k \frac{a_k}{\pi(A_k)} \leq 2$. Further,

$$\sigma^\Gamma = \sum_k a_k \pi^\Gamma_{A_k}.$$

Hence for an appropriate set $B \subseteq V$,

$$d(\sigma^\Gamma, \pi^\Gamma) = \sum_{i \in B} (\sigma^\Gamma_i - \pi^\Gamma_i) = \sum_k a_k \sum_{i \in B} (\pi^\Gamma_{A_k})_i.$$

Here we have

$$\sum_{i \in B} (\pi^\Gamma_{A_k})_i - \pi^\Gamma_i = \mathsf{P}(v^\Gamma \in B \mid v^0 \in A_k) - \mathsf{P}(v^\Gamma \in B)$$

$$= \frac{\mathsf{P}(v^0 \in A_k,\ v^\Gamma \in B) - \mathsf{P}(v^0 \in A_k)\mathsf{P}(v^\Gamma \in B)}{\mathsf{P}(v^0 \in A_k)} \leq \frac{\varepsilon}{\pi(A_k)}.$$

Thus

$$d(\sigma^\Gamma, \pi^\Gamma) \leq \varepsilon \left(\sum_{k=1}^m \frac{a_k}{\pi(A_k)} \right) \leq 2\varepsilon.$$

This proves the lemma. □

Now the first inequality in (5.1) follows from Lemma 5.22, the second, from Theorem 4.22; the next two are trivial, while the last follows easily from Lemma 4.21. Lemma 5.23 easily implies (5.2).

This completes the proof of Theorem 5.21. □

5.6. Relations between classes: proofs. Proof of Theorem 3.7. By Lemma 4.33, we have for every $k \geq 1$,

$$\mathcal{B}(\leq (1+2k)\pi, \leq (1+k)\pi) = \mathcal{B}\left(\left[\frac{2k-1}{2k}..2 \right] \pi, \left[\frac{2k-1}{2k}..\frac{3}{2} \right] \pi \right).$$

Furthermore, by Lemma 4.31,

$$\mathcal{B}\left(\left[\frac{2k-1}{2k}..2 \right] \pi, \left[\frac{2k-1}{2k}..\frac{3}{2} \right] \pi \right) \leq \mathcal{B}(\leq 2\pi, \leq (3/2)\pi) = \tilde{\mathcal{B}}_{\overline{1/2}}.$$

Let $k = \lceil \ln(1/\hat{\pi} - 1) \rceil$, then by the triangle inequality for $\mathcal{B}(\sigma, \tau)$, we have

$$\mathcal{B}(D, D_1) = \mathcal{B}(\leq (1+2^k)\pi, \leq 2\pi) = \sum_{j=0}^{k-1} \mathcal{B}(\leq (1+2^{j+1})\pi, \leq (1+2^j)\pi) \leq k\tilde{\mathcal{B}}_{\overline{1/2}}.$$

This proves the corollary. □

Proof of Theorem 3.8. Parts (a) and (c) were proved in [**LW5**]. Part (b) is straightforward linear algebra. For part (d), note that Corollary 4.12 implies that $\mathcal{H}(i,j) - \mathcal{H}(\pi,j) = \overleftarrow{\mathcal{H}}(j,i) - \overleftarrow{\mathcal{H}}(\pi,i)$, and hence the identity follows from formulas (2.9) and (2.10).

References

[AF] D.J. Aldous and J. Fill, *Reversible Markov Chains and Random Walks on Graphs* (book), to appear. URL for draft at http://www.stat.Berkeley.EDU/users/aldous/book.html.

[A1] D.J. Aldous, Some inequalities for reversible Markov chains, *J. London Math. Soc.* **25** (1982), 564–576.

[A2] D.J. Aldous, Applications of random walks on graphs, preprint (1989).

[A3] D.J. Aldous, The random walk construction for spanning trees and uniform labelled trees, *SIAM J. Discrete Math.* **3** (1990), 450–465.

[AD] D.J. Aldous and P. Diaconis, Shuffling cards and stopping times, *Amer. Math. Monthly* **93** #5 (1986), 333–348.

[ALW] D.J. Aldous, L. Lovász and P. Winkler, Mixing times for uniformly ergodic Markov chains, *Stochastic Processes and their Applications*, to appear.

[AGT] S. Asmussen, P.W. Glynn and H. Thorisson, Stationary detection in the initial transient problem, *ACM Transactions on Modeling and Computer Simulation* **2** (1992), 130–157.

[BC] J.R. Baxter and R.V. Chacon, Stopping times for recurrent Markov processes, *Illinois J. Math.* **20** (1976), 467–475.

[BD] D. Bayer and P. Diaconis, Trailing the dovetail shuffle to its lair, *Ann. Appl. Probab.* **2** (1992), 294–313.

[BL] A. Beveridge and L. Lovász, Random walks and the regeneration time, *J. Graph Theory* (to appear)

[BR] A. Broder, Generating random spanning trees, *Proc. 30th Annual Symp. on Found. of Computer Science,* IEEE Computer Soc. (1989), 442–447.

[CO] R.V. Chacon and D.S. Ornstein, A general ergodic theorem, *Illinois J. Math.* **4** (1960), 153–160.

[CRRST] A.K. Chandra, P. Raghavan, W.L. Ruzzo, R. Smolensky, and P. Tiwari, The Electrical Resistance of a Graph Captures its Commute and Cover Times, *Proceedings of the 21st Annual ACM Symposium on Theory of Computing*, May (1989).

[CTW] D. Coppersmith, P. Tetali, and P. Winkler, Collisions among Random Walks on a Graph, *SIAM J. on Discrete Mathematics* **6** #3 (1993), 363–374.

[DS] P.G. Doyle and J.L. Snell, *Random Walks and Electric Networks*, Mathematical Assoc. of America, Washington, DC 1984.

[D] L.E. Dubins, On a theorem of Skorokhod, *Ann. Math. Statist.* **39** (1968), 2094–2097.

[DFK] M. Dyer, A. Frieze and R. Kannan, A random polynomial time algorithm for estimating volumes of convex bodies, *Proc. 21st Annual ACM Symposium on the Theory of Computing* (1989), 375–381.

[JS] M. Jerrum and A. Sinclair, Conductance and the rapid mixing property for Markov chains: the approximation of the permanent resolved, *Proc. 20nd Annual ACM Symposium on Theory of Computing* (1988), 235–243.

[KLS] R. Kannan, L. Lovász and M. Simonovits: Random walks and an $O^*(n^5)$ volume algorithm, *Random Structures and Algorithms* (1997).

[LS] L. Lovász and M. Simonovits, Random walks in a convex body and an improved volume algorithm, *Random Structures and Alg.* **4** (1993), 359–412.

[LW1] L. Lovász and P. Winkler, A note on the last new vertex visited by a random walk, *J. Graph Theory* **17** (1993), 593–596.

[LW2] L. Lovász and P. Winkler, Efficient stopping rules for Markov chains, *Proc. 27th ACM Symp. on the Theory of Computing* (1995), 76–82.

[LW3] L. Lovász and P. Winkler, Exact mixing in an unknown Markov chain, *Electronic J. Comb.* **2**, (1995), Paper R15.

[LW4] L. Lovász and P. Winkler, Mixing of random walks and other diffusions on a graph, *Surveys in Combinatorics, 1995*, P. Rowlinson, ed., London Math. Soc. Lecture Note Series 218, Cambridge U. Press (1995), 119–154.

[LW5] L. Lovász and P. Winkler, Reversal of Markov chains and the forget time, *Combinatorics, Probability and Computing*, to appear.

[P] J.W. Pitman, Occupation measures for Markov chains, *Adv. Appl. Prob.* **9** (1977), 69–86.

[R] D.H. Root, The existence of certain stopping times on Brownian motion, *Ann. Math. Statist.* **40** (1969), 715–718.

[S] A. Skorokhod, *Studies in the Theory of Random Processes*, orig. pub. Addison-Wesley (1965), 2nd ed. Dover, New York 1982.

[T] P. Tetali, Random walks and effective resistance of networks, *J. Theoretical Prob.* #1 (1991), 101–109.

Dept. of Computer Science, Yale University, New Haven, CT 06510
E-mail address: lovasz@cs.yale.edu

Bell Laboratories, 700 Mountain Ave., Murray Hill, NJ 07974
E-mail address: pw@lucent.com

DIMACS Series in Discrete Mathematics
and Theoretical Computer Science
Volume 41, 1998

A Bird's-Eye View of Uniform Spanning Trees and Forests

RUSSELL LYONS

Abstract. We survey the field of uniform spanning forest measures on infinite graphs, which are weak limits of uniform spanning tree measures from finite subgraphs. These limits can be taken with free or wired boundary conditions. Among other results, Pemantle (1991) proved that in $\mathbb{Z}^d$, the free and wired spanning forests coincide and that they give a single tree iff $d \leq 4$. The theory has developed considerably since then and found further connections to random walks, potential theory, harmonic Dirichlet functions, invariant percolation and amenability. A crucial new tool is an algorithm invented by Wilson (1996) to generate random spanning trees. Uniform spanning forests also yield insights into loop-erased walks and harmonic measure from infinity.

§1. Introduction.

We begin with a brief history of this fertile and fascinating field. In subsequent sections, we will more carefully define and develop most of what we recount here. More details for much of the material surveyed here can be found in Benjamini, Lyons, Peres, and Schramm (1997), hereinafter referred to as BLPS (1997). One may also consult the book by Lyons and Peres (1997) for additional background.

A **spanning tree** of a (connected) graph is a subgraph that is connected, contains every vertex of the whole graph, and contains no cycle: see Fig. 1 for an example. The subject of random spanning trees of a graph goes back to Kirchhoff (1847), who showed its relation to electrical networks. One of these relations gives the probability that a uniformly chosen spanning tree will contain a given edge in terms of electrical current in the graph: see (3.1).

It turns out that generating spanning trees at random according to the uniform measure is of interest to computer scientists, who have developed various algorithms over the years for random generation of spanning trees. In particular, this is closely connected to generating a random state from any Markov chain. See Propp and Wilson (1996) for more on this issue.

1991 Mathematics Subject Classification. Primary 60B99. Secondary 60D05, 20F32.
Key words and phrases. Spanning trees, random walks, Cayley graphs, electric networks, harmonic Dirichlet functions, amenability, percolation, loop-erased walk.
Writing partially supported by the Institute for Advanced Studies, Jerusalem, and a Varon Visiting Professorship at the Weizmann Institute of Science.

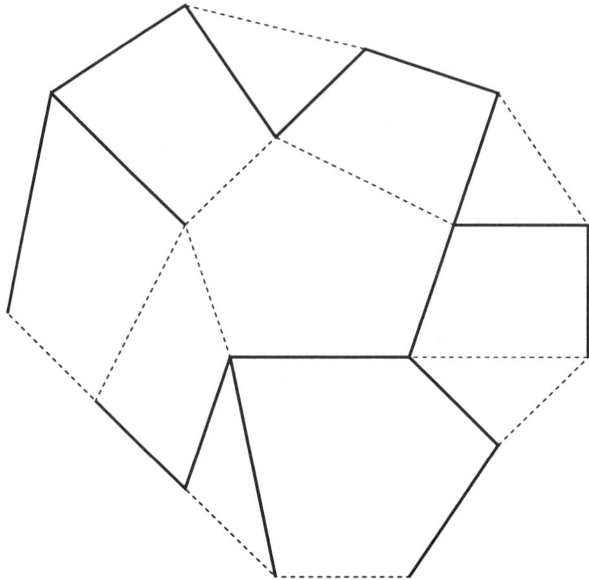

Figure 1. A spanning tree in a graph.

Early algorithms for generating a random spanning tree used the Matrix-Tree Theorem, which counts the number of spanning trees in a graph via a determinant. A better algorithm than these early ones, especially for probabilists, was introduced by Aldous (1990) and Broder (1989). It says that if you start a simple random walk at *any* vertex of a finite graph G and draw every edge it traverses except when it would complete a cycle (i.e., except when it arrives at a previously-visited vertex), then when no more edges can be added without creating a cycle, what will be drawn is a uniformly chosen spanning tree of G. This beautiful algorithm is quite efficient and useful for theoretical analysis, yet Wilson (1996) found an even better one that we'll describe later.

The study of the analogue of a uniform spanning tree on an infinite graph was begun by Pemantle (1991) at the suggestion of the author. Suppose that G is an infinite graph. Let G_n be finite subgraphs with $G_1 \subset G_2 \subset G_3 \subset \cdots$ and $\bigcup G_n = G$. In this case, we say that G is **exhausted** by the sequence $\langle G_n \rangle$. Pemantle showed that the weak limit of the uniform spanning tree measures on G_n exists. (In other words, if μ_n denotes the uniform spanning tree measure on G_n and B, B' are finite sets of edges, then $\lim_n \mu_n(B \subset T, B' \cap T = \emptyset)$ exists, where T denotes a random spanning tree.) This limit is now called the **free uniform spanning forest*** on G, denoted FUSF. Considerations of electrical networks play the dominant role in Pemantle's proof. If G is itself a tree, then this measure is trivial, namely, it is concentrated on $\{G\}$. However, Häggström (1995) drew attention to another limit,

* In graph theory, "spanning forest" usually means a maximal subgraph without cycles, i.e., a spanning tree in each connected component. We mean, instead, a subgraph without cycles that contains every vertex.

namely, the weak limit of the uniform spanning tree measures on G_n^W, where G_n^W is the graph G_n with its boundary (i.e., the vertices of G_n that are adjacent to some vertex of $G \setminus G_n$) identified ("wired") to a single vertex. In fact, Pemantle implicitly proved the existence of this limit as well on any graph; it is now called the **wired uniform spanning forest**, denoted WUSF.

Pemantle (1991) discovered the following interesting properties, among others:

• The free and the wired uniform spanning forest measures are the same on all euclidean lattices $\mathbb{Z}^d$.

• Amazingly, on $\mathbb{Z}^d$, the uniform spanning forest is a single tree a.s. if $d \leq 4$; but when $d \geq 5$, there are infinitely many trees a.s.

• If $2 \leq d \leq 4$, then the uniform spanning tree on $\mathbb{Z}^d$ has a single end a.s. (as defined in Section 6); when $d \geq 5$, each of the infinitely many trees a.s. has at most two ends. (Just how many ends they have will be seen later.)

One of Pemantle's main tools was the Aldous-Broder algorithm. In addition, the study of loop-erased random walks due to Lawler (1991) was crucial. Using Wilson's (1996) algorithm in place of the one due to Aldous and Broder, as well as other tools, BLPS (1997) redeveloped the fundamentals of the theory, originally due to Pemantle (1991) and to Burton and Pemantle (1993). As the theory has become easier to apply and the scope has broadened, new results have emerged.

For example, the general question of when the free and wired measures are the same turns out to be quite interesting: BLPS (1997) showed that the measures are the same iff there are no nonconstant harmonic Dirichlet functions on G.

Burton and Pemantle (1993) found a beautiful generalization of Kirchhoff's theorem relating to electrical networks; their theorem gives the probability that a given set of edges belongs to a uniform spanning tree in terms of an electrical determinant. Burton and Pemantle (1993) also found the entropy of the uniform spanning forest measure on $\mathbb{Z}^d$; coincidentally, it turned out to be exactly the same as the entropy of an apparently entirely different dynamical system studied by Lind, Schmidt, and Ward (1990). This remained a mystery until R. Solomyak (1997) discovered that the Matrix-Tree Theorem is responsible. In the case $d = 2$, Burton and Pemantle (1993) showed that this is also the entropy of yet another system, that of domino tilings (perfect matchings) of the square lattice $\mathbb{Z}^2$; indeed, there is a direct connection (due to Temperley) between spanning trees and domino tilings in planar graphs.

As we will see, uniform spanning forests have interesting connections not only to random walks, algorithms, domino tilings, and electrical networks, but also to planar duality, amenability, percolation, and hyperbolic spaces. Because of this, there is still an enormous number of fascinating open questions. We will present some of them in the last section of this survey.

Finally, we mention that in an entirely different context, that of euclidean functionals (such as length of the minimal spanning tree), it has also been recently discovered that comparison to a "wired" version is extremely useful: see Redmond and Yukich (1994) and Yukich (1996a,b), for example.

§2. Some Examples.

We begin with some pictures. First, consider Fig. 2. Examination of Fig. 2 reveals two spanning trees: one in white, the other in black. To be more precise, if G is a (possibly infinite) proper (connected) planar graph, meaning that G is drawn in the plane so that edges do not overlap and there are only finitely many vertices in every bounded region, then define its **planar dual** graph $G^\dagger$ to have one vertex in each face of G and one edge $e^\dagger$ for each edge e of G, with $e^\dagger$ joining the vertices in the faces on each side of e. See Fig. 3.

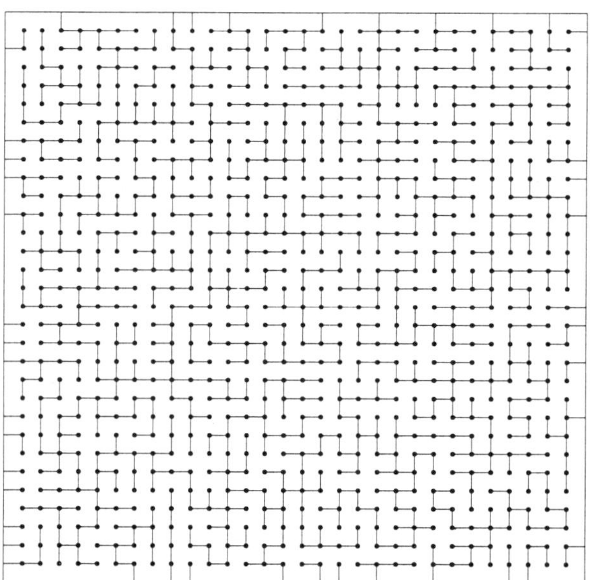

Figure 2. A uniformly chosen wired spanning tree on a subgraph of $\mathbb{Z}^2$, drawn by Wilson (see Propp and Wilson (1996)).

Thus, in Fig. 2, we may regard G to be a portion of the square lattice, with a vertex at each white square, and $G^\dagger$ a portion of the square lattice drawn in black, but with the outer boundary of the grid identified to a single vertex.

Another way of saying that the white part forms a spanning tree is to say that if we make two openings on the outer boundary, as in Fig. 4, then there is one and only one way to traverse the resulting maze without backtracking, and it is possible to get lost anywhere.

Furthermore, that the white part forms a spanning tree is equivalent to the black part forming a spanning tree: any cycle in one would disconnect the other,

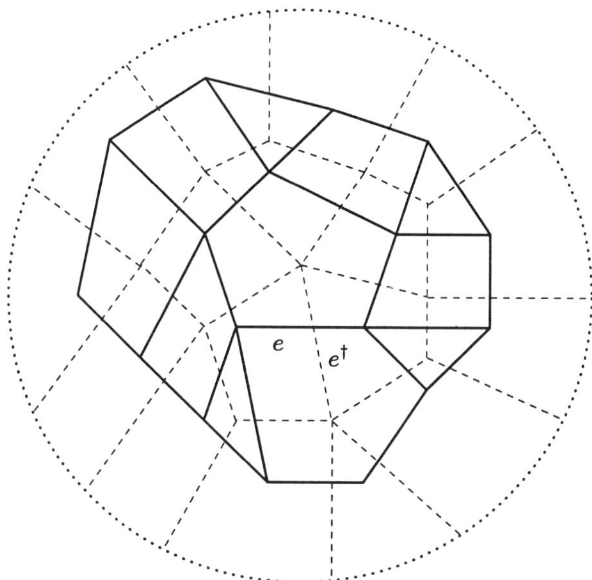

Figure 3. Planar dual graphs. The outer vertex is drawn as a dotted circle for convenience.

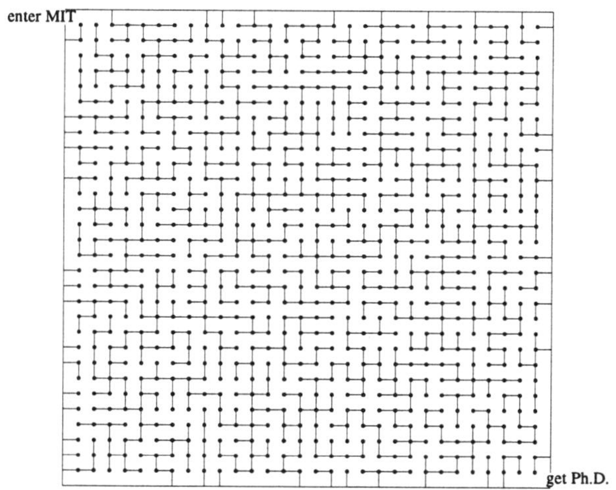

Figure 4. A maze resulting from Fig. 2, drawn by Wilson (see Propp and Wilson (1996)).

while one is disconnected only by a cycle in the other. This holds for any pair of planar dual graphs, G and $G^\dagger$: Spanning trees T of G are in one-to-one correspondence with "complementary" spanning trees $T^\dagger$ of $G^\dagger$ via

$$e \in T \iff e^\dagger \notin T^\dagger . \tag{2.1}$$

Now there are about 10^{500} spanning trees on the 31×31 portion of the square lattice shown in Fig. 2: this is an example of the entropy calculation of Burton and Pemantle (1993). This number is huge; yet the one in Fig. 2 was chosen at random according exactly to the uniform measure (or as closely as a computer can simulate randomness). How is this possible? As we mentioned in the introduction, generation of random spanning trees has drawn the attention of computer scientists. The method used for drawing Fig. 2 is the new one due to Wilson (1996). It is not only faster than all previous methods, but also turns out to be better from the theoretical viewpoint of analyzing the uniform probability measure on spanning trees and, especially, their extensions to forests on infinite graphs. We will describe Wilson's method in Section 4.

In order for the reader to have another picture in mind that is as close to the square lattice $\mathbb{Z}^2$ as possible, yet possesses a hyperbolic geometry, we present Fig. 5. This planar graph in the hyperbolic disc is also self-dual, but is transient for simple random walk. The dichotomy between transience and recurrence, as we will see in Section 5, is crucial for understanding the behavior of spanning trees and their limits on infinite graphs.

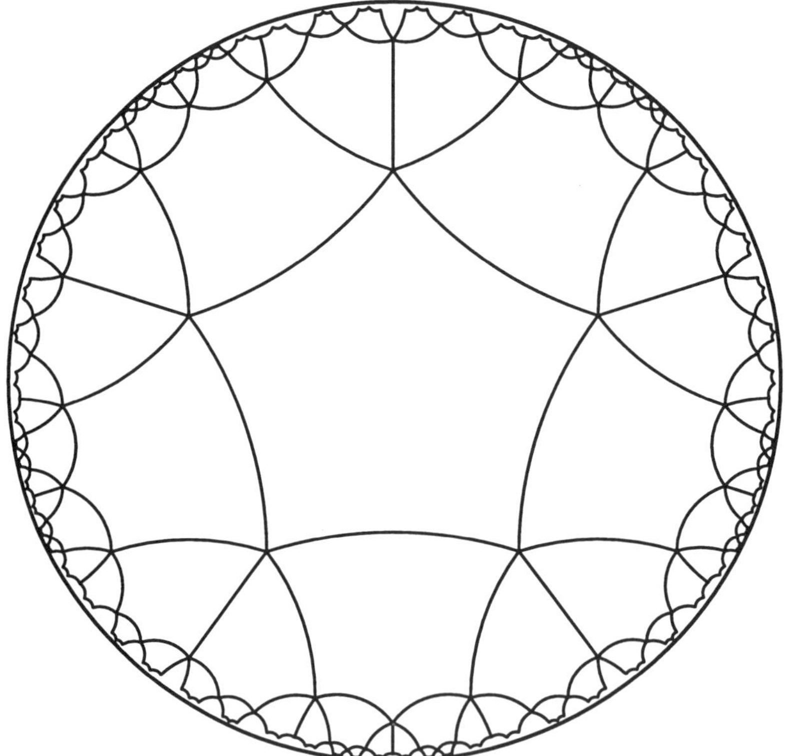

Figure 5. A self-dual graph in the hyperbolic disc.

However, we begin with a very simple finite graph on which to begin our analysis and make some preliminary probabilistic observations. Namely, consider the ladder graph of Fig. 6. Among all spanning trees of this graph, what proportion contain the bottom rung (edge)? In other words, if we were to choose at random a spanning tree, what is the chance that it would contain the bottom rung? We have illustrated the entire probability spaces for the smallest ladder graphs in Fig. 7.

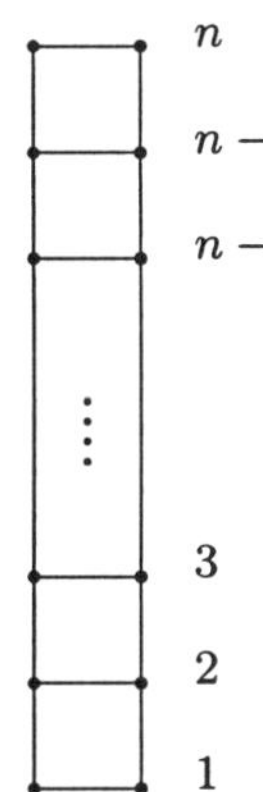

Figure 6. A ladder graph.

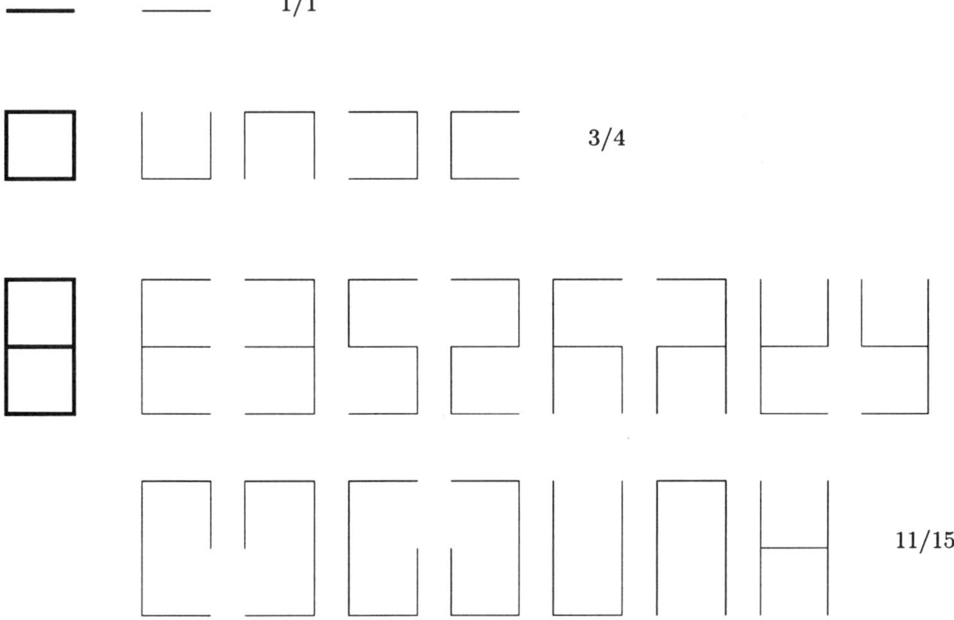

Figure 7. The ladder graphs of heights 0, 1, and 2, together with their spanning trees.

As shown, the probabilities in these cases are 1/1, 3/4, and 11/15. The next one is 41/56. Is the pattern clear? One thing that is fairly evident is that these numbers are decreasing, but hardly changing. It turns out that they come from every other term of the continued fraction expansion of $\sqrt{3}-1 = 0.73^{+}$ and, in particular, converge to $\sqrt{3}-1$. In the limit, then, the probability of using the bottom rung is $\sqrt{3}-1$ and, even before taking the limit, this gives an excellent approximation to the probability. How can we easily calculate such numbers? Of course, in this case, there is a rather easy recursion to set up and solve, but we will illustrate the more general method of electric networks. In fact, this method will show us why these probabilities are decreasing even before we calculate them.

Suppose that each edge of our graph is an electric conductor of unit conductance. Hook up a battery between the endpoints of any edge e, say the bottom rung (Fig. 8). Kirchhoff (1847) showed that the proportion of current that flows directly along e is then equal to the probability that e belongs to a randomly chosen spanning tree!

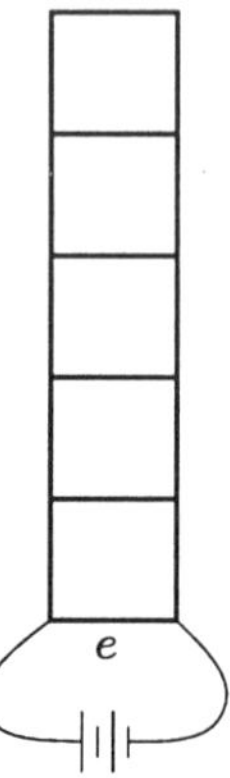

Figure 8. A battery is hooked up between the endpoints of e.

Let's review a little electric network theory before proceeding. (One source for more details is the book by Doyle and Snell (1984).) A battery produces a voltage difference across its two terminals, in this case, the endpoints of e. Let's say that the voltage of one terminal is 0 and the other is V_0. The battery also establishes voltages $V(x)$ at all the other vertices x of the network. Current then flows in the various edges in the direction of decreasing voltage. More specifically, Ohm's Law says that if x is adjacent to y, then the current that flows on the edge $[x, y]$ in the direction from x to y equals $V(x) - V(y)$ times the conductance $C(x, y)$ of the edge $[x, y]$. We are initially starting with each edge having unit conductance, as we said, but we will consider modifications that lead to other conductances. At any vertex other than the terminals, the net current flow is 0 (charge cannot build up anywhere). Consonant with Ohm's Law, we define the effective conductance of the whole network between the two terminals of the battery as I/V_0, where I is the total amount of current flowing from one terminal to the other. Moreover, suppose that the network can be decomposed into two subnetworks that intersect only at a pair of vertices. Then the total current as well as the currents flowing in one subnetwork will not change if the other subnetwork is replaced by a single wire with conductance equal to the effective conductance of the replaced subnetwork. Furthermore, the total amount flowing through each of the subnetworks is proportional to their effective conductances.

Coming back to the ladder graph and its bottom rung, e, we see that current flows in two ways: some flows directly across e and some flows through the rest of the network. From the preceding paragraph, we see that the greater the effective conductance of the rest of the network, the less (proportionally) will flow directly across e. It is intuitively clear (and justified by Rayleigh's monotonicity principle) that the higher the ladder, the greater the effective conductance of the ladder minus the bottom rung, hence, by Kirchhoff's theorem, the less the chance that a random spanning tree contains the bottom rung. This confirms our observations. Now let's continue to the actual calculation of the probabilities.

Let C_n denote the effective conductance of the ladder minus the bottom rung when the ladder has n rungs. We can calculate C_{n+1} in terms of C_n by using the familiar series-parallel laws: namely, conductances in parallel add, while resistances in series add. Here, resistance is the reciprocal of conductance; edges are said to be in parallel when they have the same endpoints and in series when they share exactly one endpoint. As Fig. 9 shows, we obtain

$$C_{n+1} = \frac{1}{2 + \frac{1}{1+C_n}}.$$

According to Kirchhoff's theorem, it follows that the probability of the bottom rung being included is

$$\frac{1}{1+C_n} = \frac{1}{1 + \frac{1}{2+\frac{1}{1+C_{n-1}}}} = \frac{1}{1 + \frac{1}{2+\frac{1}{1+\frac{1}{2+\cdots}}}},$$

as claimed.

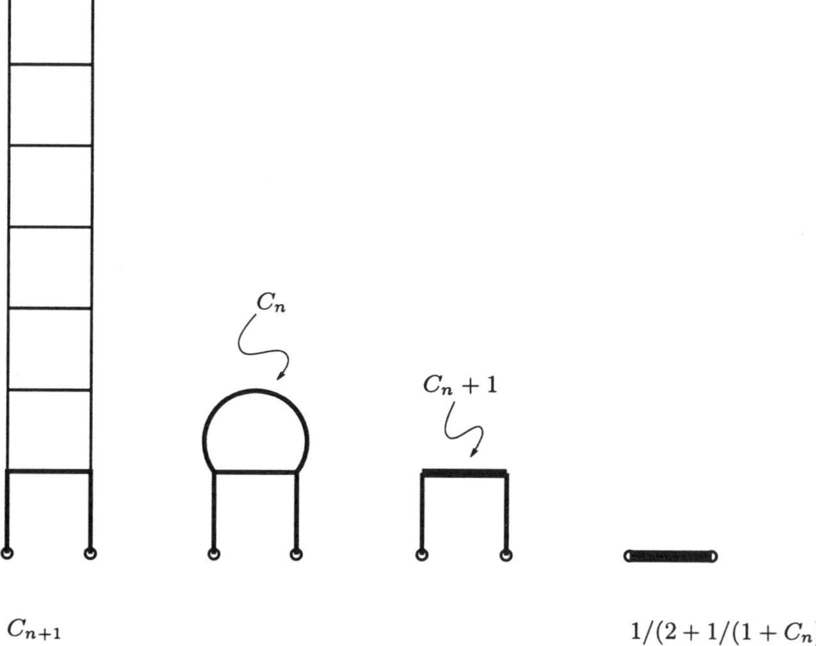

Figure 9. Calculating the effective conductance of a subnetwork.

§3. Marginal Probabilities.

We have seen that Kirchhoff gave a useful and remarkable formula for the one-edge marginal distribution of a random spanning tree. What about the other marginals? In particular, given distinct edges $e_1, \ldots, e_k$, what is $\mathbf{P}[e_1, \ldots, e_k \in T]$, where T denotes a random spanning tree? This too has an electrical answer: Let $Y(e, f)$ be the amount of current flowing along the edge f when a battery is hooked up between the endpoints of e of such voltage that in the network as a whole, unit current flows from the tail of e to the head of e. (If $e = [u, v]$ is oriented from u to v, we call u the **tail** of e and v the **head** of e.) In this notation, Kirchhoff's theorem says that

$$\mathbf{P}[e \in T] = Y(e, e) . \tag{3.1}$$

Burton and Pemantle (1993) gave the following beautiful generalization:

$$\mathbf{P}[e_1, \ldots, e_k \in T] = \det\big[Y(e_i, e_j)\big]_{1 \le i,j \le k} . \tag{3.2}$$

This is called the Transfer Current Theorem. The case $k = 2$ appeared already in Brooks, Smith, Stone, and Tutte (1940). We sketch the ideas of the proof in BLPS (1997): Write

$$\begin{aligned} \mathbf{P}[e_1, \ldots, e_k \in T] = \mathbf{P}[e_1 \in T]\mathbf{P}[e_2 \in T \mid e_1 \in T]\mathbf{P}[e_3 \in T \mid e_1, e_2 \in T] \cdots \\ \cdots \mathbf{P}[e_k \in T \mid e_1, \ldots, e_{k-1} \in T] . \end{aligned} \tag{3.3}$$

Let $\mathbf{I}_i$ denote the current vector $Y(e_i, \bullet)$ in the Hilbert space $\ell^2(\mathsf{E})$, where E is the set of edges of the graph. It turns out that $Y(e_i, e_j)$ is the inner product of $\mathbf{I}_i$ and $\mathbf{I}_j$. Thus, the matrix in (3.2) is a Gram matrix. As such, its determinant is the square of the volume of the parallelepiped spanned by the vectors $\mathbf{I}_1, \ldots, \mathbf{I}_k$. The volume can be computed by multiplying successive altitudes. It turns out that the square of the altitude of $\mathbf{I}_i$ on the space spanned by $\mathbf{I}_1, \ldots, \mathbf{I}_{i-1}$ is exactly $\mathbf{P}[e_i \in T \mid e_1, \ldots, e_{i-1} \in T]$, so that (3.2) follows from (3.3).

§4. Generating a Random Spanning Tree.

Now that we've learned a little about the marginal probabilities of random spanning trees, let's go for the whole distribution. As we said, early algorithms for generating a random spanning tree used the Matrix-Tree Theorem, which says that the total number of spanning trees of a graph equals the determinant of the following matrix, $[a_{i,j}]_{1 \le i,j \le n}$: Let the vertices of the graph be $v_0, \ldots, v_n$. For $1 \le i, j \le n$ only, let $a_{i,j}$ be -1 if v_i and v_j are adjacent, the degree of v_i if $i = j$, and 0 otherwise. (This is the same as the combinatorial Laplacian after the row and column corresponding to v_0 are removed.)

The earliest algorithms would compute the number of spanning trees containing a given edge, compute the total number of spanning trees in the graph, divide, and get the probability of putting that edge in the tree. After this random decision was made, this edge would be contracted if it were put in and deleted otherwise. Then the algorithm would proceed in the same way on the resulting smaller graph. As we said, better algorithms, especially for probabilists, were introduced by Aldous (1990) and Broder (1989) and then by Wilson (1996).

To describe Wilson's method, we define the important idea of loop erasure* of a path. If $\mathcal{P}$ is any finite path $\langle v_0, v_1, \ldots, v_l \rangle$ in G, we define the **loop erasure** of $\mathcal{P}$ by erasing cycles in $\mathcal{P}$ in the order they appear. That is, let $\mathcal{P}_0 := \mathcal{P}$. Proceed inductively to define $\mathcal{P}_k$ for $k \le l$ as follows. Given $\mathcal{P}_k = \langle u_0, \ldots, u_m \rangle$, we have two cases to consider:

(1) for some $j \le m$, we have $v_{k+1} = u_j \in \mathcal{P}_k$; in this case, let $\mathcal{P}_{k+1} := \langle u_0, \ldots, u_j \rangle$.
(2) otherwise, let $\mathcal{P}_{k+1} := \langle u_0, \ldots, u_j, v_{k+1} \rangle$.

The last path in this sequence, $\mathcal{P}_l$, is the loop erasure of $\mathcal{P}$. In case (1) above, the cycle $\langle u_j, u_{j+1}, \ldots, u_m, v_{k+1} \rangle$ also belongs to $\mathcal{P}$. We may alternatively consider this as a cycle of edges $\langle [u_j, u_{j+1}], \ldots, [u_m, v_{k+1}] \rangle$ that are removed from $\mathcal{P}$.

Now in order to generate a random spanning tree, first pick any vertex r to be the "root" of the tree. Then create a growing sequence of trees $T(i)$ $(i \ge 0)$ as follows. Let $T(0) := \{r\}$. Suppose that $T(i)$ is known. If $T(i)$ spans G, we are done. Otherwise, pick any vertex v that is not in $T(i)$ and start a simple random walk from v until it hits $T(i)$. Now create $T(i+1)$ by adding to $T(i)$ the loop erasure of this path from v to $T(i)$. The final tree in this growing sequence is the output of Wilson's algorithm (we simply forget the root). It has exactly the uniform distribution, no matter in what order we choose new starting vertices for loop-erased random walks!

How can this be? We give a brief sketch of the proof that Wilson's algorithm does indeed give the uniform distribution. However, our sketch will gloss over important subtleties. Now, when Wilson's algorithm finishes, we get a spanning tree T and a collection $C_1, \ldots, C_s$ of cycles erased by the algorithm. Consider the edges belonging to the tree or to any of the cycles (with multiplicity). The probability of the random walk having made the transitions it made along these edges is

$$\left(\prod_{v \ne r} \frac{1}{\deg v}\right) \left(\prod_{i=1}^{s} \prod_{u \in C_i} \frac{1}{\deg u}\right)$$

by the Markov property and the fact that all neighbors are equally likely transitions for simple random walk. Since we get two products, one involving only the tree and the other involving only the cycles, the tree is independent of the cycles and thus all trees have the same probability.

---

* This ought to be called "cycle erasure", but the name is fixed.

We now show how to use Wilson's algorithm to prove Kirchhoff's theorem. For those who are unfamiliar with the connection between electric networks and random walks, we present some of the basics first. (See Doyle and Snell (1984) for a superb elementary introduction.)

Consider the network shown in Fig. 10. Again, for simplicity, each edge has unit conductance. A battery joins two vertices, labelled 0 and 1 since it establishes voltages 0 and 1 at them, respectively. Note that the edge containing the battery is not part of the graph and, indeed, is shown only for physical intuition. Let x be any vertex. The fundamental relationship between electric networks and random walks is the equation

$$V(x) := \text{voltage at } x = \mathbf{P}_x[\text{visit 1 before 0}] . \tag{4.1}$$

Here, $\mathbf{P}_x$ means the probability measure for simple random walk on the graph started at x. (The random walk cannot pass through the battery.)

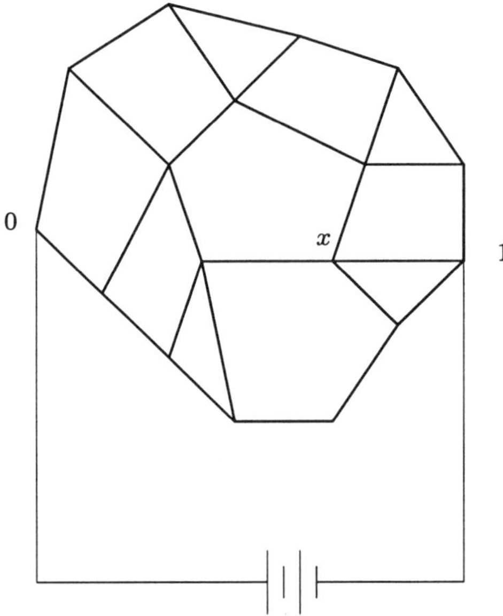

Figure 10. A battery connects vertices 0 and 1, establishing voltages 0 and 1, respectively.

The reason that (4.1) holds, as we will see, is that both sides, when viewed as functions of x, are harmonic for $x \neq 0, 1$ and, clearly, have the same "boundary" values at 0 and 1*. Here, we say that a function is **harmonic** at x if its value at x is the average of the values of neighboring vertices. Furthermore, these discrete harmonic functions obey a uniqueness principle, just as in the continuous case. All this is easy to show:

* Thus, (4.1) is the discrete analogue of the solution of the Dirichlet problem via Brownian motion.

Write $x \sim y$ to indicate that x and y are adjacent. As we saw, for $x \neq 0, 1$, the sum of the currents flowing out of x is 0:

$$0 = \sum_{y \sim x} [V(x) - V(y)] = (\deg x)\, V(x) - \sum_{y \sim x} V(y)\,,$$

i.e.,

$$V(x) = \frac{1}{\deg x} \sum_{y \sim x} V(y)\,.$$

This shows the harmonicity of the voltage function. For the hitting probabilities, we have (again, for $x \neq 0, 1$)

$$\begin{aligned} \mathbf{P}_x[\text{visit 1 before 0}] &= \sum_{y \sim x} \mathbf{P}_x[\text{first step is to } y] \cdot \mathbf{P}_y[\text{visit 1 before 0}] \\ &= \frac{1}{\deg x} \sum_{y \sim x} \mathbf{P}_y[\text{visit 1 before 0}]\,. \end{aligned}$$

Thus, the hitting probabilities are harmonic too.

Next, the Maximum Principle says that any function that is harmonic for $x \neq 0, 1$ and that achieves it maximum value at some $x \neq 0, 1$ is constant. (Here, as always, we are assuming that the network is connected.) Indeed, the values neighboring x must equal the value at x since the function is maximum there and is harmonic there. Thus, the maximum value propagates outward from x until we see that all values are the maximum.

From the Maximum Principle flows the Uniqueness Principle: Two functions that are harmonic for $x \neq 0, 1$ and that agree for $x = 0, 1$ are identically equal. The reasoning follows the usual course: subtract the two functions and apply the Maximum Principle.

This proves (4.1). Thus, we have a probabilistic interpretation of the voltage. There is also a probabilistic interpretation of the current: Use a battery of such voltage that a unit current flows from 1 to 0. Start a random walk at 1 and stop it when it reaches 0. Then

$$\begin{aligned} &\mathbf{E}\big[\#\text{ traversals of } [x, y] - \#\text{ traversals of } [y, x]\big] \\ &\qquad = \text{amount of current flowing along } [x, y] \end{aligned} \tag{4.2}$$

if $x \sim y$. It is not required that x, y be different from $0, 1$; indeed, we will use only the case when $x = 1$ and $y = 0$. We will not prove (4.2), but we use it now to establish Kirchhoff's theorem.

Let the edge e be $[x, y]$. To construct T, we use Wilson's algorithm rooted at y and started at x. (Recall that we can make these choices however we please.) As soon as this first random walk visits y, we will discover whether $e \in T$ without having to generate the rest of T; namely, $e \in T$ iff the random walk comes to y from x. Since for this random walk, $\big(\#\text{ traversals of } [x, y] - \#\text{ traversals of } [y, x]\big)$ is $1 - 0$ or $0 - 0$, the expectation in (4.2) is simply the probability that the random walk comes to y from x, and so equals $\mathbf{P}[e \in T]$.

§5. Infinite Graphs.

Wilson's method for generating spanning trees will also give a random spanning tree $T := \bigcup_{n=1}^{\infty} T(n)$ on any infinite but still recurrent graph G, provided, of course, we choose new starting vertices of the loop-erased random walks in such a way as to guarantee that T spans. How should we interpret the tree T? Suppose that G' is a finite subgraph of G and consider T_G and $T_{G'}$, the random spanning trees generated by Wilson's method on G and G', respectively. It is not hard to show that for any event B depending on only finitely many edges, we can make $\Big|\mathbf{P}[T_G \in B] - \mathbf{P}[T_{G'} \in B]\Big|$ arbitrarily small by choosing G' sufficiently large. Thus, it is reasonable to call the random spanning tree of G the **uniform spanning tree** of G. Also, this shows that Wilson's method on a recurrent graph generates a random spanning tree whose distribution again does not depend on the choice of root nor on the order in which we carry out the algorithm*.

Consider the special case $G := \mathbb{Z}^2$ (with the usual edge set). The degree of every vertex as seen in T is random, but identically distributed. As a first look at this distribution, let us find its mean. Now, the expected degree of $(0,0)$ equals the sum over all edges incident to $(0,0)$ of the probability that the edge belongs to T. By symmetry, these probabilities are all the same, so it suffices to find one of them and then multiply by 4. By taking limits, one can show that Kirchhoff's theorem (3.1) still holds. Let a unit current enter the network at $(0,0)$ and exit at $(1,0)$. We want to know how much travels directly along the edge e with these two endpoints. Here is the engineer's well-known solution: To restore symmetry, first put in a unit current at $(0,0)$ and take it out at infinity. One-quarter will flow directly along e. Second, put in unit current at infinity and take it out at $(1,0)$. Again, one-quarter will flow along e. Superposing these two currents, we obtain the current we desire and we see that one-half flows along e. Thus, the expected degree is $4 \times (1/2) = 2$.

This solution is very beautiful, but it suffers from some defects (of which the engineers are aware). The most prominent defect is that one cannot put current in at a vertex and take it out at infinity; one can do this on a network iff simple random walk on the graph is transient. Nevertheless, the answer is correct. Here are two ways to see this.

Since $\mathbb{Z}^2$ is self-dual, the "complement" $T^\dagger$ of T on the dual graph has the same law as T. By definition, $\mathbf{P}[e \in T] = \mathbf{P}[e^\dagger \notin T^\dagger]$; since, however, T and $T^\dagger$ have the same distribution in this case, we also have $\mathbf{P}[e \in T] = \mathbf{P}[e^\dagger \in T^\dagger]$. Therefore, $\mathbf{P}[e \in T] = 1/2$, as desired.

The second way to get this result is considerably more general. Let G be a graph. For a subset $\mathsf{V}' \subset \mathsf{V}$, let $\partial \mathsf{V}'$ be the set of edges that join a vertex in V' to a vertex not in V'. We say that G is **amenable** if there is an exhaustion V_n of V

---

* This can also be shown directly.

with

$$\lim_{n\to\infty} |\partial \mathsf{V}_n|/|\mathsf{V}_n| = 0\,.$$

Certainly $\mathbb{Z}^2$ is amenable. In fact, it is easy to show that the Cayley graph of every finitely generated abelian group is amenable by considering the balls V_n consisting of all vertices within distance n of the identity.

Now suppose that G is an amenable infinite graph as witnessed by the exhaustion $\langle \mathsf{V}_n \rangle$. We claim that the "average" degree of vertices in *any deterministic* spanning tree of G is 2. That is, if $\deg_T(v)$ denotes the degree of v in a spanning tree T of G, then

$$\lim_{n\to\infty} |\mathsf{V}_n|^{-1} \sum_{v\in \mathsf{V}_n} \deg_T(v) = 2\,.$$

The reader may easily deduce this from the facts that the number of components of $T \cap G_n$ is at most $|\partial \mathsf{V}_n|$ and a tree with k vertices has $k-1$ edges, where G_n is the maximal subgraph of G containing V_n. Thus, from the Bounded Convergence Theorem, we deduce that when a spanning tree is chosen at random,

$$\lim_{n\to\infty} |\mathsf{V}_n|^{-1} \sum_{v\in \mathsf{V}_n} \mathbf{E}[\deg_T(v)] = 2\,.$$

In particular, if $G = (\mathsf{V}, \mathsf{E})$ is also **transitive**, such as $\mathbb{Z}^2$, meaning that for every pair of vertices u and v, there is a bijection of V with itself that preserves adjacency and takes u to v, then every vertex has expected degree 2. (Compare Theorem 3.2 in Thomassen (1990).)

What else can we say about the distribution of the degree of a vertex in $\mathbb{Z}^2$? Again by taking limits, one may show that the Transfer Current Theorem (3.2) holds. Luckily, there are methods for calculating the entries of the transfer current matrix. One gets, for example, that the probability that the degree of $(0,0)$ is 2 is the interesting number

$$\frac{4}{\pi}\left(2 - \frac{9}{\pi} + \frac{12}{\pi^2}\right) = 0.45^-\,;$$

see Burton and Pemantle (1993).

Is there an analogue of the uniform spanning tree on a transient graph? If so, what does it look like on $\mathbb{Z}^d$ for $d \geq 3$?

It turns out that there is: as we mentioned, Pemantle (1991) showed that on any graph G, if $\langle G_n \rangle$ is an exhaustion of G, the weak limit of the uniform spanning tree measures on G_n exists and is unique, regardless of the exhaustion. This is the free uniform spanning forest, FUSF. The existence of the limit is derived from Rayleigh's monotonicity principle. What does the limit look like? First of all, the random subgraph it produces cannot have any cycles: any cycle in G is part of G_n for all sufficiently large n, but T_{G_n} has no cycles a.s. Since there are only a countable number of possible cycles, there are none a.s. in the limit. However, we cannot argue in the same way that T_G is connected. Perhaps the connections in T_{G_n} between two fixed vertices get longer and longer as $n \to \infty$; their distributions may not be tight. Indeed, this can happen: Pemantle (1991) showed that one gets a tree on $\mathbb{Z}^d$ iff $d \leq 4$. We will explain what is behind this stunning result shortly.

As we saw, there is another possibility for taking limits. In disregarding the complement of G_n, we are disregarding the possibility that a spanning tree or forest of G may connect the boundary vertices of G_n in ways that would affect the possible connections within G_n itself. An alternative approach takes the opposite view and forces all connections outside of G_n: Let G_n^W be the graph obtained from G by identifying all the vertices in G_n that are adjacent to a vertex outside G_n to a single vertex. Let μ_n^W be the random spanning tree measure on G_n^W. Again, Rayleigh's monotonicity principle proves the existence and uniqueness of a limiting probability measure μ^W, the wired uniform spanning forest, WUSF. The term "wired" comes from thinking of G_n^W as having its boundary wired together. In statistical mechanics, especially the random cluster model (see, e.g., Grimmett 1995), measures on infinite configurations are also defined using limiting procedures similar to those employed here to define FUSF and WUSF. One needs to specify boundary conditions, just as we did. The terms "free" and "wired" have analogous uses there. On a recurrent graph, it is not hard to see that FUSF = WUSF.

It turns out that the wired uniform spanning forest can be constructed by an analogue of Wilson's method: Let G be a *transient* graph. Run a loop-erased random walk from any vertex for infinite time; this is defined by transience. Then run an independent random walk from another vertex until it encounters the first path or, if it doesn't, for infinite time. Add its loop erasure to the first path. Then run another independent random walk from any vertex not already included in the union of the first two paths until it encounters the previous paths or else for infinite time, add its loop erasure, and so on. Let T be the (infinite) union of the loop-erased paths. One may show that the resulting distribution on spanning forests T is independent of the order in which we choose starting vertices, as long as the order is such as to guarantee that all vertices are eventually chosen, i.e., so that we get a spanning forest. We refer to this method of generating a random spanning forest as **Wilson's method rooted at infinity**. The good news is that it gives the wired uniform spanning forest.

Because of Wilson's method rooted at infinity, it is much easier to analyze the wired spanning forest than the free. Indeed, it follows immediately that the wired spanning forest is a tree a.s. iff random walk from any vertex x and independent loop-erased random walk from any other vertex y will intersect with probability 1. Lawler (1991) provided detailed estimates concerning intersections of this type on $\mathbb{Z}^d$; in particular, we obtain Pemantle's theorem that the uniform spanning forest on $\mathbb{Z}^d$ is a tree iff $d \leq 4$. Actually, there is one more step we can take that makes Pemantle's theorem still easier: Lyons, Peres, and Schramm (1998) showed that on any graph, random walk from any vertex x and independent loop-erased random walk from any other vertex y will intersect with probability 1 iff random walk from any vertex x and independent random walk from any other vertex y will intersect with probability 1. The latter property is much easier to analyze and quite standard. Pemantle (1991) also showed that when $d \geq 5$, the uniform spanning forest has infinitely many trees a.s. For other graphs, the number of trees is either infinite a.s. or equal a.s. to the dimension of the space of bounded harmonic functions; see BLPS (1997) for the full story.

§6. Ends of Trees.

We now ask about the topology of the trees of a random spanning forest. Call an infinite path in a tree that starts at any vertex and does not backtrack a **ray**. Call two rays **equivalent** if they have infinitely many vertices in common. An equivalence class of rays is called an **end**. How many ends do the trees of a random spanning forest have?

Let's begin by thinking about the case of the wired uniform spanning forest on a regular tree of degree $d+1$. Choose a vertex, o. Begin Wilson's method rooted at infinity from o. We obtain a ray ξ from o to start our forest. Now o has d other neighbors, $x_1, \ldots, x_d$. By beginning random walks at each of them in turn, we see that the events $A_i := \{x_i \text{ connected to } o\}$ are independent. Furthermore, it is easy to calculate that the probability that random walk started at x_i ever hits o is $1/d$. This is the probability of A_i. On the event A_i, we add only the edge (o, x_i) to the forest and then we repeat the analysis from x_i. Thus, the tree containing o includes, apart from the ray ξ, a critical Galton-Watson tree with binomial offspring distribution $(d, 1/d)$. In addition, each vertex on ξ has another random subtree attached to it; its first generation has the binomial distribution $(d-1, 1/d)$, but subsequent generations yield Galton-Watson trees with binomial distribution $(d, 1/d)$. In particular, a.s. every tree added to ξ is finite. This means that the tree containing o has only one end, the equivalence class of ξ. This analysis is easily extended to form a complete description of the entire wired spanning forest. The resulting description is due to Häggström (1997), whose work predates Wilson's algorithm (despite the publication dates) and, thus, was not this easy.

Now in general, we see that if we begin Wilson's algorithm at a vertex o in a graph, it immediately generates one end of the tree containing o. In order for this tree to have more than one end, however, we need a succession of "coincidences" to occur, building up other ends by gradually adding on finite pieces. This is possible, but it suggests that usually the wired spanning forest has trees with only one end each. Indeed, this is very often the case.

First, consider the planar recurrent case, which has an amazingly simple analysis. Let G be a proper planar graph whose planar dual $G^\dagger$ is locally finite. Given a finite connected subgraph G_n of G with connected complement $G \setminus G_n$, let $G_n^\dagger$ be its planar dual. Notice that $G_n^\dagger$ can be regarded as a finite subgraph of $G^\dagger$, but with the outer boundary vertices identified to a single vertex. Spanning trees T of G_n are in one-to-one correspondence with "complementary" spanning trees $T^\dagger$ of $G_n^\dagger$ as in (2.1). Therefore, the FUSF of G is "dual" to the WUSF of $G^\dagger$; that is, the relation (2.1) transforms one to the other. In fact, we can couple (generate simultaneously) the two spanning forests by (2.1).

THEOREM 6.1. *Suppose that G is a proper recurrent planar graph and $G^\dagger$ its planar dual. If $G^\dagger$ is locally finite and recurrent, then the uniform spanning tree on G has only one end a.s.*

Proof. Because G is recurrent, its free and wired spanning forests coincide and give a single tree a.s. Likewise for $G^\dagger$. If T had at least two ends, then the bi-infinite path in T joining two of them would separate $T^\dagger$ into at least two trees. But since $T^\dagger$ is a single tree, this is impossible. ∎

REMARK 6.2. Loop-erased random walk as a path indexed by $\mathbb{N}$ is obviously defined in any transient graph. It is not well defined on recurrent graphs in general, but this theorem provides a simple way to define loop-erased random walk from any vertex x on graphs G like $\mathbb{Z}^2$ that satisfy the hypotheses of Theorem 6.1: just take the ray from x in the spanning tree. A detailed analysis of this path on transitive graphs like $\mathbb{Z}^2$ is given by Lawler (1991).

Similar reasoning shows:

PROPOSITION 6.3. *Suppose that G is a proper planar graph and $G^\dagger$ its planar dual. Assume that $G^\dagger$ is locally finite. If each tree of the* WUSF *of G has only one end a.s., then the* FUSF *of $G^\dagger$ has only one tree a.s. If, in addition, the* WUSF *of G has infinitely many trees a.s., then the tree of the* FUSF *of $G^\dagger$ has infinitely many ends a.s.*

Going beyond the planar setting, Pemantle (1991) showed that on $\mathbb{Z}^d$, the uniform spanning forest has one end when $d \leq 4$ and at most two ends per tree when $d \geq 5$. This result was sharpened and extended in BLPS (1997):

THEOREM 6.4. *If G is a transient Cayley graph, then* WUSF-*a.s., every tree has only one end.*

Combining this with Theorem 6.1, one can show that all Cayley graphs of groups that are not finite extensions of $\mathbb{Z}$ have the property that WUSF-a.s., every tree has only one end; see BLPS (1997).

By contrast, we have:

PROPOSITION 6.5. *If G is a transitive graph with* $\mathsf{WUSF} \neq \mathsf{FUSF}$, *then* FUSF-*a.s., there is a tree with infinitely many ends.*

§7. Equality of Free and Wired Spanning Forests.

Let us return to the question of when $\mathsf{FUSF} = \mathsf{WUSF}$. In all cases, there is a simple inequality between these two probability measures, namely,

$$\forall e \in \mathsf{E} \quad \mathsf{FUSF}(e \in T) \geq \mathsf{WUSF}(e \in T) \tag{7.1}$$

by Rayleigh's monotonicity principle. In fact, it turns out that FUSF stochastically dominates WUSF by a theorem of Feder and Mihail (1992), whence by Strassen's (1965) theorem, FUSF and WUSF can be **coupled** in the sense that there is a probability measure on the set

$$\Big\{(T(1), T(2))\,;\ T(i) \text{ is a spanning forest of } G \text{ and } T(1) \subseteq T(2)\Big\}$$

that projects in the first coordinate to WUSF and in the second to FUSF.

This allows us to deduce the following result (compare Häggström 1995):

PROPOSITION 7.1. *If* $\mathbf{E}[\deg_T(v)]$ *is the same under* FUSF *and* WUSF *for every* $v \in \mathsf{V}$ *or if equality holds in (7.1), then* $\mathsf{FUSF} = \mathsf{WUSF}$.

Proof. The coupling shows that $\mathbf{E}[\deg_T(v)]$ under FUSF is at least the expectation under WUSF, with equality for all $v \in \mathsf{V}$ iff $\mathsf{FUSF} = \mathsf{WUSF}$. Since equality in (7.1) implies equality of expected degrees, the second claim follows as well. ∎

We will now see that $\mathsf{FUSF} = \mathsf{WUSF}$ for many graphs, including Cayley graphs of abelian groups such as $\mathbb{Z}^d$.

Let G be an amenable infinite graph as witnessed by the sequence $\langle \mathsf{V}_n \rangle$ (see Section 5). Let T be any deterministic spanning forest all of whose components (trees) are infinite. If k_n denotes the number of trees of $T \cap G$, then $k_n = o(|\mathsf{V}_n|)$, where G_n is the maximal subgraph of G containing V_n. Thus, the average degree of vertices is 2:

$$\lim_{n \to \infty} |\mathsf{V}_n|^{-1} \sum_{v \in \mathsf{V}_n} \deg_T(v) = 2\,.$$

In particular, if T is random, then

$$\lim_{n \to \infty} |\mathsf{V}_n|^{-1} \sum_{v \in \mathsf{V}_n} \mathbf{E}[\deg_T(v)] = 2\,. \tag{7.2}$$

Thus, we obtain the following, essentially due to Häggström (1995):

COROLLARY 7.2. *On any transitive amenable graph,* $\mathsf{FUSF} = \mathsf{WUSF}$.

Proof. By transitivity and (7.2), $\mathbf{E}[\deg_T(v)] = 2$ for every $v \in \mathsf{V}$ and for both FUSF and WUSF. ∎

In order to continue our analysis, we reformulate the electrical laws for finite graphs in terms of finite-dimensional Hilbert space.

Let $G = (\mathsf{V}, \mathsf{E})$ be a finite graph. For purposes of the electrical laws, we regard each edge $e \in \mathsf{E}$ as occurring with both possible orientations; denote the edge opposite to $e = [u, v]$ by $\check{e} := [v, u]$. Since we are interested in flows on E, it is natural to consider that what flows on e is the negative of what flows on $\check{e}$. Thus, define $\ell^2_-(\mathsf{E})$ to be the space of **antisymmetric** functions θ on E (i.e., $\theta(\check{e}) = -\theta(e)$ for each edge e) with inner product

$$(\theta, \theta') := \frac{1}{2}\sum_{e \in \mathsf{E}} \theta(e)\theta'(e) = \sum_{e \in \mathsf{E}_{1/2}} \theta(e)\theta'(e)\,,$$

where $\mathsf{E}_{1/2} \subset \mathsf{E}$ is a set of edges containing exactly one of each pair $e, \check{e}$.

We may express Kirchhoff's laws as follows. Let $\chi^e := \mathbf{1}_e - \mathbf{1}_{\check{e}}$ denote the unit flow along e represented as an antisymmetric function in $\ell^2_-(\mathsf{E})$. A function $I \in \ell^2_-(\mathsf{E})$ is called a **current** if it satisfies the following two properties:

Kirchhoff's Node Law: For any vertex v,

$$\Big(\sum_{w \sim v} \chi^{[v,w]}, I\Big) = 0$$

except if a battery is connected at v, in which case the inner product is called the amount of current flowing into the network at v.

Kirchhoff's Cycle Law: If $e_1, e_2, \ldots, e_n$ is a directed cycle in G, then

$$\Big(\sum_{i=1}^{n} \chi^{e_i}, I\Big) = 0\,.$$

Now let $\bigstar$ denote the subspace in $\ell^2_-(\mathsf{E})$ spanned by the **stars** $\sum_{w \sim v} \chi^{[v,w]}$ ($v \in \mathsf{V}$) and let $\diamondsuit$ denote the subspace spanned by the **cycles** $\sum_{i=1}^{n} \chi^{e_i}$ ($e_1, \ldots, e_n$ forming a directed cycle in G). These subspaces are clearly orthogonal. Moreover, the sum of $\bigstar$ and $\diamondsuit$ is all of $\ell^2_-(\mathsf{E})$: suppose that $\theta \in \ell^2_-(\mathsf{E})$ is orthogonal to both $\bigstar$ and $\diamondsuit$. Since θ is orthogonal to $\diamondsuit$, there is a function F such that $\theta = \nabla F$, where $\nabla F([u, v]) := F(u) - F(v)$ is called the **gradient** of F. Since θ is orthogonal to $\bigstar$, the function F is harmonic at every vertex. Therefore F is constant on G, whence $\theta = 0$, as desired.

Let I^e be the unit current from the tail of e to the head of e; this means that a unit total of current flows in the network, not that the flow directly across e is unit. According to Kirchhoff's Cycle Law, I^e is orthogonal to $\diamondsuit$, whence in $\bigstar$. Now $\chi^e - I^e$ is a sourceless flow, i.e., is orthogonal to $\bigstar$, and is thus an element of $\diamondsuit$. Therefore, the expression

$$\chi^e = I^e + (\chi^e - I^e)$$

is the orthogonal decomposition of χ^e relative to $\ell^2_-(\mathsf{E}) = \bigstar \oplus \diamondsuit$. In particular,

$$I^e = P_\bigstar \chi^e ,$$

where $P_\bigstar$ denotes the orthogonal projection onto $\bigstar$.

Recall that $\mathsf{E}_{1/2} \subset \mathsf{E}$ is a set of edges containing exactly one of each pair $e, \check{e}$. What is the matrix of $P_\bigstar$ in the orthogonal basis $\{\chi^e ;\ e \in \mathsf{E}_{1/2}\}$? We have

$$(P_\bigstar \chi^e, \chi^f) = (I^e, \chi^f) = I^e(f) = Y(e, f) .$$

In other words, the matrix coefficient at (e, f) is our familiar friend, the corresponding entry of the transfer current matrix! Since $P_\bigstar$ is self-adjoint, the transfer current matrix is symmetric*. Therefore $Y(e, f) = Y(f, e)$. (This is known as the reciprocity law.)

We now move to infinite graphs. There are two natural ways of defining currents between vertices of an infinite graph, corresponding to the two ways of defining uniform spanning forests. We will define these currents using Hilbert space; then we indicate how they correspond to limits of currents on finite subgraphs.

Considering now an infinite graph $G = (\mathsf{V}, \mathsf{E})$, we must redefine $\bigstar$ and $\diamondsuit$. Let $\bigstar$ denote the closure of the linear span of the stars and $\diamondsuit$ the closure of the linear span of the cycles of G. Since every star and every cycle are orthogonal, it is still true that $\bigstar \perp \diamondsuit$. However, it is no longer necessarily the case that $\ell^2_-(\mathsf{E}) = \bigstar \oplus \diamondsuit$; in fact, we are about to see that this is equivalent to $\mathsf{FUSF} = \mathsf{WUSF}$ on G.

Define the space of **Dirichlet** functions

$$\mathbf{D} := \{F ;\ \nabla F \in \ell^2_-(\mathsf{E})\} .$$

Our previous reasoning for finite graphs now shows that an element $\theta \in (\bigstar \oplus \diamondsuit)^\perp$, the orthocomplement of $\bigstar \oplus \diamondsuit$, is the gradient of a harmonic function $F \in \mathbf{D}$. Thus, if $\mathbf{HD}$ denotes the set of $F \in \mathbf{D}$ that are harmonic, we obtain the orthogonal decomposition

$$\ell^2_-(\mathsf{E}) = \bigstar \oplus \diamondsuit \oplus \nabla\mathbf{HD} . \tag{7.3}$$

Let $P^\perp_\diamondsuit$ denote the orthogonal projection onto $\diamondsuit^\perp$. We define two possibly different currents,

$$I^e_F := P^\perp_\diamondsuit \chi^e ,$$

the **free current** between the endpoints of e (also called the "limit current"), and

$$I^e_W := P_\bigstar \chi^e ,$$

the **wired current** between the endpoints of e (also called the "minimal current"). The names for these currents are explained by the following two well-known propositions. (See, e.g., Soardi (1994), Cor. 3.17 and Thm. 3.25.)

* Warning: when the conductances are not unit, this property does not hold; instead, the so-called transfer impedance matrix is the matrix of $P_\bigstar$ in the appropriate basis and so is symmetric. Nevertheless, it is the transfer current matrix that gives the marginals for the weighted random spanning trees corresponding to non-unit conductances.

PROPOSITION 7.3. *Let G be an infinite graph exhausted by subgraphs $\langle G_n \rangle$. Let e be an edge in G_1 and I_n be the unit current flow in G_n from the tail of e to the head of e. Then $\|I_n - I_F^e\| \to 0$ as $n \to \infty$.*

PROPOSITION 7.4. *Let G be an infinite graph exhausted by subgraphs $\langle G_n \rangle$. Let G_n^W be formed by identifying the boundary of G_n to a single vertex. Let e be an edge in G_1 and I_n be the unit current flow in G_n^W from the tail of e to the head of e. Then $\|I_n - I_W^e\| \to 0$ as $n \to \infty$.*

We now put this all together:

THEOREM 7.5. *For any graph,* $\mathsf{FUSF} = \mathsf{WUSF}$ *iff $I_W^e = I_F^e$ for every edge e iff $\ell^2_-(\mathsf{E}) = \bigstar \oplus \diamondsuit$ iff $\mathbf{HD} = \mathbb{R}$.*

Proof. From Kirchhoff's theorem (3.1) and Propositions 7.3 and 7.4, we obtain the extensions of Kirchhoff's theorem to infinite graphs: $\mathsf{FUSF}(e \in T) = I_F^e(e)$ and $\mathsf{WUSF}(e \in T) = I_W^e(e)$. Now use Proposition 7.1 to deduce the first equivalence.

For the second equivalence, note that

$$I_F^e(e) = (I_F^e, \chi^e) = (P_{\diamondsuit}^{\perp} \chi^e, \chi^e) = (P_{\diamondsuit}^{\perp} \chi^e, P_{\diamondsuit}^{\perp} \chi^e) = \|P_{\diamondsuit}^{\perp} \chi^e\|^2 ,$$

where we have used the fact that $P_{\diamondsuit}^{\perp}$ is its own square and self-adjoint. Likewise, $I_W^e(e) = \|P_{\bigstar} \chi^e\|^2$. Since $\diamondsuit^{\perp} \supseteq \bigstar$, it follows that $I_F^e(e) = I_W^e(e)$ iff $I_F^e = I_W^e$. Now $\ell^2_-(\mathsf{E}) = \bigstar \oplus \diamondsuit$ is equivalent to $P_{\bigstar} = P_{\diamondsuit}^{\perp}$. Since $\{\chi^e ;\ e \in \mathsf{E}_{1/2}\}$ is a basis for $\ell^2_-(\mathsf{E})$, this is also equivalent to $P_{\bigstar} \chi^e = P_{\diamondsuit}^{\perp} \chi^e$ (i.e., $I_F^e = I_W^e$) for all edges e, as desired.

The third equivalence follows from (7.3). ∎

This result incorporates Doyle's (1988) theorem on uniqueness of currents. Combining it with Corollary 7.2, we see that every transitive amenable graph has no nonconstant harmonic Dirichlet functions. This result is due to Medolla and Soardi (1995). See Benjamini, Lyons and Schramm (1997) for more applications of Theorem 7.5 to the study of harmonic Dirichlet functions.

If one graph is "similar" to another, how similar must their spanning forest measures be? For example, if $\mathsf{FUSF} = \mathsf{WUSF}$ on one graph, must this be the case on the other? To make the notion of "similar" precise, given two graphs $G = (\mathsf{V}, \mathsf{E})$ and $G' = (\mathsf{V}', \mathsf{E}')$, call a function $\phi : \mathsf{V} \to \mathsf{V}'$ a **rough isometry** if there are positive constants a and b such that for all $u, v \in \mathsf{V}$,

$$a^{-1} d(u,v) - b \le d'(\phi(u), \phi(v)) \le a d(u,v) + b \tag{7.4}$$

and such that every vertex in G' is within distance b of the image of V. Here, d and d' denote the usual graph distances on G and G'. Being roughly isometric is an equivalence relation. Soardi (1993) proved:

THEOREM 7.6. *Let G and G' be roughly isometric graphs with bounded degrees. Then G has nonconstant harmonic Dirichlet functions iff G' does.*

For a short proof found by Schramm, see Lyons and Peres (1997).

Thus, if G and G' are roughly isometric graphs, then the free and wired spanning forests coincide on one iff they do on the other. However, this does not mean that the geometries of the forests cannot change. Indeed, rough isometries can change the wired spanning forest from a single tree to infinitely many trees: see the example of Theorem 3.5 in Benjamini and Schramm (1996a). On the other hand, changing the generators in a Cayley graph cannot have this effect; see BLPS (1997).

Pemantle (1991) showed that the tail σ-field of the uniform spanning forest is trivial on $\mathbb{Z}^d$; his proof shows that this is a consequence of the fact that $\mathsf{FUSF} = \mathsf{WUSF}$. R. Solomyak (1998) showed a directional tail triviality of the free spanning forest on the Cayley graphs of Fuchsian groups (with the standard generators) that are *not* co-compact. Finally, using ideas about projections in Hilbert space, BLPS (1997) showed that the tail σ-fields of both the free and the wired spanning forest on *every* graph is trivial.

§8. Size and Distance of Trees.

How "close" is the WUSF on a Cayley graph G to being a tree? That is, how close is the forest to being connected? One answer comes from asking how "many" edges need to be added to connect all the trees. For example, let T be the WUSF, $0 < \epsilon < 1$, and let T_ϵ be gotten by adding to T each edge of G independently with probability ϵ. Then BLPS (1997) showed that T_ϵ is connected a.s. for all $\epsilon > 0$ iff G is amenable. Of course, since all the trees in the WUSF have only one end, were we to subtract, rather than add, edges with probability ϵ, we would obtain only finite trees a.s., no matter what $\epsilon > 0$ we used. Comparing with Bernoulli percolation, we are led to view the WUSF as a "critical" model, like percolation at p_c, the critical value for Bernoulli percolation.

Another approach to measuring how far the WUSF is from being a tree is to ask how far the trees in it are from each other. The following beautiful extension of Pemantle's (1991) theorem is due to Benjamini, Kesten, Peres, and Schramm (1998). To express the result, we use the natural metric induced from the graph structure after randomly modifying the graph.

THEOREM 8.1. *Contract the edges of each tree in the uniform spanning forest on $\mathbb{Z}^d$. The diameter of the resulting graph is a.s. $\lfloor (d-1)/4 \rfloor$.*

§9. Planar Graphs and Hyperbolic Lattices.

Let G be a graph of (uniformly) bounded degree that is roughly isometric to $\mathbb{H}^d$, hyperbolic space of dimension d. There is an interesting phase transition between dimensions 2 and 3 in hyperbolic space: there are nonconstant harmonic Dirichlet functions on G iff $d = 2$. In fact, a general result for planar graphs of Benjamini and Schramm (1996a,b) says

LEMMA 9.1. *Suppose that G is a transient planar graph with bounded degree. Then* $\mathbf{HD} \neq \mathbb{R}$.

That graphs roughly isometric to $\mathbb{H}^d$ ($d \geq 2$) are transient follows from Kanai's (1986) theorem showing that transience is preserved under rough isometries. Thus, by Lemma 9.1, they have nonconstant harmonic Dirichlet functions when $d = 2$. However, the situation is reversed in higher dimensions:

LEMMA 9.2. *If G is roughly isometric to $\mathbb{H}^d$ for some $d \geq 3$ and has bounded degree, then $\mathbf{HD} = \mathbb{R}$ on G.*

Proof. For $d \geq 3$, there are no nonconstant harmonic Dirichlet functions in $\mathbb{H}^d$; see Sario *et al.* (1977). This transfers to G by a theorem of Holopainen and Soardi (1998). ∎

Taking stock, we arrive at the following surprising conclusions:

THEOREM 9.3. *If G is a self-dual proper planar Cayley graph roughly isometric to $\mathbb{H}^2$, then the* WUSF *of G has infinitely many trees a.s., each having one end a.s., while the* FUSF *of G has one tree a.s. with infinitely many ends a.s. If G is a Cayley graph roughly isometric to $\mathbb{H}^d$ for $d \geq 3$, then the* WUSF = FUSF *of G has infinitely many trees a.s., each having one end a.s.*

Proof. From Wilson's method rooted at infinity, it follows that there are infinitely many trees in the WUSF; see BLPS (1997). Use Kanai's (1986) theorem, Theorem 6.4, and Proposition 6.3 to finish the first part. Use Lemma 9.2 and Propositions 7.5 and 6.3 for the second part. ∎

An example of a self-dual planar Cayley graph roughly isometric to $\mathbb{H}^2$ was shown in Fig. 5.

Here is a summary of the phase transitions. It is quite surprising that the free spanning forest undergoes three phase transitions as the growth rate increases. (Volume of balls in hyperbolic space grows exponentially in the radius, but only polynomially in euclidean space.)

	$\mathbb{Z}^d$		$\mathbb{H}^d$	
d	$2-4$	≥ 5	2	≥ 3
FUSF: trees	1	∞	1	∞
ends	1	1	∞	1
WUSF: trees	1	∞	∞	∞
ends	1	1	1	1

§10. Open Questions.

There are an enormous number of open questions related to uniform spanning forests. The connections to different areas continue to expand; it seems to be an excellent field for a young probabilist to enter. We will give a sample of some interesting questions here.

Question. Let G be a proper transient planar graph with bounded degree and a bounded number of sides to its faces. Is the free spanning forest a single tree a.s.? If true, this would strengthen Lemma 9.1 of Benjamini and Schramm (1996a,b), which says that FUSF $\neq$ WUSF. It follows from their work that a.s. two independent random walks intersect only finitely many times, which implies that the WUSF has infinitely many trees a.s.

Question. (This question was originally asked independently by Benjamini and Propp; see Propp (1997).) One may consider the uniform spanning tree on $\mathbb{Z}^2$ embedded in $\mathbb{R}^2$. In fact, consider it on $\epsilon\mathbb{Z}^2$ in $\mathbb{R}^2$ and let $\epsilon \to 0$. Is there some sort of limit? Does it have some sort of conformal invariance? This could be expected on the basis of conformal invariance of simple random walk. It is also supported by some simulations (Schramm, personal communication). In addition, the uniform spanning tree is intimately tied to random domino tiling of $\mathbb{Z}^2$: see, e.g., Burton and Pemantle (1993). New work of Rick Kenyon (1997) shows that domino tiling has a strong sort of conformal invariance. Finally, critical Bernoulli percolation is believed to have conformal invariance in the limit (see, e.g., Langlands, Pouliot and Saint-Aubin (1994) and Benjamini and Schramm (1996c)) and the uniform spanning tree is a sort of critical model, as explained in Section 8.

Question. Tóth and Werner (1997), §11, explain the connection between certain self-repelling random walks on $\mathbb{Z}$ and a Markov chain on $\mathbb{Z}^2$ that never visits a state more than once. They consider coalescing paths of this Markov chain; these paths form the wired random spanning tree of $\mathbb{Z}^2$ that is associated to this chain by Wilson's method rooted at infinity. The dual of this tree is considered, as well as the lattice-filling curve that "threads" between the two trees; see Fig. 11 for the lattice-filling curve of a uniform spanning tree. Most of their paper is devoted to analyzing the continuous analogue of these objects. In particular, a stochastic differential equation describes the space-filling curve. Note that the lattice-filling curve completely determines the tree and its dual. Can one give a stochastic differential equation for the space-filling curve that presumably is the limit as $\epsilon \to 0$ of the lattice-filling curves threading between the uniform spanning tree and its dual on $\epsilon\mathbb{Z}^2$?

Figure 11. The inside of this lattice-filling curve drawn by Schramm is a uniform spanning tree on a square grid, while the outside is another on the dual grid (wired).

Question. Is each tree of the wired uniform spanning forest on any graph recurrent a.s.? For each tree, is $p_c = 1$ a.s.?

Question. If a graph has positive isoperimetric constant, must each tree in its wired uniform spanning forest have one end a.s.?

Question. Does each tree in the wired uniform spanning forest on Galton-Watson trees have one end a.s.?

Question. Is there a "natural" coupling of FUSF and WUSF? For example, if G is a Cayley graph, is there a coupling that is invariant under multiplication by elements of G?

Question. Under the hypotheses of Proposition 6.5, does every tree have infinitely many ends FUSF-a.s.?

Acknowledgement. I am grateful to Oded Schramm and the referee for numerous comments on the manuscript.

REFERENCES

ALDOUS, D. (1990). The random walk construction of uniform random spanning trees and uniform labelled trees. *SIAM J. Disc. Math.* **3**, 450–465.

BENJAMINI, I., KESTEN, H., PERES, Y., and SCHRAMM, O. (1998). *Paper in preparation.*

BENJAMINI, I., LYONS, R., PERES, Y., and SCHRAMM, O. (1997). Uniform spanning forests, *preprint.*

BENJAMINI, I., LYONS, R. and SCHRAMM, O. (1997). Percolation perturbations in potential theory and random walks, *preprint.*

BENJAMINI, I. and SCHRAMM, O. (1996a). Harmonic functions on planar and almost planar graphs and manifolds, via circle packings, *Invent. Math.* **126**, 565–587.

BENJAMINI, I. and SCHRAMM, O. (1996b). Random walks and harmonic functions on infinite planar graphs using square tilings, *Ann. Probab.* **24**, 1219–1238.

BENJAMINI, I. and SCHRAMM, O. (1996c). Conformal invariance of Voronoi percolation, *preprint.*

BRODER, A. (1989). Generating random spanning trees. In *30th Annual Symp. Foundations Computer Sci.*, pp. 442-447. IEEE, New York.

BROOKS, R. L., SMITH, C. A. B., STONE, A. H., and TUTTE, W. T. (1940). The dissection of rectangles into squares, *Duke Math. J.* **7**, 213–340.

BURTON, R. and PEMANTLE, R. (1993). Local characteristics, entropy and limit theorems for spanning trees and domino tilings via transfer-impedances, *Ann. Probab.* **21**, 1329–1371.

DOYLE, P. G. (1988). Electric currents in infinite networks (preliminary version), *unpublished manuscript.*

DOYLE, P. G. and SNELL, J. L., (1984). *Random Walks and Electric Networks.* Mathematical Assoc. of America, Washington, DC.

FEDER, T. and MIHAIL, M. (1992). Balanced matroids, *Proc. 24th Annual ACM Sympos. Theory Computing (Victoria, BC, Canada)*, pp. 26–38. ACM Press, New York.

GRIMMETT, G. R. (1995). The stochastic random-cluster process and the uniqueness of random-cluster measures, *Ann. Probab.* **23**, 1461–1510.

HÄGGSTRÖM, O. (1995). Random-cluster measures and uniform spanning trees, *Stoch. Proc. Appl.* **59**, 267–275.

HÄGGSTRÖM, O. (1997). Uniform and minimal essential spanning forests on trees, *Random Struc. Alg.*, to appear.

HOLOPAINEN, I. and SOARDI, P. M. (1998). p-harmonic functions on graphs and manifolds, *to appear.*

KANAI, M. (1986). Rough isometries and the parabolicity of riemannian manifolds, *J. Math. Soc. Japan* **38**, 227–238.

KENYON, R. (1997). Conformal invariance of domino tiling, *preprint.*

KIRCHHOFF, G. (1847). Ueber die Auflösung der Gleichungen, auf welche man bei der Untersuchung der linearen Vertheilung galvanischer Ströme geführt wird, *Ann. Phys. und Chem.* **72** (12), 497–508.

LANGLANDS, R., POULIOT, P. and SAINT-AUBIN, Y. (1994). Conformal invariance in two-dimensional percolation, *Bull. Amer. Math. Soc.* (N.S.) **30**, 1–61.

LAWLER, G. F. (1991). *Intersections of Random Walks.* Birkhäuser, Boston, MA.

LIND, D., SCHMIDT, K. and WARD, T. (1990). Mahler measure and entropy for commuting automorphisms of compact groups, *Invent. Math.* **101**, 593–629.

LYONS, R. (1990). Random walks and percolation on trees, *Ann. Probab.* **18**, 931–958.

LYONS, R. (1992). Random walks, capacity, and percolation on trees, *Ann. Probab.* **20**, 2043–2088.

LYONS, R. and PERES, Y. (1997). *Probability on Trees and Networks*, Cambridge University Press, in preparation. Current version available at `http://php.indiana.edu/~rdlyons/`.

LYONS, R., PERES, Y. and SCHRAMM, O. (1998). Can a Markov chain intersect an independent copy only in its loops?, *in preparation.*

MEDOLLA, G. and SOARDI, P. M. (1995). Extension of Foster's averaging formula to infinite networks with moderate growth, *Math. Z.* **219**, 171–185.

PEMANTLE, R. (1991). Choosing a spanning tree for the integer lattice uniformly, *Ann. Probab.* **19**, 1559–1574.

PROPP, J. G. (1997). Continuum spanning trees, *these proceedings.*

PROPP, J. G. and WILSON, D. B. (1996). How to get a perfectly random sample from a generic Markov chain and generate a random spanning tree of a directed graph, *Journal of Algorithms*, to appear.

REDMOND, C. and YUKICH, J. E. (1994). Limit theorems and rates of convergence for subadditive Euclidean functionals, *Ann. Appl. Probab.* **4**, 1057–1073.

SARIO, L., NAKAI, M., WANG, C., and CHUNG, L. O. (1977). *Classification Theory of Riemannian Manifolds.* Springer, Berlin.

SOARDI, P. M. (1993). Rough isometries and Dirichlet finite harmonic functions on graphs, *Proc. Amer. Math. Soc.* **119**, 1239–1248.

SOARDI, P. M. (1994). *Potential Theory on Infinite Networks.* Springer, Berlin.

SOLOMYAK, R. (1997). On coincidence of entropies for two classes of dynamical systems, *Ergodic Theory Dynam. Systems*, to appear.

SOLOMYAK, R. (1998). Essential spanning forests processes on groups, *in preparation.*

STRASSEN, V. (1965). The existence of probability measures with given marginals, *Ann. Math. Statist.*, **36**, 423–439.

THOMASSEN, C. (1990). Resistances and currents in infinite electrical networks, *J. Comb. Theory* B **49**, 87–102.

TÓTH, B. and WERNER, W. (1997). The true self-repelling motion, *preprint.*

WILSON, D. B. (1996). Generating random spanning trees more quickly than the cover time, *1996 ACM Sympos. Theory of Computing*, pp. 296–303.

YUKICH, J. E. (1996a). Worst case growth rates for some classical optimization problems, *Combinatorica* **16**, 575–586.

YUKICH, J. E. (1996b). Ergodic theorems for some classical optimization problems, *Ann. Appl. Probab.* **6**, 1006–1023.

DEPARTMENT OF MATHEMATICS, INDIANA UNIVERSITY, BLOOMINGTON, IN 47405-5701, USA
`rdlyons@indiana.edu`
`http://php.indiana.edu/~rdlyons/`

DIMACS Series in Discrete Mathematics
and Theoretical Computer Science
Volume 41, 1998

Enumerations of trees and forests related to branching processes and random walks

Jim Pitman

ABSTRACT. In a Galton-Watson branching process with offspring distribution $(p_0, p_1, \dots)$ started with k individuals, the distribution of the total progeny is identical to the distribution of the first passage time to $-k$ for a random walk started at 0 which takes steps of size j with probability p_{j+1} for $j \geq -1$. The formula for this distribution is a probabilistic expression of the Lagrange inversion formula for the coefficients in the power series expansion of $f(z)^k$ in terms of those of $g(z)$ for $f(z)$ defined implicitly by $f(z) = zg(f(z))$. The Lagrange inversion formula is the analytic counterpart of various enumerations of trees and forests which generalize Cayley's formula kn^{n-k-1} for the number of rooted forests labeled by a set of size n whose set of roots is a particular subset of size k. These known results are derived by elementary combinatorial methods without appeal to the Lagrange formula, which is then obtained as a byproduct. This approach unifies and extends a number of known identities involving the distributions of various kinds of random trees and random forests.

1. Introduction

Let $(Z_0, Z_1, \dots)$ be a *Galton-Watson branching process with offspring distribution* $(p_i, i = 0, 1, \dots)$, where $p_i \geq 0, \sum_i p_i = 1$. Interpret p_i as the probability that an individual has i children, and Z_g as the number of individuals in the gth generation of a population starting from some initial number Z_0. For each $g \geq 0$, it is assumed that given the evolution of the population up to the gth generation, the Z_g individuals in the gth generation have independent random numbers of children distributed according to (p_i). These children are the Z_{g+1} members of the $(g+1)$th generation. See [**34, 37, 6**] for background. Let

$$S_n := X_1 + \cdots + X_n \tag{1.1}$$

be the sum of n independent random variables X_j with common distribution (p_i). From the description of $(Z_0, Z_1, \dots)$, this sequence is a Markov chain with state space $\{0, 1, 2, \dots\}$ and time-homogeneous transition probabilities

$$P(Z_{g+1} = m \,|\, Z_g = n) = P(S_n = m) \tag{1.2}$$

where for each n the sequence $(P(S_n = m), m = 0, 1, \dots)$ is the n-fold convolution of the sequence $(p_i, i = 0, 1, \dots)$ with itself. Two elementary expressions for the

1991 *Mathematics Subject Classification.* Primary 05C05, 60J80; Secondary 60J15, 60C05.

probability $P(S_n = m)$ in terms of (p_i) are recalled in formulae (1.7) and (1.8) below. Let

$$\#\mathcal{F}_k := \sum_{g=0}^{\infty} Z_g \in \{1, 2, \dots, \infty\} \tag{1.3}$$

represent the *total progeny* in the branching process given $Z_0 = k$ individuals to start with. Here $\#\mathcal{F}_k$ stands for the number of individuals in the random family $\mathcal{F}_k$ generated by the branching process. In Section 4, $\mathcal{F}_k$ will be defined as a *random family forest*, that is a collection of k *random family trees*, one for each of the k initial individuals. The following theorem presents a remarkable formula for the distribution of $\#\mathcal{F}_k$ restricted to positive integers. This formula was discovered by Otter [**48**] for $k = 1$ and extended to all $k \geq 1$ by Dwass [**20**]:

THEOREM 1.1 (Otter-Dwass formula [**48, 20**]). *For all* $n, k = 1, 2, \dots$

$$P(\#\mathcal{F}_k = n) = \frac{k}{n} P(S_n = n - k). \tag{1.4}$$

There is another setting where the same array of probabilities arises:

THEOREM 1.2 (Kemperman's formula [**38**],[**39**, (7.15)]). *Let T_{-k} be the least $n \geq 1$ such that $S_n - n = -k$, with the convention that $T_{-k} = \infty$ if there is no such n. Then for all $1 \leq k \leq n$*

$$P(T_{-k} = n) = \frac{k}{n} P(S_n = n - k). \tag{1.5}$$

The consequence of (1.4) and (1.5), that $\#\mathcal{F}_k$ has the same distribution as T_{-k} for each $k = 1, 2, \dots$, can be understood in terms of Kendall's [**40**] interpretation of the branching process (Z_n) and the random walk $(S_n - n)$ in terms of a queueing process, which is recalled in Section 5.

The first approach to these formulae was by the method of probability generating functions. Let

$$g(z) := \sum_{i=0}^{\infty} p_i z^i \tag{1.6}$$

be the probability generating function derived from the offspring distribution (p_i) Two elementary expressions for $P(S_n = m)$ are

$$P(S_n = m) = \text{coefficient of } z^m \text{ in } g(z)^n \tag{1.7}$$

$$= \sum_{\substack{\Sigma n_i = n \\ \Sigma i n_i = m}} \binom{n}{n_0, \cdots, n_m} \prod_{i \geq 0} p_i^{n_i} \tag{1.8}$$

where the sum ranges over the finite set of all sequences of non-negative integers $(n_i, i \geq 0)$ with $\Sigma n_i = n$ and $\Sigma i n_i = m$. Here

$$\binom{n}{n_0, \dots, n_m} := \frac{n!}{n_0! \cdots n_m!} \tag{1.9}$$

is the multinomial coefficient which is the number of sequences $(i_j, 1 \leq j \leq n)$ of *type* $(n_i, i \geq 0)$, that is sequences (i_j) such that $\#\{j : i_j = i\} = n_i$ for all $i \geq 0$ and hence $\sum_j i_j = \sum_i in_i$. Let

$$h_k(z) := \sum_n P(\#\mathcal{F}_k = n)z^n$$

be the generating function of the distribution of $\#\mathcal{F}_k$ restricted to finite values. Since the branching process started with k individuals can be constructed from k independent copies of the branching process started with one individual, the random variable $\#\mathcal{F}_k$ is distributed like the sum of k independent copies of $\#\mathcal{F}_1$. Similarly, $\#\mathcal{F}_1$ given $Z_1 = k$ has the same distribution as $1 + \#\mathcal{F}_k$. It follows that

$$h_k(z) = h(z)^k \text{ where } h(z) = zg(h(z)). \tag{1.10}$$

In particular, it is well known [**34**] that the *extinction probability*

$$P(\#\mathcal{F}_k < \infty) = h_k(1) = q^k$$

where $q := h_1(1)$ is the least non-negative root of $q = g(q)$; consequently $q = 1$ or $q < 1$ according as $\mu \leq 1$ or $\mu > 1$, where $\mu := \sum_i ip_i$ is the mean of the offspring distribution and it is assumed that $p_1 < 1$.

This method of determining the distribution of $\#\mathcal{F}_k$, which dates back to a 1944 report of Hawkins and Ulam [**35**], was used also by Otter [**48**] and Good [**28**]. If $\hat{h}_k(z)$ is defined in the setting of Kemperman's formula by

$$\hat{h}_k(z) := \sum_n P(T_{-k} = n)z^n$$

then the same relation (1.10) can be derived for $\hat{h}_k(z)$ instead of $h_k(z)$. According to a classical result of Lagrange the equation for $h(z)$ in (1.10) has a unique analytic solution in a neighbourhood of 0. It follows that

$$\hat{h}_k(z) = h_k(z) = h(z)^k$$

for this unique $h(z)$. The Otter-Dwass and Kemperman formulae can now be read from the power series expansion of $h(z)^k$ provided by

THEOREM 1.3 (Lagrange inversion formula [**14**]). *Let $g(z)$ be analytic in a neighbourhood of 0 with $g(0) \neq 0$. Then the equation $h(z) = zg(h(z))$ has a unique analytic solution in a neighbourhood of* 0 *such that*

$$\text{coefficient of } z^n \text{ in } h(z)^k = \frac{k}{n}\left(\text{coefficient of } z^{n-k} \text{ in } g(z)^n\right). \tag{1.11}$$

While stated here in an analytic form, it is well known [**66**, §5.4] that the Lagrange inversion formula can be formulated as an identity of formal power series. From this perspective, it is clear that both sides of (1.11) are polynomials in the first $n+1$ coefficients $p_0, \ldots, p_n$ say of $g(z) := \sum_n p_n z^n$. As observed by Wendel [**74**], to prove (1.11) it is enough to establish this identity of polynomials in $p_0, \ldots, p_n$ for $p_i \geq 0$ and $\sum_{i=1}^n p_i \leq 1$, and that is precisely the content of each of the preceding probabilistic theorems. Thus the Otter-Dwass formula and Kemperman's formula are probabilistic expressions of the Lagrange inversion formula.

It is known to combinatorialists [**60, 44, 13, 66**] that the Lagrange inversion formula is the analytic counterpart to various enumerations of trees and forests which trace back to Cayley [**11**], and that these enumerations can also be interpreted, by a suitable bijection between forests and lattice paths, as enumerations of lattice paths. This paper offers an elementary approach to this circle of ideas by development of the combinatorial and probabilistic results without appeal to the Lagrange inversion formula, which is then a byproduct as just indicated.

The term *forest* will be used here for a *finite rooted forest*, that is a directed graph with a finite number of vertices, each of whose connected components is a tree with edges directed away from its root vertex. A forest with vertex set V is said to be *labeled by* V. For vertices v and w of a forest $\mathbf{f}$ write $v \xrightarrow{\mathbf{f}} w$ to show that (v, w) is a directed edge of $\mathbf{f}$. The *number of children* or *out-degree* v in the forest $\mathbf{f}$ is $C(v, \mathbf{f}) := \#\{w : v \xrightarrow{\mathbf{f}} w\}$. In a *plane forest* $\mathbf{f}$ with k component trees, the set of roots of the tree components is ordered, as is the set of children of v for each vertex v of $\mathbf{f}$. Regard a plane forest with k root vertices as a collection of family trees, one for each of k initial individuals, with each vertex in the forest corresponding to an individual, and with the order of the roots and the orders of children corresponding to the order of birth of individuals. A plane forest is often depicted without labels as on the top panel of Figure 1, and called an *unlabeled plane forest.* However, there is a natural way to identify each vertex of a plane forest by a finite sequence of non-negative integers which indicates the location of the vertex in the forest. An individual in the gth generation of a family forest (= vertex at height g in the plane forest) is identified by a sequence of $g+1$ integers, for instance $(3, 7, 4)$ to indicate a second generation individual who is the 4th child of the 7th child of the 3rd root individual. Thus, following the convention of [**34**] for labeling family trees, the set of vertices of a plane forest will be identified as a subset of the set of all finite sequences of integers, as illustrated in the bottom panel of Figure 1.

The *type* of a forest $\mathbf{f}$ is the sequence of non-negative integers (n_i), where n_i is the number of vertices of $\mathbf{f}$ with i children. Let $1 \le k \le n$ and let (n_i) be a sequence of non-negative integers with

$$\sum_i n_i = n \text{ and } \sum_i i n_i = n - k. \tag{1.12}$$

A forest of type (n_i) has n vertices and $n-k$ non-root vertices, hence k root vertices and k tree components.

THEOREM 1.4. (Enumeration of plane forests by type [**21, 22, 48, 61**]) *For* $1 \le k \le n$ *and* (n_i) *subject to* (1.12) *the number* $N^{\text{plane}}(n_0, n_1, \ldots)$ *of plane forests of type* (n_i) *with* k *tree components and* n *vertices is*

$$N^{\text{plane}}(n_0, n_1, \ldots) = \frac{k}{n}\binom{n}{n_0, \ldots, n_n}. \tag{1.13}$$

This enumeration is due to Erdélyi and Etherington [**21, 22**] and Otter [**48**] for $k = 1$. An equivalent enumeration in terms of words instead of trees appears in Raney [**60**, Thm. 2.2] for $k \ge 1$. The corresponding result for labeled forests is:

THEOREM 1.5 (Enumeration of labeled forests by type [**66**, Cor. 3.5]). *For* $1 \le k \le n$ *and* (n_i) *subject to* (1.12), *the number* $N^{[n]}(n_0, n_1, \ldots)$ *of forests labeled by the set* $[n] := \{1, \ldots, n\}$, *of type* (n_i) *with* k *tree components and* n *vertices, is*

$$N^{[n]}(n_0, n_1, \ldots) = \frac{k}{n}\binom{n}{k}\frac{(n-k)!}{\prod_{i \ge 0}(i!)^{n_i}}\binom{n}{n_0, \ldots, n_n}. \tag{1.14}$$

The enumeration of plane forests by type is simpler than its companion for labeled forests. But the result for labeled forests has the following simpler equivalent:

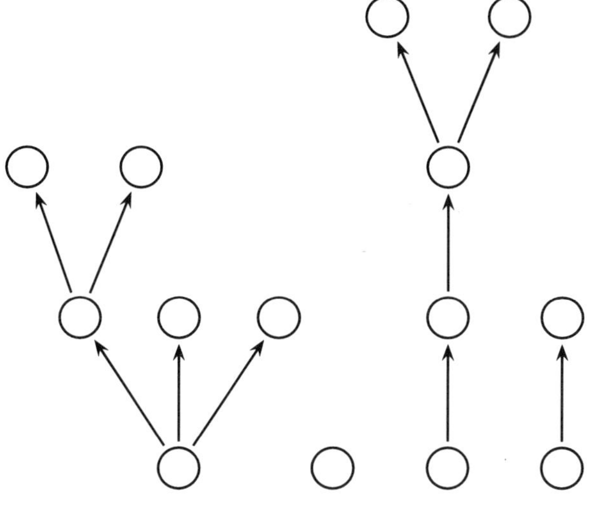

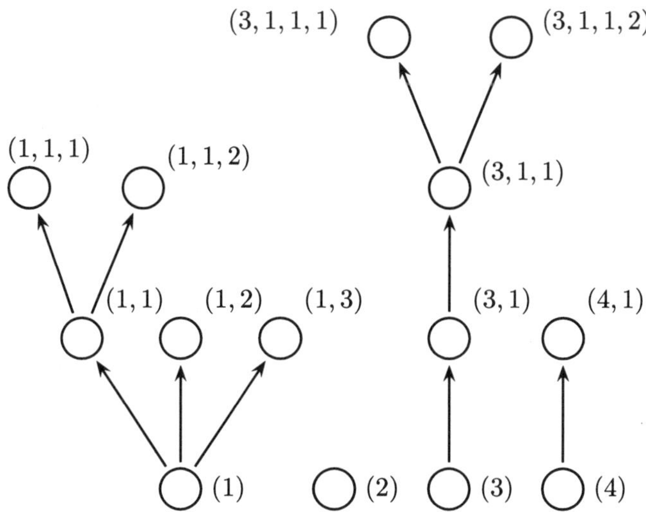

FIGURE 1

THEOREM 1.6. (Enumeration of labeled forests by out-degree sequence [**56**], [**66**, Thm. 3.4]) *For all sequences of non-negative integers* $(c_1, \dots, c_n)$ *with* $\sum_i c_i = n - k$ *the number* $N(c_1, \dots, c_n)$ *of forests* $\mathbf{f}$ *with vertex set* $[n]$ *in which vertex* i *has* c_i *children for each* $i \in [n]$ *(and hence* $\mathbf{f}$ *has* k *tree components) is*

$$N(c_1, \dots, c_n) = \frac{k}{n}\binom{n}{k}\binom{n-k}{c_1, \dots, c_n}. \tag{1.15}$$

In view of the multinomial theorem, the enumeration (1.15) amounts to the following identity of polynomials in n commuting variables $x_i, 1 \le i \le n$:

$$\sum_{\mathbf{f} \in \mathbf{F}_{k,n}} \prod_{i=1}^{n} x_i^{C(i,\mathbf{f})} = \frac{k}{n} \binom{n}{k} (x_1 + \cdots + x_n)^{n-k} \tag{1.16}$$

where the sum is over the set $\mathbf{F}_{k,n}$ of all forests with k tree components labeled by $[n]$, and $C(i, \mathbf{f})$ is the number of children of i in the forest $\mathbf{f}$. Take the x_i to be identically 1 in (1.16) to recover the well known enumeration

$$\#\mathbf{F}_{k,n} = k \binom{n}{k} n^{n-k-1} \tag{1.17}$$

which is equivalent to Cayley's [**11**] formula

$$\#\{\text{forests with root set } [k] \text{ and vertex set } [n]\} = kn^{n-k-1}. \tag{1.18}$$

In particular, for $k = 1$ the number of rooted trees labeled by $[n]$ is n^{n-1}. Equivalently, the number of unrooted trees labeled by $[n]$ is n^{n-2}. For various other approaches to these formulae of Cayley, see [**46, 56, 57, 59, 64, 66, 70**]. The multinomial expansion over rooted trees obtained from the case $k = 1$ of (1.16) is a variant of the multinomial expansion over unrooted trees indicated by Cayley [**11**] for small n, and formulated and proved for all n by Rényi [**62**]. But throughout this paper, all trees and forests are assumed to be rooted. As shown in [**57**], Cayley's multinomial expansion over trees is also equivalent to an enumeration of mappings from $[n]$ to $[n]$ with a unique cyclic point, due to Françon [**25**], whose probabilistic expression is a result of Burtin [**10, 5**].

The rest of this paper is organized as follows. Section 2 offers a simple proof of Theorem 1.6. The equivalence of Theorems 1.4, 1.5 and 1.6 is argued in Section 3. The equivalence of Theorem 1.4 and the Otter-Dwass formula is explained in Section 4. See also Kolchin [**43**, Chapter 2.1, Lemma 3] for another proof of the Otter-Dwass formula, by showing that both sides satisfy a recursion which has a unique solution. Section 5 reviews the connection between the Otter-Dwass formula for branching processes and Kemperman's formula for random walks via the interpretation of both processes in terms of queues pointed out by Kendall [**40**]. As shown by Takács [**67, 68**], Kemperman's formula is both generalized and simplified by consideration of random walks of a fixed length n whose distribution of steps is invariant under cyclic shifts. The ubiquitous factor of k/n in each of the six theorems above finds its most intuitive explanation in this context: the k/n in Kemperman's formula for random walks represents the conditional probability of the event $(T_{-k} = n)$ given $(S_n - n = -k) \cap A$ for any event A determined by the first n increments of the walk which is invariant under cyclic shifts of these increments. Section 6 reviews the well known representation of the uniform distribution on plane forests with k trees and n vertices as the distribution of a Galton-Watson forest of k trees with geometric offspring distribution conditioned on a total progeny of n. Section 7 presents a similar result which explains a number of known identities relating the uniform distribution on the set of forests of k trees labeled by $[n]$ to the distribution of the random forest generated a Galton-Watson branching process with a Poisson offspring distribution.

2. Enumeration of labeled forests by out-degree sequence.

PROOF OF THEOREM 1.6. For a forest $\mathbf{f}$ with vertex set $[n]$ and $i \in [n]$ let $J_i(\mathbf{f}) := \{j \in [n] : i \stackrel{\mathbf{f}}{\to} j\}$ be the set of children of i in $\mathbf{f}$. So $\mathbf{f}$ is determined by the sequence of disjoint sets $J_1(\mathbf{f}), \dots, J_n(\mathbf{f})$, and vice-versa. Given a sequence of disjoint subsets $J_1, \dots, J_n$ of $[n]$, for each $m \in [n]$ let $\mathbf{f}_m$ be the relation on $[m] \cup (\cup_{i=1}^m J_i)$ defined by

$$i \stackrel{\mathbf{f}_m}{\to} j \text{ iff } i \in [m] \text{ and } j \in J_i. \tag{2.1}$$

There exists a forest $\mathbf{f}$ labeled by $[n]$ such that $J_i(\mathbf{f}) = J_i$ for all $i \in [n]$ if and only if the J_i are such that $\mathbf{f}_m$ defined by (2.1) is a forest with vertex set $[m] \cup (\cup_{i=1}^m J_i)$ for every $m \in [n]$; then $\mathbf{f} = \mathbf{f}_n$ and $\mathbf{f}_m$ is the restriction of $\mathbf{f}$ to $[m] \cup (\cup_{i=1}^m J_i)$ for each $m \in [n]$. It follows that for each sequence of non-negative integers (c_i) with $\sum_i c_i = n - k$, the number

$$N(c_1, \dots, c_n) := \#\{\mathbf{f} \in \mathbf{F}_{k,n} : \#J_i(\mathbf{f}) = c_i \text{ for all } i \in [n]\} \tag{2.2}$$

is the number of ways to choose a sequence of subsets $(J_1, \dots, J_n)$ of $[n]$ such that $\#J_m = c_m$ and the relation $\mathbf{f}_m$ defined by (2.1) is a forest with vertex set $[m] \cup (\cup_{i=1}^m J_i)$ for every $m \in [n]$. Clearly, J_1 can be any of the $\binom{n-1}{c_1}$ subsets of $[n] - \{1\}$ of size c_1. For $m \in [n-1]$ make the inductive hypothesis that sets $J_1, \dots, J_m$ of sizes $c_1, \dots, c_m$ have been chosen so that $\mathbf{f}_m$ is a forest. Which choices of J_{m+1} make $\mathbf{f}_{m+1}$ a forest? There are two cases to consider. Either
(i) $m+1 \notin \cup_{i=1}^m J_i$: then J_{m+1} can be any subset of $[n] - (\cup_{i=1}^m J_i) - \{m+1\}$; or
(ii) $m+1 \in \cup_{i=1}^m J_i$: then J_{m+1} can be any subset of $[n] - (\cup_{i=1}^m J_i) - \{r_m\}$ where $r_m \notin \cup_{i=1}^m J_i$ is the root of the tree component of $\mathbf{f}_m$ which contains $m+1$. Either way, and regardless of what sets $J_1, \dots, J_m$ of sizes $c_1, \dots, c_m$ were previously chosen to make $\mathbf{f}_m$ a forest, the number of possible choices of J_{m+1} of size c_{m+1} which make $\mathbf{f}_{m+1}$ a forest is

$$\binom{n - \sum_{i=1}^m c_i - 1}{c_{m+1}}.$$

Consequently, by induction

$$N(c_1, \dots, c_n) = \binom{n-1}{c_1}\binom{n - c_1 - 1}{c_2} \cdots \binom{n - \sum_{i=1}^{n-1} c_i - 1}{c_n}$$

and this expression simplifies easily to yield (1.15). ■

3. Enumeration of forests by type

PROOF OF THEOREM 1.5. Let (n_i) subject to (1.12) be a possible type sequence for a forest of k trees with n vertices. According Theorem 1.6, for each particular sequence (c_j) of type (n_i) the number of forests in which j has c_j children for all $j \in [n]$ is

$$\frac{k}{n}\binom{n}{k}\frac{(n-k)!}{c_1! \dots c_n!} = \frac{k}{n}\binom{n}{k}\frac{(n-k)!}{\prod_{i \geq 0}(i!)^{n_i}}. \tag{3.1}$$

But the number of sequences of non-negative integers (c_j) such that (c_j) has type (n_i) is just the multinomial coefficient (1.9). The number of forests labeled by $[n]$ of type (n_i) is the product of the number in (3.1) and this multinomial coefficient. Thus Theorem 1.6 implies Theorem 1.5, and vice versa. ■

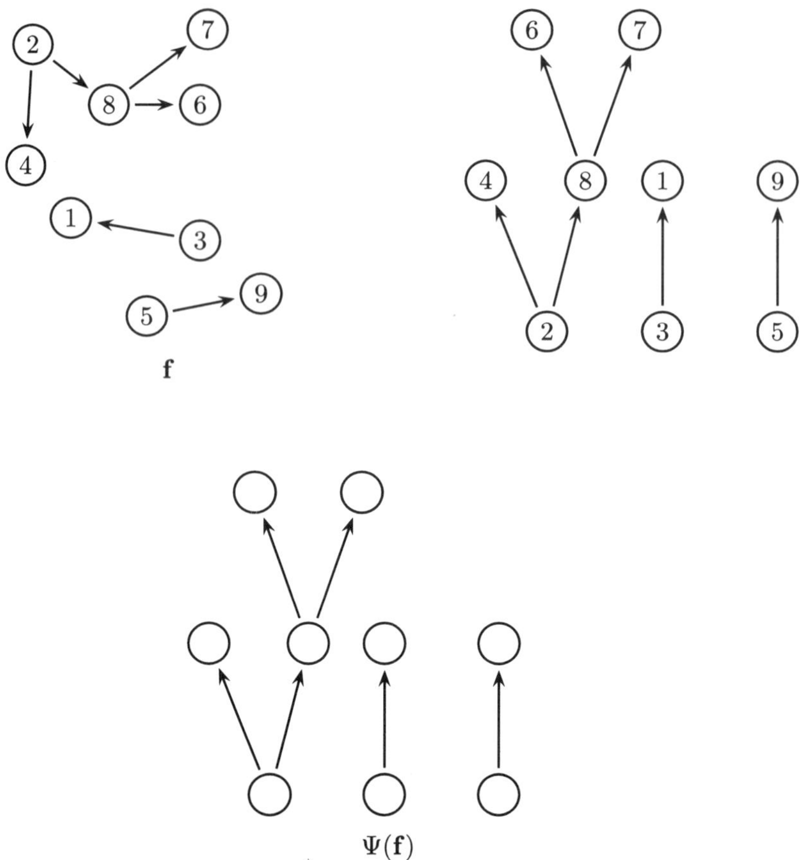

FIGURE 2

PROOF OF THEOREM 1.4. The enumeration (1.13) for plane forests corresponds to the enumeration (1.14) for labeled forests via the following map Ψ, which transforms a forest $\mathbf{f}$ labeled by $[n]$ into a plane forest $\Psi(\mathbf{f})$ of the same type. To define $\Psi(\mathbf{f})$, first place the components of $\mathbf{f}$ in order of their root labels, then recursively for each $g \geq 0$ place the children of each vertex of $\mathbf{f}$ at height g in order of their labels. Finally, delabel to obtain the plane forest $\Psi(\mathbf{f})$. See Figure 2.

For a fixed plane forest $\mathbf{f}^\circ$ with n vertices and k root vertices, and type (n_i), let $\{v_1, \dots, v_n\}$ be a listing of the vertices in some arbitrary order. Let B_0 be the set of roots of $\mathbf{f}^\circ$, and B_i the set of children of v_i for $i = 1, \dots, n$. The sets $B_i, 0 \leq i \leq n$, some of which must be empty, are disjoint with union $V(\mathbf{f}^\circ)$, the set of vertices of $\mathbf{f}^\circ$. As discussed earlier, $V(\mathbf{f}^\circ)$ is regarded as a subset of the set of finite sequences of positive integers. Each non-empty B_i has a linear ordering, and each $\mathbf{f} \in \Psi^{-1}(\mathbf{f}^\circ)$ corresponds to a unique bijection from $V(\mathbf{f}^\circ)$ to $[n]$ which is increasing on each non-empty B_i. It follows that $\#\Psi^{-1}(\mathbf{f}^\circ)$, the number of forests

$\mathbf{f}$ labeled by $[n]$ such that $\Psi(\mathbf{f}) = \mathbf{f}^\circ$ is the multinomial coefficient

$$(3.2) \qquad \#\Psi^{-1}(\mathbf{f}^\circ) = \binom{n!}{k, \#B_1, \dots, \#B_n} = \frac{n!}{k! \prod_{v \in V(\mathbf{f}^\circ)} C(v, \mathbf{f}^\circ)!} = \frac{n!}{k! \prod_{i \geq 0} (i!)^{n_i}}$$

where $C(v, \mathbf{f}^\circ)$ is the number of children of v in the forest $\mathbf{f}^\circ$, and (n_i) is the type of $\mathbf{f}^\circ$ and of every $\mathbf{f} \in \Psi^{-1}(\mathbf{f}^\circ)$. It follows that the number $N^{\text{plane}}(n_0, n_1, \dots)$ of plane forests of type (n_i) and the corresponding number $N^{[n]}(n_0, n_1, \dots)$ of forests of type (n_i) labeled by $[n]$ are related by

$$(3.3) \qquad \frac{N^{[n]}(n_0, n_1, \dots)}{N^{\text{plane}}(n_0, n_1, \dots)} = \frac{n!}{k! \prod_{i \geq 0} (i!)^{n_i}}.$$

Hence the equivalence of Theorems 1.4 and 1.5. ∎

Alternative Proof of Theorem 1.4. Stanley [**66**, Thm. 3.10 of Ch. 5] gives two proofs of Theorem 1.4 based on a correspondence between forests and lattice paths, and enumerations of lattice paths by consideration of cyclic shifts as in Section 5 of this paper. The following similar argument was suggested by a referee. First, label the vertices of the plane forest $\mathbf{f}$ by $[n]$ according the order they are visited in a depth-first search, as follows: start at the root of the left-most tree, then follow a path around the vertices of that tree as indicated in Figure 3, then go to the root of the next tree to the right, follow the path around that tree, and so on. For a plane forest $\mathbf{f}$ consider the lattice path starting at $(0,0)$ with increments

$$c_m, c_{m+1}, \dots, c_n, c_1, \dots, c_{m-1}$$

where c_i is the number of children in $\mathbf{f}$ of the vertex of $\mathbf{f}$ labeled i by the depth-first search, and m is the label of some arbitrarily chosen vertex of $\mathbf{f}$. Since $\sum_i c_i = n-k$ the path goes from $(0,0)$ to $(n, n-k)$. The multinomial coefficient in (1.13) is the number of possible paths for the given type (n_i). Because the path remains the same if the trees of $\mathbf{f}$ are interchanged cyclically along with the choice of vertex, for each given type (n_i) the map from pairs $(m, \mathbf{f})$, with $m \in [n]$ and $\mathbf{f}$ of the given type, to paths with increments of the corresponding type, is k to 1. The enumeration (1.13) follows. ∎

4. The Otter-Dwass formula

Proof of the equivalence of Theorems 1.1 and 1.4. Following the approach of Otter [**48**] and subsequent authors [**34, 41, 16, 47**], regard a Galton-Watson process started with $Z_0 = k$ individuals as generating a collection of k family trees, which combine to form a *random family forest* $\mathcal{F}_k$. On the event $(\#\mathcal{F}_k < \infty)$ the random family forest $\mathcal{F}_k$ can be defined in an elementary way as a random element with values in the countable set $\mathbf{F}$ of all plane forests. The *distribution of* $\mathcal{F}_k$ is then the sub-probability distribution on $\mathbf{F}$ defined by the formula [**48**]

$$(4.1) \qquad P(\mathcal{F}_k = \mathbf{f}) = \prod_{v \in V(\mathbf{f})} p_{C(v, \mathbf{f})} = \prod_{i \geq 0} p_i^{n_i(\mathbf{f})} \quad \text{for } k \geq 1, \ \mathbf{f} \in \mathbf{F}_k^{\text{plane}}$$

where $V(\mathbf{f})$ is the set of vertices of $\mathbf{f}$, the number of vertices of $\mathbf{f}$ with i children is denoted $n_i(\mathbf{f})$, and $\mathbf{F}_k^{\text{plane}}$ is the set of plane forests with k root vertices. This distribution on $\mathbf{F}_k^{\text{plane}}$ has total mass $P(\#\mathcal{F}_k < \infty) \leq 1$. For each $k = 1, 2, \dots$ the probability of the event $(\#\mathcal{F}_k = n)$ is obtained by summing the expression (4.1) over all plane forests $\mathbf{f}$ of k trees with a total of n vertices. The terms in this sum can be classified by the type (n_i) of the forest $\mathbf{f}$. Since the number of vertices in a forest of type (n_i) is $\sum_i n_i$ and the number of root vertices is $n - \sum_i in_i$, the result is

$$P(\#\mathcal{F}_k = n) = \sum_{\substack{\Sigma n_i = n \\ \Sigma in_i = n-k}} N^{\text{plane}}(n_0, n_1, \dots) \prod_{i \geq 0} p_i^{n_i} \tag{4.2}$$

where the sum is over all possible types (n_i) of a sequence with length $\sum_i n_i = n$ and sum $\sum_i n_i = n - k$. Now fix n and k and regard the probability displayed in (4.2), and the probability $P(S_n = m)$ displayed in (1.8) for $m = n - k$, as functions of the sequence $(p_0, \dots, p_n)$. Since each probability is a polynomial in $(p_0, \dots, p_n)$, the Otter-Dwass formula (1.4) is an identity of polynomials whose coefficient identity is the enumeration (1.13) of plane forests by type. ∎

Lagrangian distributions. Recall that $h(z)$ determined by the offspring generating function $g(z)$ via (1.10) or (1.11) is the probability generating function of the total progeny of a Galton-Watson branching process started with one individual. For Z_0 with an arbitrary distribution with probability generating function $f(z) := \sum_n P(Z_0 = n)z^n$ it is evident by conditioning on Z_0 that the unconditional distribution of the total progeny is that determined by the generating function $f(h(z))$. This distribution is known as the *Lagrangian distribution* derived from the distribution of Z_0 and the offspring distribution. See [**19, 20, 49, 12, 65**] regarding Lagrangian distributions and their applications, and [**29, 30, 31, 27**] for the multivariate extension of Lagrange's expansion and its relation to multi-type branching processes.

5. Random walks, queues, and branching processes.

The following interpretation of the random walk $(S_n - n)$ and the branching process (Z_n) in terms of queuing theory is due to Kendall [**40**]. See also [**29, 67, 69, 71, 26**]. Suppose customers arrive and wait for service in a queue with a single server. The time the server is working consists of alternating busy and idle periods, each busy period consisting of one or more service periods, one per customer. Let X_j denote the number of customers arriving during the jth service period, so S_n represents the number of customers arriving by the end of the nth service period. Suppose at time zero there are $k \geq 1$ customers already present in the queue. For $0 \leq n \leq T_{-k}$ the number of customers in the queue just after the end of the end of the nth service period is $k + S_n - n$. So T_{-k} represents the number of customers served during the first busy period, and $P(T_{-k} = n)$ is the probability that the server has to deal with n customers before taking a break, given k customers in the queue to start with. The queuing process defines a random family $\mathcal{F}_k$, starting with k individuals, such that $\#\mathcal{F}_k = T_{-k}$. Each individual j in $\mathcal{F}_k$ represents a different customer, with j' the child of j if j' arrives during the service period of j. Assuming that the numbers X_j are independent with common distribution (p_i), the random family derived from the queue defines a Galton-Watson branching process (Z_n) with offspring distribution (p_i). This argument explains why the total progeny of the branching process (Z_n) started with k individuals has the same distribution as the first passage time T_{-k} of the random walk $(S_n - n)$. The argument can be made more precise by setting up an appropriate bijection between the following two sets: the set of walk paths starting at $(0,0)$ and first reaching $-k$ at time n by a sequence of integer increments with no increment less than -1, and the set of plane forests of k trees and n vertices. See [**66, 26, 7**] for details and further developments.

In the setting of Theorem 1.2, by definition

$$(T_{-k} = n) := (\forall_{m=1}^{n-1} S_m - m \neq -k, S_n - n = -k) \subseteq (S_n - n = -k) \tag{5.1}$$

where $S_n - n = Y_1 + \cdots + Y_n$ for $Y_j := X_j - 1$, derived from independent X_j with common distribution (p_i). So Kemperman's formula (1.5) can be restated as follows:

$$P(T_{-k} = n \,|\, S_n - n = -k) = \frac{k}{n}. \tag{5.2}$$

That is to say, given that the random walk $(S_m - m, m = 1, 2, \dots)$ is at $-k$ at time $m = n$, the chance that the walk first reached $-k$ at time n is k/n. The work of Takács [**67, 68**] shows that this form of Kemperman's result can be generalized as follows. See [**67**, Thm 1 of §4 and Thm 5 of §28] for two other essentially equivalent formulations related to the classical ballot theorem. The basic idea of considering cyclic shifts traces back to Dvoretsky and Motzkin [**17**]. See also [**24, 23, 18, 15**] for closely related results and further references.

For a sequence of integers $\mathbf{y} := (y_1, \dots, y_n)$ and $i \in [n]$ let $\mathbf{y}^{(i)}$ denote the ith *cyclic shift* of $\mathbf{y}$, that is the sequence whose jth term is y_{i+j} with addition modulo n. Call a set of sequences A *cyclically invariant* iff $\mathbf{y} \in A$ implies $\mathbf{y}^{(1)} \in A$, in which case $\mathbf{y}^{(i)} \in A$ for all $i \in [n]$. Call a sequence of random variables $\mathbf{Y}_n := (Y_1, \dots, Y_n)$ *cyclically exchangeable* if $(Y_2, \dots, Y_n, Y_1)$ has the same distribution as $(Y_1, \dots, Y_n)$. For integer valued Y_i this is equivalent to

$$P(\mathbf{Y}_n = \mathbf{y}) = P(\mathbf{Y}_n = \mathbf{y}^{(i)}) \tag{5.3}$$

for each sequence of integers $\mathbf{y}$ and all $i \in [n]$.

THEOREM 5.1 (Takacs [**67, 68**]). *Suppose that $\mathbf{Y}_n := (Y_1, \dots, Y_n)$ is a cyclically exchangeable sequence of random variables, let $S_m^- = Y_1 + \cdots + Y_m$, and let $(T_{-k} = n)$ be the event that the walk $(S_m^-, 1 \le m \le n)$ first reaches $-k$ at time n. Let $\mathbb{N}_- := \{-1, 0, 1, 2, \dots\}$. Then for every cyclically invariant subset A of $\mathbb{N}_-^n$,*

$$P(T_{-k} = n \,|\, S_n^- = -k, \mathbf{Y}_n \in A) = \frac{k}{n}. \tag{5.4}$$

PROOF. Every cyclically invariant A decomposes as the union of some collection of cyclic orbits, where for a sequence $\mathbf{y}$ the *cyclic orbit of* $\mathbf{y}$ is the set $A_{\mathbf{y}} := \{\mathbf{z} : \mathbf{z} = \mathbf{y}^{(i)} \text{ for some } i \in [n]\}$. So it suffices to prove (5.4) in the case $A = A_{\mathbf{y}}$ for an arbitrary $\mathbf{y} \in \mathbb{N}_-^n$. This special case is a consequence of following elementary lemma. ■

For a sequence $\mathbf{y}$, let $t_m = y_1 + \cdots + y_m$ and call the sequence of partial sums $(t_1, \dots, t_n)$ the *walk with steps* $\mathbf{y}$. Say the walk *first reaches b at time n* if $t_i \neq b$ for $i < n$ and $t_n = b$.

LEMMA 5.2. (Wendel [**74**, §3]) *Let $\mathbf{y} \in \mathbb{N}_-^n$ be such that $y_1 + \cdots + y_n = -k$ for some $1 \le k \le n$, and let $\mathbf{y}^{(i)}$ be the ith cyclic shift of $\mathbf{y}$. Then there are exactly k distinct $i \in [n]$ such that the walk with steps $\mathbf{y}^{(i)}$ first reaches $-k$ at time n.*

For different formulations of the lemma, which show how it generalizes the classical ballot theorem, see [**67**, Thm. 3 and Thm. 4 of §2]. While Theorem 5.1 was stated in a probabilistic way, the lemma reveals its combinatorial essence. Thus Theorem 5.1 can be formulated in purely combinatorial terms as follows:

COROLLARY 5.3. *Let A be a cyclically invariant subset of $\mathbb{N}_-^n$ such that*

$$y_1 + \cdots + y_n = -k \text{ for all } \mathbf{y} \in A.$$

Let A^* be the set of all $\mathbf{y} \in A$ such that the walk with steps $\mathbf{y}$ first reaches $-k$ at time n. Then the fraction of elements of A that are elements of A^* equals k/n.

To illustrate the corollary, let $I := \{a, a+1, \ldots, b\}$ be a finite set of consecutive integers with $a \leq -1$. Given a sequence $(n_i, i \in I)$ of non-negative integers with $\sum_i n_i = n$ and $\sum_i i n_i = -k$, consider the set A of all sequences $\mathbf{y} \in I^n$ of type (n_i), so

$$\#A = \binom{n}{n_a, \ldots, n_b}. \tag{5.5}$$

For $a \leq -2$ there is no simple formula for $\#A^*$, the number of sequences $\mathbf{y}$ of type (n_i) such that the walk with steps $\mathbf{y}$ first reaches $-k$ at time n. But in the special case $a = -1$, Corollary 5.3 implies

$$\#A^* = \frac{k}{n}\binom{n}{n_{-1}, \ldots, n_b}. \tag{5.6}$$

Implicit in the previous description of the branching process derived from a queuing process is a bijection between the set of walks with step sequence in A^*, as enumerated by (5.6), and the set of plane forests of k trees in which n_{i+1} vertices have i children, as enumerated by Theorem 1.4.

6. Uniform random plane forests

For $1 \leq k \leq n$ let $\mathbf{F}_{k,n}^{\text{plane}}$ denote the set of all plane forests of k trees with a total of n vertices, and let $\mathcal{F}_{k,n}^{\text{plane}}$ be a uniformly distributed random element of $\mathbf{F}_{k,n}^{\text{plane}}$. For $k = 1, 2, \ldots$ and $0 < p < 1$ let $\mathcal{G}_{k,p}$ be a Galton-Watson forest of k trees with the geometric(p) offspring distribution $p_i := p(1-p)^i$. The general product formula (4.1) gives

$$P(\mathcal{G}_{k,p} = \mathbf{f}) = \prod_{v \in V(\mathbf{f})} p(1-p)^{C(v,\mathbf{f})} = p^n (1-p)^{n-k} \text{ for } \mathbf{f} \in \mathbf{F}_{k,n}^{\text{plane}}. \tag{6.1}$$

Since this probability is the same for all $\mathbf{f} \in \mathbf{F}_{k,n}^{\text{plane}}$, the distribution of $\mathcal{G}_{k,p}$ given $(\#\mathcal{G}_{k,p} = n)$ is uniform on $\mathbf{F}_{k,n}^{\text{plane}}$, as observed in [**33, 53**]. Symbolically

$$\mathcal{F}_{k,n}^{\text{plane}} \stackrel{d}{=} (\mathcal{G}_{k,p} \,|\, \#\mathcal{G}_{k,p} = n) \tag{6.2}$$

where $\stackrel{d}{=}$ denotes equality of distributions. The Otter-Dwass formula implies

$$P(\#\mathcal{G}_{k,p} = n) = \frac{k}{n} P(S_{n,p} = n-k) = \frac{k}{n}\binom{2n-k-1}{n-k} p^n (1-p)^{n-k} \tag{6.3}$$

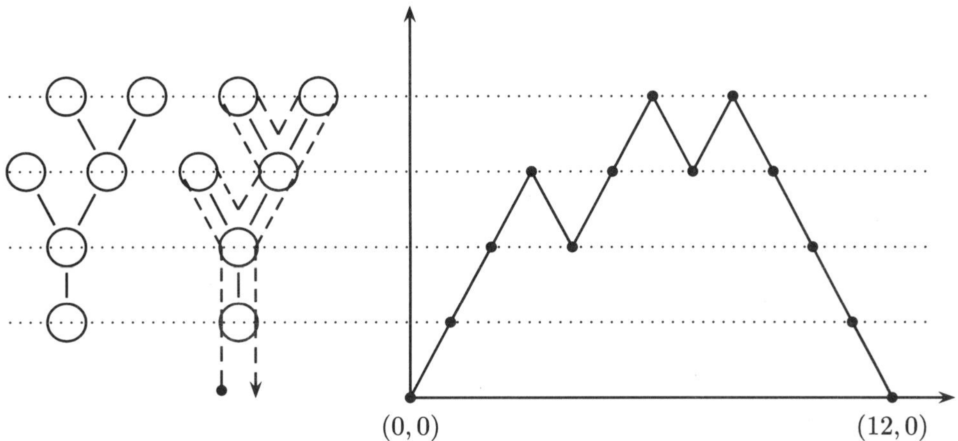

FIGURE 3

where $S_{n,p}$ is the sum of n independent geometric(p) random variables, and the second equality in (6.3) is read from the negative binomial formula [**24**, VI.8] for the distribution of $S_{n,p}$. Compare (6.1) and (6.3) to obtain

$$\#\mathbf{F}_{k,n}^{\text{plane}} = \frac{k}{n}\binom{2n-k-1}{n-k}. \tag{6.4}$$

This argument simplifies and corrects a similar derivation of Pavlov [**53**]. In particular the number of plane trees with n vertices is

$$\#\mathbf{F}_{1,n}^{\text{plane}} = \frac{1}{n}\binom{2n-2}{n-1}$$

which is the $(n-1)$th Catalan number [**9, 36**]. This is also the number of *lattice excursions of length* $2n$, that is sequences $(s_j, 0 \le j \le 2n)$ where $s_0 = s_{2n} = 0$, and $s_j > 0$ and $s_{j+1} - s_j \in \{-1, +1\}$ for all $0 \le j \le 2n-1$. As observed by Harris [**33**], there is a natural bijection between $\mathbf{F}_{1,n}^{\text{plane}}$ and the set of lattice excursions of length $2n$. See Figure 3.

This bijection extends to a bijection between $\mathbf{F}_{k,n}^{\text{plane}}$ and the set of non-negative lattice walk paths from $(0,0)$ to $(0,2n)$ with increments of ± 1 and exactly k returns to 0. By another bijection, the number of plane forests with n vertices equals the number of plane trees with $n+1$ vertices:

$$\# \bigcup_{k=1}^{n} \mathbf{F}_{k,n}^{\text{plane}} = \#\mathbf{F}_{1,n+1}^{\text{plane}}.$$

In view of (6.4) this yields the identity

$$\sum_{k=1}^{n} \frac{k}{n}\binom{2n-k-1}{n-k} = \frac{1}{n+1}\binom{2n}{n}. \tag{6.5}$$

The sum is the nth Catalan number, which is the the number of non-negative lattice walk paths from $(0,0)$ to $(0,2n)$ with increments of ± 1. The kth term of the sum is the number of such paths with k returns to zero.

Similarly, the number of ways to pick a plane forest with n vertices and assign each tree component a sign ± 1 equals the number of lattice paths from $(0,0)$ to $(0,2n)$ with increments of ± 1, that is

$$\sum_{k=1}^{n} \frac{k}{n} \binom{2n-k-1}{n-k} 2^k = \binom{2n}{n}. \tag{6.6}$$

In agreement with the result of Feller [**24**, Thm. 4 of §III.7], the kth term in the sum (6.6) divided by 2^{2n} is the probability that a simple symmetric random walk, started at 0 and moving with increments of ± 1, returns to 0 for the kth time after $2n$ steps.

The bijection between plane trees and lattice excursions implies that the large n asymptotic distribution of many functionals of a random tree $\mathcal{T}_n$ with uniform distribution on $\mathbf{F}_{1,n}^{\text{plane}}$ can be read from the asymptotic distribution of a functional of a uniformly distributed random lattice excursion of length $2n$, which is typically the distribution of a corresponding functional of a Brownian excursion [**2, 16**]. As shown by Aldous [**2, 3**], the same holds for any random plane tree $\mathcal{T}_n$ with $\mathcal{T}_n$ distributed like $\mathcal{T}$ given $\#\mathcal{T} = n$ where $\mathcal{T}$ is a Galton-Watson tree whose offspring distribution has mean 1 and finite variance. In particular, due to the result of the next section, this conclusion applies to $\mathcal{T}_n$ derived from a random tree with uniform distribution on the set $\mathbf{T}_n$ of all n^{n-1} rooted trees labeled by $[n]$. The general class of distributions for a planar tree of size n obtained by conditioning a Galton-Watson tree to be of size n is the class of distributions of "simply generated trees" studied by Meir and Moon [**45**]. See [**2, 3, 1, 16, 32, 58**] for further developments.

7. The plane forest derived from a uniform labeled forest

Recall from around (3.2) the map $\Psi : \mathbf{F}_{k,n} \to \mathbf{F}_{k,n}^{\text{plane}}$, where $\mathbf{F}_{k,n}$ is the set of forests with k tree components labeled by $[n]$, and $\mathbf{F}_{k,n}^{\text{plane}}$ is the set of plane forests with k tree components and n vertices. For a lighter notation, write $\mathbf{f}^\circ$ instead of $\Psi(\mathbf{f})$, and call $\mathbf{f}^\circ$ the *plane forest derived from* $\mathbf{f}$. So $\mathbf{f}^\circ$ is just $\mathbf{f}$ regarded as a plane forest by giving the set of roots of $\mathbf{f}$ and the sets of children of various vertices of $\mathbf{f}$ the order these sets acquire from the usual ordering of $[n]$. The following theorem strengthens connections discovered Kolchin [**42**] and Pavlov[**51, 52**] between the uniform distribution on $\mathbf{F}_{k,n}$ and the distribution of a Galton-Watson forest with the Poisson(μ) offspring distribution $p_i := e^{-\mu}\mu^i/i!$. Kolchin and Pavlov [**42, 43, 50, 51, 52, 55, 53, 54**] exploited these connections to derive the asymptotic distributions of functionals of a uniform random forest of k trees labeled by $[n]$, such as the numbers of trees of various sizes and the maximum tree size, as $n \to \infty$ for various ranges of k. The case $k = 1$ of the theorem is implicit in the discussion of Aldous [**2, 3**].

THEOREM 7.1. *For $\mu \in (0,\infty)$ let $\mathcal{P}_{k,\mu}$ be a Galton-Watson forest with the Poisson(μ) offspring distribution, for $1 \le k \le n$ let $\mathcal{F}_{k,n}$ have uniform distribution on the set of forests of k trees labeled by $[n]$, and let $\mathcal{F}_{k,n}^\circ$ be $\mathcal{F}_{k,n}$ regarded as a plane forest. Then $\mathcal{F}_{k,n}^\circ$ has the same distribution as $\mathcal{P}_{k,\mu}$ given $(\#\mathcal{P}_{k,\mu} = n)$:*

$$\mathcal{F}_{k,n}^\circ \stackrel{d}{=} (\mathcal{P}_{k,\mu} \,|\, \#\mathcal{P}_{k,\mu} = n). \tag{7.1}$$

PROOF. To be more explicit, there is the following formula. For all plane forests $\mathbf{f}$ of k trees with n vertices

$$P(\mathcal{F}^{\circ}_{k,n} = \mathbf{f}) = P(\mathcal{P}_{k,\mu} = \mathbf{f} \mid \#\mathcal{P}_{k,\mu} = n) = \frac{n(n-k)!}{kn^{n-k}} \prod_{v \in V(\mathbf{f})} \frac{1}{C(v,\mathbf{f})!}. \tag{7.2}$$

The first probability in (7.2) is the number displayed in (3.2) divided by $\#\mathbf{F}_{k,n}$ in (1.17), which reduces to the last expression in (7.2) by cancellation. The second probability reduces similarly, by application of (4.1) and the consequence of the Otter-Dwass formula (1.4) that the total progeny in a Poisson-Galton-Watson family forest of k trees has the distribution

$$P(\#\mathcal{P}_{k,\mu} = n) = \frac{k}{n} \frac{(\mu n)^{n-k}}{(n-k)!} e^{-\mu n} \text{ for } n = k, k+1, \dots \tag{7.3}$$

which is known as the *Borel-Tanner distribution* [**8, 48, 72, 73**]. ∎

Call a function Φ of forests $\mathbf{f}$ an *invariant* if $\Phi(\mathbf{f}) = \Phi(\mathbf{f}')$ whenever $\mathbf{f}'$ is a relabeling of $\mathbf{f}$, meaning $v \xrightarrow{\mathbf{f}'} w$ iff $\ell(v) \xrightarrow{\mathbf{f}} \ell(w)$ for some bijection ℓ from the vertices of $\mathbf{f}$ to the vertices of $\mathbf{f}'$. For example, the number $Z_h\mathbf{f}$ of vertices of $\mathbf{f}$ at height h is an invariant. So is the matrix $M(\mathbf{f}) := (M_{h,c}(\mathbf{f}), h \geq 0, c \geq 0)$ where $M_{h,c}(\mathbf{f})$ is the number of individuals in generation h of $\mathbf{f}$ that have c children. Since the plane forest $\mathcal{F}^{\circ}_{k,n}$ is by definition a relabeling of the uniform labeled forest $\mathcal{F}_{k,n}$, the identity (7.1) implies

$$\Phi(\mathcal{F}_{k,n}) \stackrel{d}{=} (\Phi(\mathcal{P}_{k,\mu}) \mid \#\mathcal{P}_{k,\mu} = n) \text{ for all invariant } \Phi. \tag{7.4}$$

For $\Phi = M$ this result is due to Kolchin [**42**] and Pavlov [**52**]. The identity (7.4) is expressed more intuitively by the following construction, suggested by Aldous [**2**] for $k = 1$. Fix $\mu > 0$ and generate a Poisson-Galton-Watson family forest $\mathcal{P}_{k,\mu}$ starting from k root individuals. Given that $\mathcal{P}_{k,\mu}$ has vertex set V with $\#V = n$, let $\mathcal{P}^*_{k,\mu} \in \mathbf{F}_{k,n}$ be $\mathcal{P}_{k,\mu}$ relabeled by a uniform random permutation $\sigma : V \to [n]$. Then

$$\mathcal{F}_{k,n} \stackrel{d}{=} (\mathcal{P}^*_{k,\mu} \mid \#\mathcal{P}_{k,\mu} = n). \tag{7.5}$$

That is, given that $\mathcal{P}_{k,\mu}$ has n vertices, a random relabeling of $\mathcal{P}_{k,\mu}$ has uniform distribution over the set of all forests of k trees labeled by $[n]$. For some recent applications of this relation between uniform random trees and Poisson-Galton-Watson trees see [**56, 63, 4**].

Acknowledgments

Thanks to Vlada Limic for preparation of the graphics, to Richard Stanley for providing a draft of [**66**], to Anthony Pakes for help with the literature, to Steven Evans for comments on a preliminary draft, and to an anonymous referee for helpful suggestions.

References

[1] D. Aldous and J. Pitman. Brownian bridge asymptotics for random mappings. *Random Structures and Algorithms*, 5:487–512, 1994.

[2] D.J. Aldous. The continuum random tree I. *Ann. Probab.*, 19:1–28, 1991.

[3] D.J. Aldous. The continuum random tree II: an overview. In M.T. Barlow and N.H. Bingham, editors, *Stochastic Analysis*, pages 23–70. Cambridge University Press, 1991.

[4] D.J. Aldous and J. Pitman. Tree-valued Markov chains derived from Galton-Watson processes. Technical Report 481, Dept. Statistics, U.C. Berkeley, 1997.

[5] S. Anoulova, J. Bennies, J. Lenhard, D. Metzler, Y. Sung, and A. Weber. Six ways of looking at Burtin's lemma. Preprint, 1998.

[6] K.B. Athreya and P. Ney. *Branching Processes*. Springer, 1972.

[7] J. Bennies and G. Kersting. A random walk approach to Galton-Watson trees. Preprint, 1996.

[8] E. Borel. Sur l'emploi du théorème de Bernoulli pour faciliter le calcul d'un infinité de coefficients. Application au probleme de l'attente á un guichet. *C.R. Acad. Sci. Paris*, 214:452–456, 1942.

[9] W.G. Brown. Historical note on a recurrent combinatorial problem. *Amer. Math. Monthly*, 72:973–977, 1965.

[10] Y. D. Burtin. On a simple formula for random mappings and its applications. *J. Appl. Probab.*, 17:403 - 414, 1980.

[11] A. Cayley. A theorem on trees. *Quarterly Journal of Pure and Applied Mathematics*, 23:376–378, 1889. (Also in *The Collected Mathematical Papers of Arthur Cayley. Vol XIII*, 26-28, Cambridge University Press, 1897).

[12] P.C. Consul. *Generalized Poisson Distributions*. Dekker, 1989.

[13] R. Cori. Words and Trees. In M. Lothaire, editor, *Combinatorics on Words*, volume 17 of *Encyclopedia of Mathematics and its Applications*, pages 215–229. Addison-Wesley, Reading, Mass., 1983.

[14] L. de Lagrange. Nouvelle méthode pour résoudre des équations littérales par le moyen des séries. *Mém. Acad. Roy. Sci. Belles-Lettres de Berlin*, 24, 1770.

[15] N. Dershowitz and S. Zaks. The cycle lemma and its applications. *Europ. J. Combinatorics*, 11:35–40, 1990.

[16] R. Durrett, H. Kesten, and E. Waymire. On weighted heights of random trees. *J. Theoret. Probab.*, 4:223–237, 1991.

[17] A. Dvoretsky and Th. Motzkin. A problem of arrangements. *Duke Math. J.*, 14:305–313, 1947.

[18] M. Dwass. A fluctuation theorem for cyclic random variables. *Ann. Math. Stat.*, 33:1450–1453, 1962.

[19] M. Dwass. A theorem about infinitely divisible distributions. *Z. Wahrsch. Verw. Gebiete*, 9:287–289, 1967.

[20] M. Dwass. The total progeny in a branching process. *J. Appl. Probab.*, 6:682–686, 1969.

[21] A. Erdélyi and I.M.H. Etherington. Some problems of non-associative combinations (2). *Edinburgh Math. Notes*, 32:7–12, 1940.

[22] I.M.H. Etherington. Some problems of non-associative combinations (1). *Edinburgh Math. Notes*, 32:1–6, 1940.

[23] W. Feller. *An Introduction to Probability Theory and its Applications, Vol 2.* Wiley, 1966.

[24] W. Feller. *An Introduction to Probability Theory and its Applications,* Vol 1,3rd ed. Wiley, New York, 1968.

[25] J. Françon. Preuves combinatoires des identités d'Abel. *Discrete Mathematics*, 8:331–343, 1974.

[26] J-F. Le Gall and Y. Le Jan. Branching processes in Lévy processes: the exploration process. To appear in *Ann. Probab.*, 1997.

[27] I. Gessel. A combinatorial proof of the multivariable Lagrange inversion formula. *J. Combinatorial Theory A*, 45:178–195, 1987.

[28] I.J. Good. The number of individuals in a cascade process. *Proc. Camb. Phil. Soc.*, 45:360–363, 1949.

[29] I.J. Good. Generalizations in several variables of Lagrange's expansion, with applications to stochastic processes. *Proc. Camb. Phil. Soc.*, 56:366–380, 1963.

[30] I.J. Good. The generalization of Lagrange's expansion and the enumeration of trees. *Proc. Camb. Phil. Soc.*, 61:499–517, 1965.

[31] I.J. Good. The Lagrange distributions and branching processes. *SIAM Journal on Applied Mathematics*, 28:270–275, 1975.

[32] W. Gutjahr. Expectation transfer between branching processes and random trees. *Random Structures and Algorithms*, 4:447–467, 1993.

[33] T. E. Harris. First passage and recurrence distributions. *Trans. Amer. Math. Soc.*, 73:471–486, 1952.

[34] T.E. Harris. *The Theory of Branching Processes.* Springer-Verlag, New York, 1963.

[35] D. Hawkins and S.M. Ulam. Theory of Multiplicative Processes, 1. Technical Report LA-171, Los Alamos Scientific Laboratory, 1944. (reprinted in *Analogies between analogies: the mathematical reports of S.M. Ulam and his collaborators*, A.R. Bednarek and F. Ulam editors, University of California Press, Berkeley(1990)).

[36] P. Hilton and J. Pedersen. Catalan numbers, their generalization, and their uses. *Math. Intelligencer*, 13:64–75, 1991.

[37] P. Jagers. *Branching Processes with Biological Applications.* Wiley, 1975.

[38] J.H.B. Kemperman. The general one-dimensional random walk with absorbing barriers. Thesis, Excelsior, The Hague, 1950.

[39] J.H.B. Kemperman. *The Passage Problem for a Stationary Markov Chain.* University of Chicago Press, 1961.

[40] D.G. Kendall. Some problems in the theory of queues. *J.R.S.S. B*, 13:151–185, 1951.

[41] H. Kesten. Subdiffusive behavior of random walk on a random cluster. *Ann. Inst. H. Poincaré Probab. Statist.*, 22:425–487, 1987.

[42] V.F. Kolchin. Branching processes, random trees, and a generalized scheme of arrangements of particles. *Mathematical Notes of the Acad. Sci. USSR*, 21:386–394, 1977.

[43] V.F. Kolchin. *Random Mappings.* Optimization Software, New York, 1986. (Translation of Russian original).

[44] G. Labelle. Une nouvelle démonstration combinatoire des formules d'inversion de Lagrange. *Adv. in Math.*, 42:217–247, 1981.

[45] A. Meir and J.W. Moon. On the altitude of nodes in random trees. *Canad. J. Math.*, 30:997–1015, 1978.

[46] J.W. Moon. Various proofs of Cayley's formula for counting trees. In F. Harary, editor, *A Seminar on Graph Theory*, pages 70–78. Holt, Rineharte and Winston, New York, 1967.

[47] J. Neveu. Arbres et processus de Galton-Watson. *Ann. Inst. H. Poincaré Probab. Statist.*, 22:199 - 207, 1986.

[48] R. Otter. The multiplicative process. *Ann. Math. Statist.*, 20:206–224, 1949.

[49] A. G. Pakes and T. P. Speed. Lagrange distributions and their limit theorems. *SIAM Journal on Applied Mathematics*, 32:745–754, 1977.

[50] Yu. L. Pavlov. Limit theorems for the number of trees of a given size in a random forest. *Math. USSR Subornik*, 32:335–345, 1977.

[51] Yu. L. Pavlov. The asymptotic distribution of maximum tree size in a random forest. *Theory of Probability and its Applications*, 22:509–520, 1977.

[52] Yu. L. Pavlov. Limit distributions of the height of a random forest. *Theory of Probability and its Applications*, 28:471 - 480, 1983.

[53] Yu. L. Pavlov. Limit distributions of the height of a random forest of plane rooted trees. *Discrete Math. Appl.*, 4:73–88, 1994.

[54] Yu. L. Pavlov. Limit distribution of the number of trees of a given size in a random forest. *Discrete Math. Appl.*, 6:117–133, 1996.

[55] Yu.L. Pavlov. A case of the limit distribution of the maximal volume on a tree in a random forest. *Mathematical Notes of the Acad. Sci. USSR*, 25:387–392, 1979.

[56] J. Pitman. Coalescent random forests. Technical Report 457, Dept. Statistics, U.C. Berkeley, 1996. Available via http://www.stat.berkeley.edu/users/pitman.

[57] J. Pitman. Abel-Cayley-Hurwitz multinomial expansions associated with random mappings, forests and subsets. Technical Report 498, Dept. Statistics, U.C. Berkeley, 1997. Available via http://www.stat.berkeley.edu/users/pitman.

[58] J. Pitman. The SDE solved by local times of a Brownian excursion or bridge derived from the height profile of a random tree or forest. Technical Report 498, Dept. Statistics, U.C. Berkeley, 1997. Available via http://www.stat.berkeley.edu/users/pitman.

[59] H. Prüfer. Neuer Beweis eines Satzes über Permutationen. *Archiv für Mathematik und Physik*, 27:142–144, 1918.

[60] G.N Raney. Functional composition patterns and power series reversion. *Trans. Amer. Math. Soc.*, 94:441–451, 1960.

[61] G.N. Raney. A formal solution of $\sum_{i=1}^{\infty} a_i e^{B_i X} = x$. *Canad. J. Math.*, 16:755–762, 1964.

[62] A. Rényi. On the enumeration of trees. In R. Guy, H. Hanani, N. Sauer, and J. Schonheim, editors, *Combinatorial Structures and their Applications*, pages 355–360. Gordon and Breach, New York, 1970.

[63] R.K. Sheth and J. Pitman. Coagulation and branching process models of gravitational clustering. *Mon. Not. R. Astron. Soc.*, 289:66–80, 1997.

[64] P.W. Shor. A new proof of Cayley's formula for counting labelled trees. *J. Combinatorial Theory A.*, 71:154–158, 1995.

[65] M. Sibuya, N. Miyawaki, and U. Sumita. Aspects of Lagrangian probability distributions. *Studies in Applied Probability. Essays in Honour of Lajos Takács (J. Appl. Probab.)*, 31A:185–197, 1994.

[66] R. Stanley. Enumerative combinatorics, vol. 2. Book in preparation, to be published by Cambridge University Press, 1996.

[67] L. Takács. A generalization of the ballot problem and its application to the theory of queues. *J. Amer. Stat. Assoc.*, 57:154–158, 1962.

[68] L. Takács. *Combinatorial Methods in the Theory of Stochastic Processes.* Robert E. Kreiger Publ. Co., Huntington, New York, 1977.

[69] L. Takács. Ballots, queues and random graphs. *J. Appl. Probab.*, 26:103–112, 1989.

[70] L. Takács. Counting forests. *Discrete Mathematics*, 84:323–326, 1990.

[71] L. Takács. Limit distributions for queues and random rooted trees. *J. Applied Mathematics and Stochastic Analysis*, 6:189–216, 1993.

[72] J.C. Tanner. A problem of interference between two queues. *Biometrika*, 40:58–69, 1953.

[73] J.C. Tanner. A derivation of the Borel distribution. *Biometrika*, 1961:222–224, 1961.

[74] J.G. Wendel. Left continuous random walk and the Lagrange expansion. *Amer. Math. Monthly*, 82:494–498, 1975.

Department of Statistics, University of California, 367 Evans Hall # 3860, Berkeley, CA 94720-3860

DIMACS Series in Discrete Mathematics
and Theoretical Computer Science
Volume 41, 1998

Coupling from the Past: a User's Guide

James Propp and David Wilson

ABSTRACT. The Markov chain Monte Carlo method is a general technique for obtaining samples from a probability distribution. In earlier work, we showed that for many applications one can modify the Markov chain Monte Carlo method so as to remove all bias in the output resulting from the biased choice of an initial state for the chain; we have called this method Coupling From The Past (CFTP). Here we describe this method in a fashion that should make our ideas accessible to researchers from diverse areas. Our expository strategy is to avoid proofs and focus on sample applications.

1. Introduction

In Markov chain Monte Carlo studies, one attempts to sample from a distribution π by running a Markov chain whose unique steady-state distribution is π. Ideally, one has proved a theorem that guarantees that the time for which one plans to run the chain is substantially greater than the mixing time of the chain, so that the distribution $\tilde{\pi}$ that one's procedure actually samples from is known to be close to the desired π in variation distance. More often, one merely hopes that this is the case, and the possibility that one's samples are contaminated with substantial initialization bias cannot be ruled out with complete confidence.

The "coupling from the past" procedure introduced in [**PW1**] provides one way of getting around this problem. Where it is applicable, this method delivers samples that are governed by π itself, rather than $\tilde{\pi}$. In the past two years, many researchers have found ways to apply the basic idea in a wide variety of settings. To paraphrase Wilfrid Kendall, one of the first to jump into the fray: "We must now be more ambitious about our simulation objectives. We can no longer be content merely with long-run approximations to equilibrium distributions but instead must strive after perfection" [**K**].

Coupling from the past (hereafter CFTP) is based on what Kendall calls *stochastic flows* (in discrete or continuous time). We restrict attention for now to the discrete-time, finite-state version. Given a finite state-space $\mathbf{X}$, a discrete-time stochastic flow on $\mathbf{X}$ is a random process indexed by discrete time taking

---

1991 *Mathematics Subject Classification.* Primary 11K45.

The first author was supported by NSA grant MDA904-92-H-3060, NSF grant DMS92-06374, and a career development grant from the M.I.T. Class of 1922.

The second author was supported by an NSF Postdoctoral Fellowship.

its values in the set of maps from $\mathbf{X}$ to itself. If one lets f_i denote the (random) map at time i, then the stochastic flow determines "streamlines" of the form $(x, f_i(x), f_{i+1}(f_i(x)), \dots)$. Just as the Markov chain Monte Carlo method is predicated on the availability of rapidly-mixing Markov chains, CFTP is predicated on the availability of rapidly-coalescent stochastic flows, in which there is a strong tendency for the streamlines to converge. (Note that once two streamlines have converged, they coincide forever after.)

It is worth stressing at the outset that CFTP is especially valuable as an alternative to standard Markov chain Monte Carlo when one is working with stochastic flows for which one suspects, but has not proved, that rapid coalescence occurs. In such cases, the availability of CFTP makes it less urgent that theoreticians obtain bounds on time-to-coalescence, since CFTP (unlike Markov chain Monte Carlo) cleanly separates the issue of efficiency from the issue of quality of output. That is to say, one's samples are guaranteed to be uncontaminated by initialization bias, regardless of how quickly or slowly they are generated. (See section 7 for a discussion of how to avoid subtle sources of error arising from the variability of the running time of the procedure.)

Even if one has rigorously proved that rapid mixing occurs (as Randall et al. have done [**LRS**] and [**MR**] for the particular application for which CFTP was originally invented), CFTP may be preferable to the usual Markov chain Monte Carlo strategy of running the system until it is "well-mixed" (e.g., running the system until the variation distance between $\tilde{\pi}$ and π is less than 10^{-6}). For one thing, the bounds that have been obtained may be unduly pessimistic. Even if the bounds are tight, CFTP may be the preferred option. Consider, for instance, a situation in which a rigorous upper bound on mixing time has been obtained via coupling methods. On the one hand, the standard Monte Carlo approach would require running the Markov chain for some multiple of the estimated mixing time, and this multiple grows without bound as the desired bound on variation distance shrinks. On the other hand, the fact that the bound on mixing time was proved using coupling techniques is likely to be symptomatic of the existence of a natural way of realizing the Markov chain as part of a stochastic flow whose time-to-coalescence is equal (or comparable) to the mixing time of the chain; if this is the case, one can reduce the initialization bias to zero by running the Markov chain under CFTP for a (random) duration whose expected value is bounded by a fixed multiple of the mixing time.

As an historical aside, we mention that the conceptual ingredients of CFTP were in the air even before the versatility of the method was made clear in [**PW1**]. Precursors include Letac [**Le**], Thorisson [**T**], and Borovkov and Foss [**BF**]. Even back in the 1970's, one can find foreshadowings in the work of Ted Harris (on the contact process, the exclusion model, random stirrings, and coalescing and annihilating random walks), David Griffeath (on additive and cancellative interacting particle systems), and Richard Arratia (on coalescing Brownian motion). One can even see traces of the idea in the work of Loynes [**Lo**] thirty-five years ago.

2. Coupling from the past

Computationally, one needs three things in order to be able to implement the CFTP strategy: a way of generating (and representing) certain maps from $\mathbf{X}$ to itself; a way of composing these maps (so as to be able to simulate the flow for many time-steps); and a way of ascertaining whether total coalescence has occurred, or equivalently, a way of ascertaining whether a certain composite map (obtained by composing many random f's) collapses all of $\mathbf{X}$ to a single element.

The first component is what we call the random map procedure; we model it as an oracle that on successive calls returns independent, identically distributed functions f from $\mathbf{X}$ to $\mathbf{X}$, governed by some selected probability distribution $\mathbf{P}$ (typically supported on a very small subset of the set of all maps from $\mathbf{X}$ to itself). We use the oracle to choose independent, identically distributed maps f_{-1}, f_{-2}, f_{-3}, ..., f_{-N}, where how far into the past we have to go (N steps) is determined during run-time itself. The defining property that N must have is that the composite map

$$F^0_{-N} \stackrel{\text{def}}{=} f_{-1} \circ f_{-2} \circ f_{-3} \circ \cdots \circ f_{-N}$$

must be collapsing. Finding such an N thus requires that we have both a way of composing f's and a way of testing when such a composition is collapsing. (Having the test enables one to find such an N, since one can iteratively test ever-larger values of N, say by successive doubling, until one finds an N that works. Such an N will be a random variable that is measurable with respect to $f_{-N}, f_{-N+1}, \ldots, f_{-1}$.)

Once a suitable N has been found, the algorithm outputs $F^0_{-N}(x)$ for any $x \in \mathbf{X}$ (the result will not depend on x, since F^0_{-N} is collapsing). We call this output the CFTP sample. It must be stressed that when one is attempting to determine a usable N by guessing successively larger values and testing them in turn, one must use the *same* respective maps f_i during each test. That is, if we have just tried starting the chain from time $-N_1$ and failed to achieve coalescence, then, as we proceed to try starting the chain from time $-N_2 < -N_1$, we must use the same maps f_{-N_1}, f_{-N_1+1}, ..., f_{-1} as in the preceding attempt. Failure to abide by this principle "voids the warranty" of our algorithm.

We showed in [**PW1**] that, as long as the nature of $\mathbf{P}$ guarantees (almost sure) eventual coalescence, and as long as $\mathbf{P}$ bears a suitable relationship to the distribution π, the CFTP sample will be distributed according to π. Specifically, it is required that $\mathbf{P}$ preserve π in the sense that if a random state x is chosen in accordance with π and a random map f is chosen in accordance with $\mathbf{P}$, then the state $f(x)$ will be distributed in accordance with π.

We will not repeat here the proof, whose simplicity makes it seem rather amazing that the idea was not exploited sooner. Our goal is to show that designing good stochastic flows for CFTP is not much harder than designing good Markov chains for Monte Carlo, and to argue that the Monte Carlo community should be looking for opportunities to apply these ideas. We will give several applications of the CFTP philosophy, and conclude with some words of warning about the proper use of CFTP algorithms.

The article [**PW1**] gives many examples of situations in which CFTP works by virtue of monotonicity. In particular, there is a "top state" $\hat{1}$ and a "bottom state" $\hat{0}$, and the random maps f that have positive probability under $\mathbf{P}$ have the property that a composite map F is collapsing if and only if $F(\hat{0}) = F(\hat{1})$. There are many situations in which this is the case for non-obvious reasons, e.g., the hard-core model on a bipartite graph [**KSW**]. However, in this article we will focus on situations in which this simple state of affairs does not prevail.

For an interesting attempt to embed CFTP in a general probabilistic framework, see the article by Foss and Tweedie [**FT**]. For a discussion of the many uses of the notion of backwards composition of random maps, see the forthcoming survey article by Diaconis [**D**]. For an on-line bibliography on perfect random sampling using Markov chains, see `http://dimacs.rutgers.edu/~dbwilson/exact.html`.

3. The hard-core model

The states of this model are given by subsets of the vertex-set of a finite graph G, or equivalently, by $0,1$-valued functions on the vertex-set. We think of 1 and 0 as respectively denoting the presence or absence of a particle. In a legal state, no two adjacent vertices may both be occupied by particles. The probability of a particular legal state is proportional to λ^m, where m is the number of particles (which depends on the choice of state) and λ is some fixed parameter-value. We denote this probability distribution by π. That is, $\pi(S) = \lambda^{|S|}/Z$ where S is a state, $|S|$ is the number of particles in that state, and $Z = \sum_S \lambda^{|S|}$.

Luby and Vigoda [**LV**] provide a simple Markov chain Monte Carlo procedure for randomizing an initial hard-core state. The random moves they consider are determined by a pair of adjacent vertices u, v and a pair of numbers i, j with (i, j) equal to $(0,0)$, $(0,1)$, or $(1,0)$. They assume that the pair u, v is chosen uniformly from the set of pairs of adjacent vertices in G, and that (i, j) is $(0,0)$ with probability $\frac{1}{1+2\lambda}$, $(0,1)$ with probability $\frac{\lambda}{1+2\lambda}$, and $(1,0)$ with probability $\frac{\lambda}{1+2\lambda}$. Once such a quadruple u, v, i, j is chosen, the algorithm proposes to put a vacancy (resp. particle) at vertex u if i is 0 (resp. 1), and similarly for v and j; if the proposed move would lead to an illegal state, it is rejected, otherwise it is accepted. It is not hard to show that this randomization procedure has π as its unique steady-state distribution.

Luby and Vigoda show that as long as $\lambda \leq \frac{1}{\Delta-3}$, where $\Delta \geq 4$ is the maximum degree of G, this Markov chain is rapidly mixing. They do this by using a coupling argument: two initially distinct states, evolved in tandem, tend to coalesce over time. That is, the authors implicitly embed the Markov chain in a stochastic flow. As such, the method cries out to be turned into a perfect sampling scheme via CFTP.

This is easy to do. One can associate with each *set* of hard-core states a three-valued function on the vertex-set, where the value "1" means that all states in the set are known to have a particle at that vertex, the value "0" means that all states in the set are known to have a vacancy at that vertex, and the value "?" means that it is possible that some of the states in the set have a particle there while others have a vacancy. We can operate directly on this three-valued state-model by means of simple rules that mimic the Luby-Vigoda algorithm on the original two-valued model.

More specifically, we start with a three-valued configuration in which the adjacencies 0–0, 0–?, and ?–? are permitted but in which a 1 can only be adjacent to 0's. Proposals are still of the form $(0,0)$, $(0,1)$, $(1,0)$, and they still have respective probabilities $\frac{1}{2\lambda+1}$, $\frac{\lambda}{2\lambda+1}$, and $\frac{\lambda}{2\lambda+1}$, but proposals are implemented differently. When it is proposed to put 0's at u and v, the proposal is always accepted. When it is proposed to put 0 at u and 1 at v, there are three cases. If all the vertices adjacent to v (other than u) have a 0, the proposal is accepted. If any vertex adjacent to v (other than u) has a 1, the proposal is simply rejected and nothing happens. However, if vertex v has a neighbor (other than u) that has a ? but no neighbor (other than u) that has a 1, then v gets marked with ? and u also gets marked with ? (unless u was already 0, in which case the marking of u does not change). When it is proposed to put 1 at u and 0 at v, the same procedure is followed, but with the roles of u and v reversed.

In short, we can take the work of Luby and Vigoda and, without adding any new ideas, check that their way of coupling two copies of the Luby-Vigoda Markov chain extends to a stochastic flow on the whole state-space. Moreover, this flow can be simulated in such a way that coalescence is easily detected: it is not hard to show that if the 0,1,? Markov chain, starting from the all-?'s state, ever reaches a state in which there are no ?'s, then the Luby-Vigoda chain, using the same random proposals, maps all initial states into the same final state. Hence we might want to call the 0,1,? Markov chain the "certification chain", for it tells us when the stochastic flow of primary interest has achieved coalescence.

One might fear that it would take exponentially long for the certification chain to certify coalescence, but the proof that Luby and Vigoda give carries over straightforwardly to the three-valued setting, and shows that the number of ?'s tends to shrink to zero in polynomial time (relative to the size of the system).

It should be mentioned that Häggström and Nelander [**HN**] have another approach to the hard-core model, in which only one vertex at a time needs to be modified, and that their approach makes use of the very interesting notion of antimonotone CFTP which has its roots in the work of Kendall on repulsive point-processes. Work of Randall and Tetali [**RT**], in conjunction with the Luby-Vigoda result, implies that the single-site heat-bath Markov chain studied by Häggström and Nelander is also rapidly mixing for $\lambda \leq \frac{1}{\Delta-3}$.

4. Point-processes with attractive or repulsive interaction

The states of this model are given by discrete subsets of a finite or infinite region of a Euclidean space; for simplicity, we will focus on finite regions. Typically, such processes are defined via a Radon-Nikodym density with respect to the unit-rate Poisson process on the region, and the density rewards or penalizes configurations in which points are close together. (One typical way of measuring clustering is to take the volume of a union of balls of fixed radius, centered on the points in the discrete set; the more the points cluster together, the smaller this volume will be.)

This might seem like an unpromising area for CFTP, since the state space is uncountable. However, there is a kind of monotonicity in the model, and Kendall [**K**] figured out a way to exploit this. (Strictly speaking, the system is monotonic in the attractive case and anti-monotonic in the repulsive case; for now, we focus on the attractive case.)

Kendall effectively constructs a continuous-time stochastic flow. This flow can be modeled as a censored version of a birth/death process whose state at any instant is a point process, and for which the behavior in time is that points get added to the current state, remain alive for an exponentially-distributed random time, and then die. Births are proposed in accordance with Poisson distribution in space-time, but not all proposed births are accepted; proposals are accepted/rejected randomly, with the probability of acceptance being higher when there are one or more "alive" points nearby.

To apply CFTP, one picks an integer N and evolves the system from time $-N$ to time 0, using two different starting states. In one of these states, no particles are present at time $-N$; the other state is generated by an interaction-free birth/death process (call it P) conceived of as running from time $-\infty$ to time 0 with censorship turned off up until time $-N$. The two initial states are evolved in tandem, in accordance with the rules for accepting or rejecting births, using the same random proposals for birth and death events for both cases. If the two simulations agree at time 0, then coalescence occurs over the time-interval $[-N, 0]$ and the state at time 0 can be delivered as the unbiased CFTP sample. If the two simulations disagree, then a larger window $[-N', 0]$ must be tried, using the same realization of the Poisson birth/death process P (this time via its state at time $-N'$) and using the same birth/death proposals on the interval $[-N, 0]$ as before. In practice the realization of the Poisson birth/death process on the window $[-N', 0]$ is built from the one on the window $[-N, 0]$ by simulating the process starting from the state at time $-N$ back into the past until time $-N'$.

Note that in effect one is doing monotone CFTP; the interesting wrinkle is that the "top" state is itself a random variable. It is worth pointing out that in the anti-monotone, repulsive case, a clever variant works, in which one makes a decision about a birth for the *top* simulation in accordance with the environment that the proposed birth faces in the *bottom* simulation, and vice versa. This approach of Kendall was further developed by Häggström and Nelander [**HN**]. Also, Häggström, Van Lieshout, and Møller [**HLM**] have proposed an algorithm for the special case of penetrable spheres that is simpler than Kendall's and works faster, though it is not as amenable to generalization.

It should also be mentioned that CFTP may have theoretical as well as practical significance for such models. For, it gives a way of coupling together different point-processes (by having them hitch a ride on the same Poisson realization). Indeed, what Kendall actually does is associate with each proposed birth a random number chosen uniformly in $[0, 1]$, and the proposal is accepted if and only if the random number falls between 0 and a configuration-specific probability threshold. To get an informative coupling between two models, one should use the same numbers in $[0, 1]$ (called "marks" by Kendall). Such joint realizations of two processes can allow one to prove comparative assertions that might otherwise be quite difficult to establish.

Finally, we mention that the article of Murdoch and Greene [**MG**] has other interesting applications of CFTP to continuous state-spaces.

5. Spanning arborescences with specified root

Let G be a strongly connected directed finite graph with positive weights associated with its arcs. (These assumptions are convenient pedagogically, but they are more stringent than the best theorems require.) A (spanning) arborescence in G with root r is an acyclic directed subgraph of G in which r has out-degree 0 and every other vertex has out-degree 1. We define the weight of an arborescence as the product of the weights of its constituent arcs. This determines a unique probability distribution π on the set of arborescences for which the probability of each arborescence is proportional to its weight. Propp and Wilson [**PW2**] consider the problem of sampling from this distribution π.

A much-studied Markov chain for arborescences can be defined in terms of biased random walk on G, where the transition-probabilities from a vertex v to each of its neighbors are proportional to the weights of the respective arcs connecting v to those vertices. If the current arborescence is rooted at r, one can get a new arborescence by adding an arc from r to s (where s is a random successor of r

under the random walk on G) and deleting the arc that used to emanate from s. It is well known that π is the unique stationary probability measure for this Markov chain on the set of arborescences of G. (This bit of folklore was first published by Anantharam and Tsoucas [**AT**].)

If two arborescences have the same root, there is an obvious way to simulate their future jointly, by having the root take the same random walk in G. This coupling has the property that the two arborescences will continue to have the same root. Moreover, this stochastic flow is rapidly coalescent if the cover-time of the random walk on G is not too large. To see this, note that at any instant, the unique arc emanating from any vertex v other than the current root is the arc that the random walk took on its most recent departure from v. Hence when the random walk has visited all vertices, the two arborescences must have coalesced.

To cash in on this, we define a stochastic flow whose states are arborescences of G rooted at a fixed vertex r. To advance this flow, we perform a random walk on G that starts at r and walks until it next encounters r; we call the walk a random excursion from r to r. While this excursion is going on, we repeatedly update each of the arborescences in the fashion already described, in accordance with the latest step of the random walk. If the excursion happens to cover G, then all the updated arborescences will be identical; this is, the random map associated with this excursion (a map from the set of arborescences rooted at r to itself) is a collapsing map. More generally, if during a succession of excursions the root visits every vertex of G, the corresponding composition of random maps is collapsing.

There is a simple way of describing the effect that an excursion E has on any spanning arborescence A rooted at r. The effect of the excursion can be summarized by an array f indexed by the vertices of the graph. We write $f[v]$ when we think of f as an array and wish to refer to the entry indexed by vertex v, and we write $f(A)$ when we think of f as the "tree-map" associated with the excursion E and wish to refer to the arborescence A' that arises from A as a result of the excursion E. For all vertices v other than r that were visited during excursion E, let $f[v]$ be the vertex of G that was visited immediately following the last visit to v made during the excursion; for all other vertices v, let $f[v]$ be undefined. If $f[v]$ is undefined (and v is not the root), then the excursion never visited v, so the arc out of v in A' is the same as the arc out of v in A. If $f[v] = w$, then the last time the excursion visited v, the next vertex was w, so the arc out of v in A' leads to w. So the array f does indeed contain all the information about the tree-map defined by the excursion from the root to itself. The tree-map defined by any finite sequence of excursions is similarly representable by an array indexed by the vertices of the graph.

To obtain a procedure for randomly generating spanning arborescences, one needs not only a way to represent and randomly generate tree-maps, but also a way to compose them, and a way to detect when a tree-map is a collapsing map. Composing tree-maps is easy: The combined effect of two excursions with respective tree-maps f and g (with the g-excursion preceding the f-excursion in time) is the tree-map $f \circ g$, where the array entry $(f \circ g)[v]$ is equal to $f[v]$ unless $f[v]$ was undefined (corresponding to the situation in which the excursion associated with f did not visit v), in which case $(f \circ g)[v]$ is equal to $g[v]$ (which may itself be undefined). Finally, detecting when coalescence has occurred is simple: one merely checks that $f[v]$ is defined for all v other than the root. Hence we can use CFTP to generate random rooted spanning trees, via prepended excursions; further details can be found in [**PW2**].

If one wants to use CFTP to generate a random spanning arborescence of a graph *without* specifying a root, then one might think one needs a way to couple histories in which two arborescences start out with different roots. That is, one might try to do away with the notion of excursions and simply let the stochastic flow advance one step at a time, so that eventually the roots coincide. Such an approach might be workable, but we have proposed a different way. Specifically, in [**PW2**] we suggest that one first choose a random vertex r to serve as the root, where r is governed by the steady-state distribution for the random walk on G, and then apply the procedure for finding a random arborescence with root r. A CFTP-based procedure for choosing such an r is described in the next section (although the non-CFTP-based method called cycle-popping, described in [**PW2**], is likely to be superior in most applications).

We also mention that in the case where one knows the transition probabilities for the time-reversal of the random walk on G, the procedure for determining the latest starting time $-N$ that leads to coalescence by time 0 is simple: one simply runs the walk backwards from time 0 into the past until all vertices have been visited. Two such cases are (unweighted) random walk on an undirected graph and (unweighted) random walk on an Eulerian directed graph (a directed graph in which each vertex has its in-degree equal to its out-degree). In the first case, the CFTP algorithm is effectively equivalent to the random walk method of Aldous [**A1**] and Broder [**B**]; in the second case, the CFTP algorithm is effectively equivalent to the random walk method of Kandel, Matias, Unger, and Winkler [**KMUW**].

Finally, we stress that CFTP is only one of many schemes for generating random spanning arborescences; see [**PW2**] for others.

6. Random state of an unknown Markov chain

Now we come to a problem that in a sense encompasses all the cases we have discussed so far: the problem of sampling from the steady-state distribution $\pi(\cdot)$ of a general Markov chain. Of course, in the absence of further strictures this problem admits a trivial "solution": just solve for the steady-state distribution analytically! In the case of the systems studied in sections 3 through 5, this is not practical, since the state spaces are large. We now consider what happens if the state space is small but the analytic method of simulation is barred by imposing the constraint that the transition probabilities of the Markov chain are unknown: one merely has access to a black box that simulates the transitions.

It might seem that, under this stipulation, no solution to the problem is possible, but in fact a solution was found by Asmussen, Glynn, and Thorisson [**AGT**]. However, their algorithm was not very efficient. Subsequently Aldous [**A2**] and Lovász and Winkler [**LW**] found faster procedures (although the algorithm of Aldous involves controlled but non-zero error). The CFTP-based solution given below is even faster than that of Lovász and Winkler.

For pictorial concreteness, we envision the Markov chain as biased random walk on some directed graph G whose arcs are labeled with weights, where the transition probabilities from a given vertex are proportional to the weights of the associated arcs (as in the preceding section). We denote the vertex set of G by $\mathbf{X}$, and denote the steady-state distribution on $\mathbf{X}$ by π. Propp and Wilson [**PW2**] give a CFTP-based algorithm that lets one sample from this distribution π.

Our goal is to define suitable random maps from $\mathbf{X}$ to $\mathbf{X}$ in which many states are mapped into a single state. We might therefore define a random map from $\mathbf{X}$ to itself by starting at some fixed vertex r, walking randomly for some large number N of steps, and mapping all states in $\mathbf{X}$ to the particular state v that one has landed in after N steps. However, v is subject to initialization bias, so this random map procedure typically does not preserve π in the sense defined in section 2.

What actually works is a multi-phase scheme of the following sort: Start at some vertex r and take a random walk for a *random* amount of time T_1, ending at some state v; then map every state that has been visited during that walk to v. In the second phase, continue walking from v for a further random amount of time T_2, ending at some new state v'; then map every state that was visited during the second phase but not the first to v'. In the third phase, walk from v' for a random time to a new state v'', and map every hitherto-unvisited state that was visited during that phase to the state v''. And so on. Eventually, every state gets visited, and every state gets mapped to some state. Such maps, like tree-maps, are easy to compose, and it is easy to recognize when such a composition is coalescent (it maps every state to one particular state).

There are two constraints that our random durations T_1, T_2, ... must satisfy if we are planning to use this scheme for CFTP. (For convenience we will assume henceforth that they T_i's are i.i.d.) First, the distribution of each T_i should have the property that, at any point during the walk, the (conditional) expected time until the walk terminates does not depend on where one is or how one got there. This ensures that the stochastic flow determined by these random maps preserves π. Second, the time for the walk should be neither so short that only a few states get visited by the time the walk ends nor so long that generating even a single random map takes more time than an experimenter is willing to wait. Ideally, the expected duration of the walk should be on the order of the cover-time for the random walk. Propp and Wilson [**PW2**] show that by using the random walk itself to estimate its own cover-time, one gets an algorithm that generates a random state distributed according to π in expected time at most 15 times the cover time.

At the beginning of this section, we said that one has access to a black box that simulates the transitions. This is, strictly speaking, ambiguous: Does the black box have an "input port" so that we can ask it for a random transition from a specified state? Or are we merely passively observing a Markov chain in which we have no power to intervene? This ambiguity gives rise to two different versions of the problem, of separate interest. Our CFTP algorithm works for both of them.

For the "passive" version of the problem, it is not hard to show that no scheme can work in expected time less than the expected cover time of the walk, so in this setting our algorithm runs in time that is within a constant factor of optimal. It is possible to do better in the active setting, but no good lower bounds are currently known for this case.

7. Caveats

Of course a major warning that should come on the "packaging" of all algorithms that purport to generate random objects is that we do not know of any source of truly random bits, and that if we found one we would not be able to prove that it was random. But leaving that deep issue aside, there are other matters that need to be addressed.

Any time one uses a sampling scheme in which the running time is random, one runs the risk of introducing bias if one is not careful. CFTP is no different in this regard than most other schemes for randomly generating combinatorial or geometric objects. Clearly if the current run of the algorithm has been running for 1000 CPU hours and shows no sign of terminating, the experimenter would be likely to abort the run; and, in effectively discarding the output that the procedure *would* have generated, had it been allowed to run through to completion, the experimenter would risk biasing his set of samples from $\mathbf{X}$ in favor of those for which the procedure tends to finish more quickly.

One way of dealing with this problem would be to devise algorithms for which the (random) running-time is independent (or nearly independent) of the running time; interrupting such an algorithm and discarding the current output would then contaminate one's collection of samples with no (or hardly any) bias. For a significant step in this direction, see the article of Fill [**F**].

Another way to address the problem of bias caused by user impatience is given by work of Glynn and Heidelberger [**GH**]. Under their approach, an experimenter generates as many samples $x_1, x_2, \ldots, x_n$ as she can during some pre-selected time-interval, subject to the condition that if no samples at all were generated during that interval, she must continue her simulation until one sample is generated (giving $n = 1$). This approach avoids bias in the sense that, for any function f on the space from which one is sampling, the sample-average $(f(x_1) + f(x_2) + \cdots + f(x_n))/n$ is an unbiased estimate for the mean value of $f(x)$.

We suggest that the thing to do when using algorithms in which the correlation between output and running time is unknown is to handle the premature termination problem preventatively: one should perform preliminary assays that enable one to design an experiment that is likely to finish tolerably soon. The probability that the duration of the experiment will exceed one's estimated running-time by a factor of k falls off rapidly in k. Thus the experimenter can confidently commit herself to running the experiment through to completion. Of course, there is always a chance that the experiment will take so long to finish that the experimenter, despite her intentions, will go back on her word and abort (and re-design) the experiment. However, this probability can be made as minuscule as the experimenter desires if she takes the time to pre-test the design of the experiment.

Even in the case where a run is interrupted, through deliberate user impatience or technical mishap, all is not lost. For instance, suppose one is using perfect simulation as a way of estimating the mean value of some quantity $f(x)$ via the sample-average $(f(x_1) + f(x_2) + \cdots + f(x_n))/n$. If one is willing to replace the sample-average by an interval estimate, one can replace the unknown value $f(x_n)$ (corresponding to the run that was terminated prematurely) by the interval whose endpoints are upper and lower bounds on this quantity; in many cases, good two-sided bounds on $f(x_n)$ may be available from the terminated run. We suggest that a user of coupling from the past take the point of view that a prematurely terminated run is not a failed run but a run that determined partial information about a particular random object, namely, the random object that she would have obtained if the run had gone through to completion. Indeed, coupling from the past can be envisioned as a process in which more and more information about a random object is obtained until one finally has determined it completely; it is natural to extend this to situations in which one is free to decide, adaptively, that one's partial information is sufficient for the purpose at hand.

References

[A1] D. Aldous, *A random walk construction of uniform spanning trees and uniform labelled trees*, SIAM Journal on Discrete Mathematics **3** (1990), 450–465.

[A2] D. Aldous, *On simulating a Markov chain stationary distribution when transition probabilities are unknown*, in: D. Aldous, P. Diaconis, J. Spencer, and J. M. Steele, editors, "Discrete Probability and Algorithms", volume 72 of IMA Volumes in Mathematics and its Applications, 1–9, Springer-Verlag 1995.

[AGT] S. Asmussen, P. Glynn, and H. Thorisson, *Stationary detection in the initial transient problem*, ACM Transactions on Modeling and Computer Simulation **2** (1992), 130–157.

[AT] V. Anantharam and P. Tsoucas, A proof of the Markov chain tree theorem, *Statistics and Probability Letters* **8** (1989), 189–192.

[BF] A. A. Borovkov and S. G. Foss, *Stochastically recursive sequences and their generalizations*, Siberian Advances in Mathematics **2** (1992), 16–81.

[B] A. Broder, *Generating random spanning trees*, in: "30th Annual Symposium on Foundations of Computer Science" (1989), 442–447.

[D] P. Diaconis, *Backwards composition of random maps*, survey article in preparation.

[F] J. A. Fill, *An interruptible algorithm for perfect sampling via Markov chains*, preprint (1997); to appear in Annals of Applied Probability.

[FT] S. G. Foss and R. L. Tweedie, *Perfect simulation and backward coupling*, to appear in Stochastic Models.

[GH] P. W. Glynn and P. Heidelberger, *Bias properties of budget constrained simulations*, Operations Research **38** (1990), 801–814.

[HLM] O. Häggström, M. N. M. Van Lieshout, and J. Møller, *Characterisation results and Markov chain Monte Carlo algorithms including exact simulation for some spatial point processes*, Technical Report R-96-2040, Aalborg University (1996).

[HN] O. Häggström and K. Nelander, *Exact sampling from anti-monotone systems*, preprint (1997).

[KMUW] D. Kandel, Y. Matias, R. Unger, and P. Winkler, *Shuffling biological sequences*, Discrete Applied Mathematics **71** (1996), 171–185.

[K] W. S. Kendall, *Perfect simulation for the area-interaction point process*, to appear in Probability Perspective.

[KSW] J. H. Kim, P. Shor, and P. Winkler, *Random independent sets*, article in preparation.

[Le] G. Letac, *A contraction principle for certain Markov chains and its applications*, Contemporary Mathematics **50** (1986), 263–273.

[LW] L. Lovász and P. Winkler, *Exact mixing in an unknown Markov chain*, Electronic Journal of Combinatorics **2** (1995), paper #R15.

[Lo] R. M. Loynes, *The stability of a queue with non-independent inter-arrival and service times*, Proceedings of the Cambridge Philosophical Society **58** (1962), 497–520.

[LRS] M. Luby, D. Randall, and A. Sinclair, *Markov chain algorithms for planar lattice structures*, in: Proceedings of the 36th IEEE Symposium on Foundations of Computing (1995), 150–159.

[LV] M. Luby and E. Vigoda, *Approximately counting up to four* (extended abstract), in: Proceedings of the Twenty-Ninth Annual ACM Symposium on Theory of Computing (1995), 150–159.

[MR] N. Madras and D. Randall, *Factoring graphs to bound mixing rates*, in: Proceedings of the 37th IEEE Symposium on Foundations of Computing (1996), 194–203.

[MG] D. Murdoch and P. Green, *Exact sampling from a continuous state space*, to appear in the Scandinavian Journal of Statistics.

[PW1] J. Propp and D. Wilson, *Exact sampling with coupled Markov chains and applications to statistical mechanics*, Random Structure and Algorithms **9** (1996), 223–252.

[PW2] J. Propp and D. Wilson, *How to get a perfectly random sample from a generic Markov chain and generate a random spanning tree of a directed graph*, to appear in the Journal of Algorithms (SODA '96 special issue).

[RT] D. Randall and P. Tetali, *Analyzing Glauber dynamics by comparison of Markov chains* (extended abstract), preprint (1997).

[T] H. Thorisson, *Backward limits*, Annals of Probability **16** (1988), 914–924.

Department of Mathematics, Massachusetts Institute of Technology, Cambridge, Massachusetts 02139
E-mail address: propp@math.mit.edu

Institute for Advanced Study, Princeton, New Jersey 08540
E-mail address: dbwilson@math.ias.edu

DIMACS Series in Discrete Mathematics
and Theoretical Computer Science
Volume **41**, 1998

Couplings for Normal Approximations with Stein's Method

Gesine Reinert

ABSTRACT. Coupling constructions for Poisson approximation using the Chen-Stein method is now a standard technique; a systematic study of related couplings for normal approximation using Stein's method has begun only a few years ago. This small survey of coupling methods for normal approximations includes size-bias couplings, which are natural for nonnegative random variables such as counts, and zero-bias couplings, which may be applied to mean zero variables and are especially useful for random variables with vanishing third moments.

1. Introduction

In 1972, Stein [**47**] published a very elegant method to prove normal approximations. It is based on the fact that a random variable Z is standard normal if and only if for all smooth, real-valued functions f,

$$E\{Zf'(Z) - f''(Z)\} = 0.$$

(This is easily seen using dominated convergence and integration by parts.) Stein [**47**] then showed that for any smooth, real-valued function h there is a function f solving the now-called "Stein equation"

$$xf'(x) - f''(x) = h(x) - \Phi h, \tag{1.1}$$

Φh denoting the expectation of h with respect to the standard normal density. Moreover, there is a solution f of the Stein equation (1.1) satisfying

$$\|f'\| \leq \sqrt{\frac{\pi}{2}}\|h - \Phi h\|; \quad \|f''\| \leq (\sup h - \inf h); \quad \|f^{(3)}\| \leq 2\|h'\|; \tag{1.2}$$

where $\|\cdot\|$ denotes the supremum norm (see Stein [**48**], p.25 and Baldi *et al.* [**5**]). Now, for any random variable W, taking expectations in (1.1) gives

$$\mathbb{E}h(W) - \Phi h = \mathbb{E}Wf'(W) - \mathbb{E}f''(W). \tag{1.3}$$

Thus the distance of W from the normal, in terms of a test function h, can be bounded by bounding the right-hand side of (1.3); the immediate bound on the distance is one of the key advantages of Stein's method compared to moment-generating functions or characteristic functions. Typical classes of test functions h are $C^4(\mathbb{R})$ (for weak convergence), or the class of indicator functions of half-lines (giving the Kolmogorov distance).

1991 *Mathematics Subject Classification.* 60F05, 60E10, 62D05.

The author was supported in part by NSF Grant DMS-9505075.

Nearly twenty years later, Barbour [**7**] and Götze [**32**] proved similar results for more general Gaussian approximations; Barbour [**7**] considered diffusion approximations, Götze [**32**] multivariate normal approximations, with the bound

$$\|f^{(j)}\| \leq j^{-1}\|h^{(j)}\|, j = 1, 2, \ldots \tag{1.4}$$

A key tool for solving the higher-dimensional case is the generator method developed by Barbour [**6**], [**7**], [**8**]; the left-hand side of (1.1) can be written as $Af(x)$, where A is the generator of an Ornstein-Uhlenbeck process. Thus semigroup theory can be applied to solve the generator equation. Note that the target distribution, here the standard normal, is the stationary distribution of this Markov process.

Stein's method has been generalized to many other distributions, foremost the Poisson distribution (see Chen [**21**], Arratia *et al.* [**1**], Barbour *et al.* [**15**], Aldous [**2**], to cite but a few). Other distributions include the uniform distribution (Diaconis [**26**]), the binomial distribution (Ehm [**28**]), the compound Poisson distribution (Barbour *et al.* [**9**], Barbour and Utev [**17**], Roos [**46**]), the multinomial distribution (Loh [**36**]), the gamma distribution (Luk [**38**]; for the χ^2 distribution see also Mann [**39**]), the geometric distribution (Peköz [**40**]) and, more generally, Pearson curves (Diaconis and Zabell [**27**], Loh [**37**]).

The most obvious advantage of Stein's method is that it yields immediate bounds on the distance. Moreover in many situations where dependence comes into play the application is straightforward; many examples are of combinatorial nature. An early success of Stein's method is the work by Bolthausen [**18**] for a combinatorial central limit theorem; he was the first to obtain the correct order for this approximation. In examples from random graph theory, where the method of moments used to be the most popular technique, Stein's method allowed not only to provide rates of convergence for the first time, but also to weaken conditions; see, for instance, Barbour *et al.* [**16**]. Another advantage of Stein's method is that it can also be used to derive lower bounds for the approximations; Hall and Barbour [**34**] applied it to give lower bounds for the rate of convergence in the central limit theorem for independent random variables.

Unfortunately such a straightforward application of Stein's method may not yield the correct order for the rate of convergence; additional work may be needed to sharpen the bounds. An example are dependency graphs; first, Baldi and Rinott [**3**] proved a normal approximation using the method of moments, without any result on the rate of convergence. Next, Baldi and Rinott [**4**] employed Stein's method to obtain a bound of the order $n^{-1/4}$; Baldi *et al.* [**5**] derived related results for the number of local maxima in a graph whose vertices are randomly ranked. About five years later, Rinott [**43**] improved the bounds considerably to the correct order $n^{-1/2}$, for bounded random variables. Shortly after, Dembo and Rinott [**25**] simplified this bound, whereas Goldstein and Rinott [**31**] and Rinott and Rotar [**44**] provide multivariate extensions. Another, earlier example that illustrates the effort in obtaining optimal bounds is Bolthausen's [**18**] proof of the Berry-Esséen theorem, the first to yield the correct order when applying Stein's method.

Evidently, Stein's method is in place when the right-hand side of (1.3) is easier to bound than the left-hand side of (1.3). A typical approach for evaluating the right-hand side of (1.3) is to employ couplings, and often the success of the method is connected with finding an effective coupling for the right-hand side of (1.3).

In what follows, we will only consider random variables W that are the sum of n random variables $X_1, \ldots X_n$; that is, $W = \sum_{i=1}^n X_i$. For convenience we will assume throughout that $\mathrm{Var}(W) = 1$. Moreover we will only discuss smooth test functions, because the treatment of nonsmooth test functions is slightly more technical, and the purpose of this paper is to lay out the basic methods. References for nonsmooth test functions will be given. As our main two examples, we will firstly consider that $X_1, \dots X_n$ are independent, and secondly that $X_1, \dots X_n$ is a simple random sample from a finite population. Moreover, for $\|f''\|$ in (1.2), Stein **[48]** proved the bound $2\|h - \Phi h\|$; Baldi *et al.* **[5]** showed the improved inequality in (1.2). In what follows, we will use the improved inequality for theorems we cite, even if the theorems used the first inequality.

In Section 2 we describe the perhaps most common coupling, the "local" approach. It is very effective if each X_i depends only on a small number of the other $X_j, j \neq i$. Typical examples are m-dependent sequences.

In case of global but weak dependence between the $X_1, \dots X_n$, exchangeable pair couplings are usually more natural. This approach is discussed in Section 3.

Section 4 gives a coupling that is particularly adapted to describe counts; the size bias coupling. In Barbour *et al.* **[15]** it has been developed as a major tool for proving Poisson approximations; its importance for normal approximations has been described in Baldi *et al.* **[5]**, Goldstein and Rinott **[31]**, and Stein **[49]**. A drawback in the context of normal approximations is that it requires W to be nonnegative, with positive mean.

For mean zero W, and in particular for symmetric W, the zero bias coupling discussed in Section 5 might give better results, especially when the test functions are smooth. The zero-bias coupling is a "second-order" refinement of exchangeable pair ideas .

Finally, Section 6 collects other couplings that work well in special cases. It illustrates that specific problems may benefit from constructing couplings that do not fit into the above classes, and thus illustrates the dynamical structure of the field.

2. The local approach

Suppose $X_1, \dots X_n$ are independent, mean zero, and let $\sigma_i^2 = \mathrm{Var}(X_i)$. Put $W = \sum_{i=1}^n X_i$, and let $\mathrm{Var}(W) = \sum_{i=1}^n \sigma_i^2 = 1$. For each $i = 1, \dots, n$ put

$$W_i = W - X_i = \sum_{j \neq i} X_j. \tag{2.1}$$

Thus W and W_i are defined on the same probability space. For any smooth function f we have, by Taylor expansion

$$\begin{aligned} \mathbb{E} W f'(W) &= \sum_{i=1}^n \mathbb{E} X_i f'(W) \\ &= \sum_{i=1}^n \mathbb{E} X_i f'(W_i) + \sum_{i=1}^n \mathbb{E} X_i^2 f''(W_i) + R, \end{aligned}$$

where

$$|R| \leq \frac{1}{2} \|f^{(3)}\| \sum_{i=1}^n \mathbb{E}|X_i^3|.$$

Using the independence we obtain for the right-hand side of (1.3), with $\mu = 0$, that

$$\begin{aligned} \mathbb{E}Wf'(W) - \mathbb{E}f''(W) &= \sum_{i=1}^{n} \mathbb{E}X_i^2 \mathbb{E}f''(W_i) - \mathbb{E}f''(W) + R \\ &= \sum_{i=1}^{n} \sigma_i^2 \mathbb{E}\{f''(W_i) - f''(W)\} + R. \end{aligned}$$

Taylor expansion and the bounds (1.2) now give the following theorem.

THEOREM 2.1. *Suppose* $X_1, \dots X_n$ *are independent, mean zero, and let* $\sigma_i^2 = \mathrm{Var}(X_i)$. *Put* $W = \sum_{i=1}^n X_i$, *and let* $\mathrm{Var}(W) = 1$. *For any continuous, bounded function* h *with piecewise continuous, bounded first derivative, we have*

$$|\mathbb{E}h(W) - \Phi h| \leq \|h'\| \left(2\sum_{i=1}^{n} \sigma_i^3 + \sum_{i=1}^{n} \mathbb{E}|X_i^3| \right).$$

This approach, first used by Stein [**47**], was employed by Bolthausen [**18**] to obtain sharp rates in the Berry-Esséen Theorem, by Barbour and Hall [**13**] for smoother metrics, by Barbour and Hall [**14**] for the non-identically distributed case, and by Hall and Barbour [**34**] for reversing the Berry-Esséen Theorem. For more general Gaussian approximations of sums of independent random elements, it was employed by Barbour [**7**] for a functional CLT, by Götze [**32**] for a multivariate CLT, and by Reinert [**41**] for a Gaussian approximation of empirical measures.

The local approach might well be the most widely applied coupling approach for Stein's method, being both effective and easy to construct in situations of local dependence; the amount of literature where it is used is large, and listing it all would be beyond the scope of this paper. As briefly mentioned in the introduction, the correct bound on the rate of convergence for bounded variates has been obtained by Rinott [**43**]; Goldstein and Rinott [**31**] and Rinott and Rotar [**44**] give multivariate extensions. Important applications include m-dependent sequences, where the generalization using Taylor expansion is straightforward, dependency graphs (Rinott [**43**]), and sums of dissociated random variables (Chen [**22**], Barbour and Eagleson [**10**]). Refinements with different types of neighborhoods are derived by Chen [**23**], and by Barbour *et al.* [**16**] for decomposable random variables; typical applications are graph-related statistics, see also Goldstein and Rinott [**31**].

Moreover the local approach can be seen as a special case of a conditional expectation coupling used by Stein [**47**], [**48**]. Let $(\tilde{\Omega}, \tilde{B}, \tilde{\mathbb{P}})$ be a probability space, let B and C be sub-σ-algebras of $\tilde{B}$, let G be a $\tilde{B}$-measurable random variable such that $\mathbb{E}|G| < \infty$, and let W^* be C-measurable. Put

$$W = \mathbb{E}(G|B).$$

Assume $\mathbb{E}W = 0$ and $\mathrm{Var}(W) = 1$. Then (see Stein [**48**], p.106, Theorem 1)

THEOREM 2.2. *Let* $W = \mathbb{E}(G|B)$ *be constructed as above. For any continuous, bounded function* $h : \mathbb{R} \to \mathbb{R}$ *with bounded, piecewise continuous derivatives, we have*

$$\begin{aligned} |\mathbb{E}h(W) - \Phi h| &\leq (\sup h - \inf h)\sqrt{\mathbb{E}\{1 - \mathbb{E}(G(W - W^*)|B)\}^2} \\ &\quad + \sqrt{\frac{\pi}{2}}\|h - \Phi h\| \mathbb{E}|\mathbb{E}(G|C)| + \|h'\| \mathbb{E}|G|(W - W^*)^2. \end{aligned}$$

In the independent example, an index I would be chosen uniformly from $\{1,\dots,n\}$; we would put $\tilde{B} = \sigma\{X_1,\dots,X_n,I\}$, $B = \sigma\{X_1,\dots,X_n\}$, and $C = \sigma\{X_j, j \neq I; I\}$. Furthermore put $G = nX_I$ and $W^* = W - X_I$. Then we have $W = \mathbb{E}(G|B) = \sum_{i=1}^n X_i$. The conditional expectation formulation also extends to mixing sequences (Stein [**47**], Chen [**23**]), and can be applied whenever the dependence between any X_i and $X_j, j \neq i$ is strong only for a few indices j, and very weak for the other indices.

However, this approach does not work well for global weak dependence structures, such as given by the example of simple random sampling.

3. Exchangeable pair couplings

Note that the local coupling can be described in a more abstract setting as follows. Let I be chosen randomly, uniformly from $\{1,\dots,n\}$ and put $W' = W_I$; recall (2.1). Assume as usual that $\mathrm{Var}(W) = 1$. Then we have for all smooth functions f that

$$\mathbb{E}Wf'(W) = n\mathbb{E}(W - W')(f'(W) - f'(W')). \tag{3.1}$$

Taylor expansion gives

$$\mathbb{E}Wf'(W) - \mathbb{E}f''(W) \quad = \quad \mathbb{E}\{1 - n(W - W')^2\}f''(W) + R,$$

where R is a remainder term that can be bounded by

$$|R| \leq \frac{n}{2}\|f^{(3)}\|\mathbb{E}|W - W'|^3.$$

Using the Cauchy-Schwarz inequality gives

$$\begin{aligned}
&|\mathbb{E}Wf'(W) - \mathbb{E}f''(W)| \\
&\leq \|f''\|\sqrt{\mathbb{E}(1 - n\mathbb{E}((W - W')^2|W)} + \frac{n}{2}\|f^{(3)}\|\mathbb{E}|W - W'|^3 \\
&\leq (\sup h - \inf h)\sqrt{\mathbb{E}(1 - n\mathbb{E}((W - W')^2|W)} + \frac{n}{2}\|h'\|\mathbb{E}|W - W'|^3.
\end{aligned}$$

Indeed, all that is required for this derivation is an equation of the type (3.1). Another method to achieve this type of equation is the method of exchangeable pairs. A pair (W, W') of random variables defined on the same probability space is called *exchangeable* if for all measurable sets B and B',

$$\mathbb{P}(W \in B, W' \in B') = \mathbb{P}(W \in B', W' \in B).$$

Following Stein [**48**] we assume that W is mean zero, variance 1, and that there is a $0 < \lambda < 1$ such that

$$\mathbb{E}(W'|W) = (1 - \lambda)W. \tag{3.2}$$

This assumption can be related to regression; if (W, W') is bivariate normal with correlation ρ, then $1 - \lambda = \rho$. Under (3.2) it is easy to see that

$$\mathbb{E}Wf'(W) = \frac{1}{2\lambda}\mathbb{E}(W - W')(f'(W) - f'(W')),$$

so that equation (3.1) is satisfied with n replaced by $\frac{1}{2\lambda}$. The same reasoning as above gives the following result (see Stein [**48**]).

THEOREM 3.1. *Let (W, W') be an exchangeable pair satisfying (3.2) and assume $\mathbb{E}W = 0, \mathrm{Var}(W) = 1$. For any continuous, bounded function $h : \mathbb{R} \to \mathbb{R}$ with bounded, piecewise continuous derivatives, we have*

$$\begin{aligned} &|\mathbb{E}h(W) - \Phi h| \\ &\leq \ (\sup h - \inf h)\sqrt{\mathbb{E}\left(1 - \frac{1}{2\lambda}\mathbb{E}((W - W')^2|W)\right)} + \frac{1}{4\lambda}\|h'\|\mathbb{E}|W - W'|^3. \end{aligned}$$

EXAMPLE 3.2. Let us assume that $X_1, \ldots, X_n$ are independent, mean zero; let $\sigma_i^2 = \mathrm{Var}(X_i)$ and $\sum_{i=1}^n \sigma_i^2 = 1$. As usual, we consider $W = \sum_{i=1}^n X_i$. To construct W' such that (W, W') is an exchangeable pair, pick an index I uniformly from $\{1, \ldots, n\}$. If $I = i$, we replace X_i by an independent copy X_i^*, and we put

$$W' = W - X_I + X_I^*. \tag{3.3}$$

Then (W, W') is exchangeable, and

$$\mathbb{E}(W'|W) = W - \frac{1}{n}W + \mathbb{E}X_I^* = \left(1 - \frac{1}{n}\right)W,$$

so (3.2) is satisfied with $\lambda = \frac{1}{n}$. Theorem 3.1 thus gives

$$\begin{aligned} &|\mathbb{E}h(W) - \Phi h| \\ &\leq \ (\sup h - \inf h)\sqrt{\mathbb{E}\left(1 - \frac{1}{2\lambda}\mathbb{E}((W - W')^2|W)\right)} + \frac{1}{4\lambda}\|h'\|\mathbb{E}|W - W'|^3. \end{aligned}$$

Bounding the expectations gives (see [**48**], p.37)

$$|\mathbb{E}h(W) - \Phi h| \ \leq \ (\sup h - \inf h)\sqrt{\sum_{i=1}^n \mathbb{E}X_i^4 - \sigma_i^4} + \frac{1}{2}\|h'\|\sum_{i=1}^n \left(\mathbb{E}|X_i|^3 + 3\sigma_i^3\right).$$

The construction used in Example 3.2 is typical for the exchangeable pair approach.

CONSTRUCTION 3.3. Let $X_1, \ldots, X_n$ be possibly dependent, mean zero variates with existing variances, and let $W = \sum_{i=1}^n X_i$. Pick an index I uniformly from $\{1, \ldots, n\}$. If $I = i$, replace X_i by an independent copy X_i^*. If $X_i^* = x$, construct $\hat{X}_j, j \neq i$ such that

$$\mathcal{L}(\hat{X}_j, j \neq i) = \mathcal{L}(X_j, j \neq i | X_i = x). \tag{3.4}$$

Then put

$$W' = \sum_{j \neq I} \hat{X}_j + X_I^*. \tag{3.5}$$

The pair (W, W') is exchangeable. If, in addition, (3.2) holds, then Theorem 3.1 can be applied. However, not always will the exchangeable pair (W, W') constructed above satisfy (3.2); see, for example, Rinott and Rotar [**45**].

Compared to Theorem 2.1 the bound obtained in Example 3.2 is worse. However, a similar construction leads to an exchangeable pair for the simple random sampling example, which caused problems in the local approach; see Stein [**48**].

EXAMPLE 3.4. Let $X_1, \ldots, X_n$ be a simple random sample of size n from a finite population A. Assume for convenience that all elements of A are distinct. We follow Construction 3.3. To construct $\hat{X}_j, j \neq i$ satisfying (3.4), we choose $\hat{X}_j, j \neq i$ as a simple random sample of size $n-1$ from $A \setminus \{X_I^*\}$. In particular, if $X_I^* \notin \{X_j, j \neq i\}$, then we may choose $\hat{X}_j = X_j, j \neq i$. If $X_I^* \in \{X_j, j \neq i\}$, so that $X_I^* = X_J$, say, then let $\hat{X}_j = X_j, j \neq i, J$, and choose $\hat{X}_j$ uniformly from $A \setminus \{X_I^*, \hat{X}_j, j \neq i, J\}$. Then (3.2) is satisfied with $\lambda = \frac{2}{n-1}$. Note that W and W' differ for at most two summands, so that the coupling is efficient.

Other examples where Construction 3.3 works well include random permutations (Stein [**48**], Fulman [**29**], for example), random allocations (Stein [**48**]) and combinatorial central limit theorems (Bolthausen [**18**], Bolthausen and Götze [**19**]). Note that Construction 3.3, if repeated, yields a Markov chain - this relates to the generator approach developed by Barbour [**6**], [**7**], [**8**]. Indeed, this construction can be used to derive a generator associated with a target distribution; see, e.g., Reinert [**41**].

Conversely, an exchangeable pair can be constructed from a reversible Markov chain; see Rinott and Rotar [**45**]. Let $X_1, \ldots, X_n$ be random variables, and suppose that $\mathcal{L}(X_1, \ldots, X_n)$ is the stationary distribution of a reversible, ergodic Markov chain $(X_1(t), \ldots, X_n(t))_{t=0,1,\ldots}$. Let $W = W(X_1, \ldots, X_n)$ be the quantity of interest. Put

$$\begin{aligned} W &= W(X_1(t), \ldots, X_n(t)) \\ W' &= W(X_1(t+1), \ldots, X_n(t+1)), \end{aligned}$$

then (W, W') is an exchangeable pair.

Moreover, the approach is not restricted to requiring that Condition (3.2) is satisfied. Following Rinott and Rotar [**45**], assume that (W, W') is an exchangeable pair such that $\mathbb{E}W = 0, \mathbb{E}W^2 = 1$, and let $R = R(W)$ be such that

$$\mathbb{E}(W'|W) = (1-\lambda)W + R \tag{3.6}$$

for some $0 < \lambda < 1$. Similarly as for Theorem 3.1, we can show that, under the above setting, for W real-valued,

THEOREM 3.5. *Let (W, W') be an exchangeable pair such that Condition (3.6) is satisfied, and assume $\mathbb{E}W = 0, \mathrm{Var}(W) = 1$. For any continuous, bounded function h with piecewise continuous, bounded first derivative, we have*

$$\begin{aligned} |\mathbb{E}h(W) - \Phi h| &\leq (\sup h - \inf h)\sqrt{\mathbb{E}(1 - \frac{1}{2\lambda}\mathbb{E}((W-W')^2|W)} \\ &\quad + \frac{1}{4\lambda}\|h'\|\mathbb{E}|W-W'|^3 + \sqrt{\frac{\pi}{2}}\frac{1}{\lambda}\|h - \Phi h\|\mathbb{E}|R|. \end{aligned}$$

Thus, if R is small, the normal approximation will be good. Examples include the anti-voter model and weighted U-statistics; see Rinott and Rotar [**45**].

Finally it must be remarked that the method of exchangeable pairs has a much wider range than normal approximations; in particular it can be used to estimate ratios of probabilities, see Stein [**49**]. Moreover Diaconis [**26**] employed it for for the uniform distribution, and more generally, for the convergence of a Markov chain to its stationary distribution. Recently, Mann [**39**] has applied it to yield a χ^2 approximation for statistics based on the multinomial distribution. This illustrated the vast potential of the method of exchangeable pairs.

In contrast to the local approach, where we used a coupling that reduced the variability (by leaving out a summand, or by taking conditional expectations), the exchangeable pair coupling introduces additional randomness. This is a first example of what, following Stein [**49**], might be called the method of auxiliary randomization. Section 4 and Section 5 provide more examples for this method.

4. Size-bias couplings

There are situations where couplings other than exchangeable pairs seem more natural. A wide class of examples is provided in the context of Poisson approximations, see Barbour *et al.* [**15**], where counts are considered. Then size bias couplings seem to be more adapted. In the context of Stein's method for normal approximations, they have been explored by Baldi *et al.* [**5**], Stein [**49**], Goldstein and Rinott [**31**], Dembo and Rinott [**25**], and Reinert [**42**]. In Baldi *et al.* [**5**], and Stein [**49**], sums of $0-1$ random variables are considered. Goldstein and Rinott [**31**] generalize it to multivariate normal approximations of any sums, Reinert [**42**] uses it for empirical processes, and Dembo and Rinott [**25**] prove approximations for nonsmooth functions.

Let W be a nonnegative random variable, $\mathrm{Var}(W)=1$, and $\mathbb{E}W=\mu>0$. W^* is said to have the W-size biased distribution if, for all functions g for which the expectation exists,

$$\mathbb{E}Wg(W)=\mu\mathbb{E}g(W^*).$$

Thus, if w is discrete, say, then, for all w we have $\mathbb{P}(W^*=w)=\frac{w}{\mu}\mathbb{P}(W=w)$. This illustrates that size biasing corresponds to sampling proportional to size; the larger a subpopulation, the more likely it is to be in the sample.

If (W,W^*) are defined on the same probability space, with W^* having the W-size biased distribution, then the right-hand side of equation (1.3) becomes

$$\begin{aligned}\mathbb{E}(W-\mu)f('W-\mu) &= \mu\mathbb{E}(f'(W^*-\mu)-f'(W-\mu))\\ &= \mu\mathbb{E}f''(W-\mu)(W^*-W)+R,\end{aligned}$$

using Taylor expansion, where

$$|R|\le\mu\|f^{(3)}\|\mathbb{E}(W^*-W)^2.$$

Note that $\mu\mathbb{E}(W^*-W)=\mathbb{E}W^2-\mu^2=\mathrm{Var}(W)=1$. Thus

$$\begin{aligned}\mu\mathbb{E}f''(W-\mu)(W^*-W)-\mathbb{E}f''(W-\mu) &= \mathbb{E}f''(W-\mu)(\mu\mathbb{E}(W^*-W|W)-1)\\ &\le \|f''\|\sqrt{\mathbb{E}(\mu\mathbb{E}(W^*-W|W)-1)^2}\\ &= \|f''\|\mu\sqrt{\mathrm{Var}\mathbb{E}(W^*-W|W)}.\end{aligned}$$

This gives (see Goldstein and Rinott [**31**])

THEOREM 4.1. *Let W be nonnegative, $\mathbb{E}W=\mu>0$, and $\mathrm{Var}(W)=1$. Let W^* have the W-size biased distribution. For any continuous, bounded function h with piecewise continuous, bounded first derivative, we have*

$$\begin{aligned}&|\mathbb{E}h(W-\mu)-\Phi h|\\ &\le (\sup h-\inf h)\mu\sqrt{\mathrm{Var}\mathbb{E}(W^*-W|W)}+\|h'\|\mu\mathbb{E}(W^*-W)^2.\end{aligned}$$

EXAMPLE 4.2. let $X_1, \dots X_n$ be independent, nonnegative, $\mathbb{E}X_i = \mu_i > 0$, $W = \sum_{i=1}^n X_i$, $\mathbb{E}W = \mu$, $\mathrm{Var}(W) = 1$. Choose an index I from $\{1, \dots, n\}$ according to

$$\mathbb{P}(I = i) = \frac{\mu_i}{\mu}, \tag{4.1}$$

that is, choose index i proportionally to its expectation. If $I = i$, replace X_i by X_i^* having the X_i-size bias distribution, and put

$$W^* = W - X_I + X_I^*. \tag{4.2}$$

Then W^* has the W-size biased distribution. (Note the similarity to (3.3)). Using Theorem 4.1 gives

$$|\mathbb{E}h(W - \mu) - \Phi h| \quad \leq \quad (\sup h - \inf h)\frac{\mu}{n} + \|h'\| \sum_{i=1}^n \mathbb{E}|X_i|^3.$$

In the usual scaling , $X_i \asymp \frac{1}{\sqrt{n}}$, in which case $\mu \asymp \sqrt{n}$ and the bound is of order $\frac{1}{\sqrt{n}}$. Depending on $\sigma_i, i = 1, \dots, n$, this bound may be better than the one obtained in Theorem 2.1 by the local approach.

Goldstein and Rinott [**31**] also give a general construction.

CONSTRUCTION 4.3. Let $X_1, \dots X_n$ be nonnegative, $\mathbb{E}X_i = \mu_i > 0$, $W = \sum_{i=1}^n X_i$, $\mathbb{E}W - \mu$, $\mathrm{Var}(W) - 1$. Choose an index I from $\{1, \dots, n\}$ according to (4.1). If $I = i$, replace X_i by a variate X_i^* having the X_i-size bias distribution. If $X_i^* = x$, construct $\hat{X}_j, j \neq i$ such that

$$\mathcal{L}(\hat{X}_j, j \neq i) = \mathcal{L}(X_j, j \neq i | X_i = x).$$

Then

$$W^* = \sum_{j \neq I} \hat{X}_j + X_I^*$$

has the W-size biased distribution. (The difference from Construction 3.3 are the choice of I and the distribution of X_i^*.)

The above construction can also be applied to the example of simple random sampling.

EXAMPLE 4.4. As in Example 3.4, let $X_1, \dots, X_n$ be a simple random sample of size n from a finite population A, where all elements of A are distinct. We follow Construction 4.3. Once X_I^* is constructed, we may continue as in Example 3.4; if $X_I^* \notin \{X_j, j \neq I\}$, then we may choose $\hat{X}_j = X_j, j \neq I$, whereas if $X_I^* = X_J$ for some $J \in \{j, j \neq I\}$, then let $\hat{X}_j = X_j, j \neq I, J$, and choose $\hat{X}_J$ uniformly from $A \setminus \{X_I^*, \hat{X}_j, j \neq I, J\}$. This procedure is known as Midzuno's procedure, see Luk [**38**], and is used to obtain unbiased ratio estimators. Note that, as again W and W^* differ for at most two summands, the coupling is efficient.

Size bias couplings for functions of variables are described in Dembo and Rinott [**25**]. Typical examples are counts, as occurring in random graphs (the number of vertices of a fixed degree, for example) and in random allocations.

However, a necessary ingredient is that W is nonnegative. For bounded variables W one might shift the distribution to the positive axis, but this is not very natural, in particular when the distribution of W is symmetric around zero, because W should be closer to normal than any shifted version of it. This drawback motivated the introduction of the zero bias coupling described in the next section.

5. Zero-bias couplings

Let W be mean zero, variance one. We say that a random variable W^* has the W-zero biased distribution if, for all g for which the expectation exists,

$$\mathbb{E}Wg(W) = \mathbb{E}g'(W^*). \tag{5.1}$$

This notion was introduced in Goldstein and Reinert [**30**]. Related analytical ideas, without coupling constructions, appear in Ho and Chen [**35**], Chen [**24**], and in Cacoullos *et al.* [**20**]. Bolthausen [**18**] uses a related coupling, without formalizing it. As the standard normal distribution is the unique fixed point of (5.1), it seems to be another natural approach for normal approximations.

Using (5.1), the right-hand side of Equation (1.3) can be written as

$$\begin{aligned}\mathbb{E}Wf'(W) - \mathbb{E}f'(W) &= \mathbb{E}(f''(W^*) - f''(W)) \\ &\leq R,\end{aligned}$$

where

$$|R| \leq \|f^{(3)}\|\mathbb{E}|W^* - W|.$$

Thus we save a step in the Taylor expansion. Expanding one step further, though, gives the following, sharper result, see Goldstein and Reinert [**30**], where we use the bounds (1.4), as derivatives higher than second order occur.

THEOREM 5.1. *Let W be mean zero, variance 1, and let W^* have the W-zero biased distribution. For any bounded, continuous function h with bounded derivatives up to order 4, we have*

$$|\mathbb{E}h(W) - \Phi h| \leq \frac{1}{3}\|h^{(3)}\|\sqrt{\mathbb{E}(\mathbb{E}(W^* - W|W))^2} + \frac{1}{8}\|h^{(4)}\|\mathbb{E}(W - W^*)^2.$$

EXAMPLE 5.2. Let $X_1, \dots X_n$ be independent, mean zero, $\mathrm{Var}(X_i) = \sigma_i^2$, $W = \sum_{i=1}^n X_i$, and assume $\mathrm{Var}(W) = 1$. Choose an index I from $\{1, \dots, n\}$ according to

$$\mathbb{P}(I = i) = \sigma_i^2,$$

that is, choose index i proportionally to its variance. (This resembles (4.1), where i is drawn proportionally to its expectation.) If $I = i$, replace X_i by X_i^* having the X_i-zero biased distribution, and put

$$W^* = W - X_I + X_I^*. \tag{5.2}$$

Then W^* has the W-zero biased distribution; note the similarity to (3.3) and (4.2). Using Theorem 5.1 gives

$$|\mathbb{E}h(W - \mu) - \Phi h| \leq \frac{1}{3}\|h^{(3)}\|\sqrt{\mathbb{E}\left(\mathbb{E}X_I^3 - \frac{W}{n}\right)^2} + \frac{1}{8}\|h^{(4)}\|\left(\frac{\mathbb{E}X_I^4}{3} + \mathbb{E}X_I^2\right).$$

In the special case that $\mathbb{E}X_i^3 = 0$ we obtain

$$|\mathbb{E}h(W - \mu) - \Phi h| \leq \frac{1}{3}\|h^{(3)}\|\frac{1}{n} + \frac{1}{8}\|h^{(4)}\|\left(\frac{\mathbb{E}X_I^4}{3} + \mathbb{E}X_I^2\right). \tag{5.3}$$

With the scaling $X_i \asymp \frac{1}{\sqrt{n}}$ we thus obtain a $\frac{1}{n}$ - bound on the rate of convergence for smooth test functions.

The feature of a $\frac{1}{n}$ - bound on the rate of convergence for smooth test functions under vanishing third moment conditions is a main advantage of the zero bias coupling. (It could also be derived using Edgeworth expansions, but those assume some smoothness of the density, see for instance Hall [**33**], Chapter 2.8). For nonsmooth test functions it is easy to see that the rate of $\frac{1}{\sqrt{n}}$ is unimprovable - consider the sum of n independent centered binomials and, as test function, the indicator of the negative half axis.

Despite the similarity of the construction for the independent case in Example 5.2 to Construction 4.3 and Construction 3.3, a general construction is much more involved, unfortunately; see Goldstein and Reinert [**30**]. This more complicated construction, Construction 5.3 below, makes it seem advisable to mainly consider applications with vanishing third moments, as then there is hope for the better rate for smooth test functions.

CONSTRUCTION 5.3. Let $X_1, \dots X_n$ be mean zero, $W = \sum_{i=1}^n X_i$, and assume $\mathrm{Var}(W) = 1$. Denote the distribution of $X_1, \dots X_n$ by dF_n. Suppose that for each $i = 1, \dots, n$ there exists a distribution $dF_{n,i}(x_1, \dots, x_{i-1}, x_i', x_i'', x_{i+1}, \dots, x_n)$ on $n+1$ random variables $X_1, \dots, X_{i-1}, X_i', X_i'', X_{i+1}, \dots, X_n$ such that

$$(X_1, \dots, X_{i-1}, X_i', X_i'', X_{i+1}, \dots, X_n) \stackrel{d}{=} (X_1, \dots, X_{i-1}, X_i'', X_i', X_{i+1}, \dots, X_n),$$

and

$$(X_1, \dots, X_{i-1}, X_i, X_{i+1}, \dots, X_n) \stackrel{d}{=} (X_1, \dots, X_{i-1}, X_i', X_{i+1}, \dots, X_n).$$

Suppose that there is a ρ such that for all f for which $\mathbb{E}Wf(W)$ exists,

$$\sum_{i=1}^n \mathbb{E}X_i' f(W_i + X_i'') = \rho \mathbb{E}Wf(W), \tag{5.4}$$

where $W_i = W - X_i$ as in (2.1). Let

$$\sum_{i=1}^n v_i^2 > 0 \quad \text{where} \quad v_i^2 = \mathbb{E}(X_i' - X_i'')^2,$$

and let I be a random index independent of the $X's$ such that

$$\mathbb{P}(I = i) = v_i^2 / \sum_{j=1}^n v_j^2.$$

Further, for i such that $v_i > 0$, let $\hat{X}_1, \dots, \hat{X}_{i-1}, \hat{X}_i', \hat{X}_i'', \hat{X}_{i+1}, \dots, \hat{X}_n$ be chosen according to the distribution

$$\begin{aligned} &d\hat{F}_{n,i}(\hat{x}_1, \dots, \hat{x}_{i-1}, \hat{x}_i', \hat{x}_i'', \hat{x}_{i+1}, \dots, \hat{x}_n) \\ &= \frac{(\hat{x}_i' - \hat{x}_i'')^2}{v_i^2} dF_{n,i}(\hat{x}_1, \dots, \hat{x}_{i-1}, \hat{x}_i', \hat{x}_i'', \hat{x}_{i+1}, \dots, \hat{x}_n). \end{aligned}$$

Then, with U a uniform $U[0,1]$ variate which is independent of the X's and the index I,

$$W^* = U\hat{X}_I' + (1-U)\hat{X}_I'' + \sum_{j \neq I} \hat{X}_j$$

has the W-zero biased distribution.

Although this construction is rather involved, it is not so difficult to apply it to the simple random sampling example.

EXAMPLE 5.4. As in Example 3.4, let $X_1, \ldots, X_n$ be a simple random sample of size n from a finite population A, where all elements of A are distinct. Assume that $\sum_{a \in A} a^3 = 0$. In Construction 5.3, because of exchangeability we may choose $I = 1$. Independently of $X_1, \ldots, X_n$, pick a pair $(\hat{X}_1', \hat{X}_1'')$ from the distribution

$$q(u,v) = \frac{(u-v)^2}{2N} \mathbf{1}(\{u,v\} \subset A).$$

Firstly then, independently of the chosen sample $\mathbf{X}$, pick $(\hat{X}_1', \hat{X}_1'')$ from the distribution $q(u,v)$. The random variables $(\hat{X}_1', \hat{X}_1'')$ are now placed as the first two components in the vector $\hat{\mathbf{X}}$. The remaining $n-1$ random variables $\hat{\mathbf{X}}$ are sampled by rejection. If the two sets $\{X_2, \ldots, X_n\}$ and $\{\hat{X}_1', \hat{X}_1''\}$ do not intersect, fill in the remaining $n-1$ components of $\hat{\mathbf{X}}$ with $(X_2, \ldots, X_n)$. If the sets have an intersection, remove from the vector $(X_2, \ldots, X_n)$ the two random variables (or single random variable) that intersect and replace them (or it) with values obtained by a simple random sample of size two (one) from $A \setminus \{\hat{X}_1', \hat{X}_1'', X_2, \ldots, X_n\}$. This new vector now fills in the remaining $n-1$ positions in $\hat{\mathbf{X}}$. In Goldstein and Reinert [**30**] it is shown that this construction satisfies Condition (5.4) with $\rho = -n/(N-n)$, and that it yields a bound of order $\frac{1}{n}$ for the normal approximation of W, provided the elements of A are scaled to be $\asymp \frac{1}{\sqrt{n}}$.

The zero-bias coupling also displays an interesting connection with the method of exchangeable pairs. Let (W, W') be an exchangeable pair with distribution function dF such that Condition 3.2 is satisfied. Pick a pair $(\hat{W}, \hat{W}')$ from the distribution

$$\frac{(\hat{w} - \hat{w}')^2}{\mathbb{E}(W - W')^2} dF(\hat{w}, \hat{w}').$$

Let U be an independent $U(0,1)$ variable. Then

$$W^* = U\hat{W} + (1-U)\hat{W}'$$

has the W-zero biased distribution (see Goldstein and Reinert [**30**]).

Other examples where the zero bias coupling might be useful include the anti-voter model, U-statistics, and permutation statistics.

6. Other couplings

There are many other couplings that work well in specific situations. One example is symmetric arrays as treated by Barbour and Eagleson [**10**]. Assume $a_1, \ldots, a_n$ and $b_1, \ldots, b_n$ are real numbers with $\sum_{i=1}^n a_i = \sum_{j=1}^n b_j = 0$. Let π be a random permutation of $\{1, \ldots, n\}$, chosen uniformly, and put

$$X_i = a_i b_{\pi(i)}.$$

Let, as usual, $W = \sum_{i=1}^n X_i$. Assume that $\mathrm{Var}(W) = (n-1)^{-1} \sum_{i=1}^n a_i^2 \sum_{j=1}^n b_j^2 = 1$. Now, if $\pi(i)$ is known, we automatically know X_i. This can be put to use as follows.

$$\begin{aligned} \mathbb{E}Wf'(W) &= \frac{1}{n}\sum_{i,j=1}^{n} a_i b_j \mathbb{E}(f'(W)|\pi(i)=j) \\ &= \frac{1}{n}\sum_{i,j=1}^{n} a_i b_j \mathbb{E}f'(W+D_{i,j}), \end{aligned}$$

where

$$D_{i,j} = (a_i - a_{\pi^{-1}(j)})(b_j - b_{\pi(i)}).$$

Now

$$\begin{aligned} &\frac{1}{n}\sum_{i,j=1}^{n} a_i b_j \mathbb{E}f'(W+D_{i,j}) \\ &\approx \frac{1}{n}\sum_{i,j=1}^{n} a_i b_j \mathbb{E}D_{i,j}f'(W) \\ &= \frac{1}{n}\sum_{i,j=1}^{n} a_i b_j \frac{1}{n(n-1)}\sum_{k\neq j}\sum_{l\neq i}(a_i-a_l)(b_j-b_k)\mathbb{E}f''(W+\tilde{D}_{i,k;l,j}), \end{aligned}$$

where $\tilde{D}_{i,k;l,j} = D_{i,k} + D_{l,j}$ whenever $\{\pi(i),\pi(j)\}\cap\{k,j\}$ is empty, and when the intersection is nonempty, the expressions are slightly modified. After some work this yields a normal approximation. This approach can be generalized to obtain a Wald-Wolfowitz Theorem for processes, see Barbour [**7**], [**8**] and Barbour and Eagleson [**11**].

Furthermore, sometimes couplings for normal approximations are used in a different sense. A variate T is first coupled to a variate T_0 using the structure inherent of the problem, and then a normal approximation is shown or known to hold for T_0. This is applied by Bolthausen and Götze [**19**] to multivariate sampling statistics, and by Barbour *et al.* [**12**] to iterations of expanding maps.

Finally it should be emphasized that the above is a collection of techniques. Depending on the problem that is to be solved, they might provide useful tools. Yet there is always the possibility that a coupling of a different nature might yield better results. Moreover, a concentration inequality approach has been proved to be another powerful tool when using Stein's method for normal approximations; see Chen [**24**] for an overview.

Acknowledgements. I would like to thank the organizers for the opportunity to present the paper, and the referee for many useful suggestions. Larry Goldstein has provided many useful discussions over the last years that certainly influenced my perspective. Finally, many thanks to Andrew Barbour and Charles Stein for teaching me on this subject.

References

[1] Arratia, R., Goldstein, L., and Gordon, L. (1989). Two moments suffice for Poisson approximations: the Chen-Stein method. *Ann. Probab.* **17** 9–25.

[2] Aldous, D. (1989). Stein's method in a two-dimensional coverage problem. *Statist. Probab. Lett.* **8**, 307 - 314.

[3] Baldi, P., and Rinott, Y. (1989). Asymptotic normality of some graph-related statistics. *J. Appl. Prob.* **26**, 171–175.

[4] Baldi, P., and Rinott, Y. (1989). On normal approximations of distributions in terms of dependency graphs. *Ann. Probab.* **17**, 1646–1650.
[5] Baldi, P., Rinott, Y., and Stein, C. (1989). A normal approximation for the number of local maxima of a random function on a graph. In *Probability, Statistics and Mathematics, Papers in Honor of Samuel Karlin.* T. W. Anderson, K.B. Athreya and D. L. Iglehart, eds., Academic Press, 59–81.
[6] Barbour, A.D. (1988). Stein's method and Poisson process convergence. *J. Appl. Probab.* **25 (A)**, 175–184.
[7] Barbour, A.D. (1990). Stein's method for diffusion approximations. *Probab. Theory Related Fields* **84**, 297–322.
[8] Barbour, A.D. (1997). Stein's method. *Encyclopaedia of Statistical Science,* Update, Vol.1, Wiley, New York, 513–521.
[9] Barbour, A.D., Chen, L.H.Y., and Loh, W.-L. (1992) Compound Poisson approximation for nonnegative random variables via Stein's method. *Ann. Probab.* **20**, 1843 - 1866.
[10] Barbour, A.D., and Eagleson, G.K. (1985). Multiple comparisons and sums of dissociated random variables. *Adv. Appl. Prob.* **17**, 147–162.
[11] Barbour, A.D., and Eagleson, G.K. (1986). Random association of symmetric assays. *Stoch. Analysis Applics* **4**, 239–281.
[12] Barbour, A.D., Gerrard, R., and Reinert, G. (1997). Iterates of expanding maps. Preprint.
[13] Barbour, A.D., and Hall, P. (1983). On bounds to the rate of convergence in the central limit theorem. *Bull. London Math. Soc.* **17**, 151–156.
[14] Barbour, A.D., and Hall, P. (1984). Stein's method and the Berry-Esséen theorem. *Austral. J. Statist.* **26 (1)**, 8–15.
[15] Barbour, A.D., Holst, L., and Janson, S. (1992). *Poisson Approximation.* Oxford Science Publications.
[16] Barbour, A.D., Karoński, M., and Ruciński, A. (1989). A central limit theorem for decomposable random variables with applications to random graphs. *J. Comb. Theory,* Ser. B, **47**, 125–145.
[17] Barbour, A.D., and Utev, S. (1997). Compound Poisson approximation in total variation. Preprint.
[18] Bolthausen, E. (1984). An estimate of the remainder in a combinatorial central limit theorem. *Z. Wahrscheinlichkeitstheor. Verw. Geb.* **66**, 379–386.
[19] Bolthausen, E., and Götze, F. (1993). The rate of convergence for multivariate sampling statistics. *Ann. Statist.* **21**, 1692–1710.
[20] Cacoullos, T., Papathanasiou, V., and Utev, S. (1994). Variational inequalities with examples and an application to the central limit theorem. *Ann. Probab.* **22**, 1607–1618.
[21] Chen, L.H.Y. (1975). Poisson approximation for dependent trials. *Ann. Probab.* **3**, 534–545.
[22] Chen, L.H.Y. (1978). Two central limit problems for dependent random variables. *Z. Wahrscheinlichkeitstheor. Verw. Geb.* **43**, 223–243.
[23] Chen, L.H.Y. (1986). The rate of convergence in a central limit theorem for dependent random variables with arbitrary index set. *IMA Preprint Series* **243**, Univ. Minnesota.
[24] Chen, L.H.Y. (1996). Stein's method: some perspectives with applications. Preprint.
[25] Dembo, A., and Rinott, Y. (1996). Some examples of normal approximations by Stein's method. In *Random Discrete Structures*, IMA volume 76, Aldous, D. and Pemantle, R., eds., Springer-Verlag, 25–44.
[26] Diaconis, P. (1989). An example for Stein's method. Stanford Stat. Dept. Tech. Rep.
[27] Diaconis, P., and Zabell, S. (1991). Closed form summation for classical distributions: variations on a theme of de Moivre. *Statistical Science* **6**, 284-302.
[28] Ehm, W. (1991). Binomial approximation to the Poisson binomial distribution. *Statist. Probab. Lett.* **11**, 7–16.
[29] Fulman, J. (1997). A Stein's method proof for the asymptotic normality of descents and inversions in the symmetric group. Preprint.
[30] Goldstein, L., and Reinert, G. (1997) Stein's method and the zero bias transformation with application to simple random sampling. *Ann. Appl. Probab.,* in print.
[31] Goldstein, L., and Rinott, Y. (1996). On multivariate normal approximations by Stein's method and size bias couplings. *J. Appl. Prob.* **33**, 1–17.
[32] Götze, F. (1991). On the rate of convergence in the multivariate CLT. *Ann. Probab.* **19**, 724–739.

[33] Hall, P. (1992). *The Bootstrap and Edgeworth Expansion*. Springer-Verlag.
[34] Hall, P., and Barbour, A.D. (1984). Reversing the Berry-Esséen theorem. *Proc. AMS* **90**, 107–110.
[35] Ho, S.-T., and Chen, L.H.Y. (1978). An L_p bound for the remainder in a combinatorial central limit theorem. *Ann. Probab.* **6**, 231–249.
[36] Loh, W.-L. (1992). Stein's method and multinomial approximation. *Ann. Appl. Probab.* **2**, 536–554.
[37] Loh, W.-L. (1997). Stein's method and Pearson Type 4 approximations. Research report.
[38] Luk, H.M. (1994). Stein's method for the gamma distribution and related statistical applications. Ph.D. thesis. University of Southern California, Los Angeles, USA.
[39] Mann, B. (1997). Stein's method for χ^2 of a multinomial. Preprint.
[40] Peköz, E. (1995). Stein's method for geometric approximation. Tech. Report #225, Stat. Dept. U. California. Berkeley.
[41] Reinert, G. (1994). A weak law of large numbers for empirical measures via Stein's method, and applications. Ph.D. thesis. University of Zurich, Switzerland.
[42] Reinert, G. (1997). A Gaussian approximation for random measures via Stein's method. Preprint.
[43] Rinott, Y. (1994). On normal approximation rates for certain sums of dependent random variables. *J. Comp. Appl. Math.* **55**, 135–143.
[44] Rinott, Y., and Rotar, V. (1996). A multivariate CLT for local dependence with $n^{-1/2}\log n$ rate and applications to multivariate graph related statistics. *J. Multivariate Analysis.* **56**, 333–350.
[45] Rinott, Y., and Rotar, V. (1997). On coupling constructions and rates in the CLT for dependent summands with applications to the anti-voter model and weighted U-statistics. Preprint.
[46] Roos, M. (1994). Stein's method for compound Poisson approximation: the local approach. *Ann. Appl. Probab.* **4**, 1177–1187.
[47] Stein, C. (1972). A bound for the error in the normal approximation to the distribution of a sum of dependent random variables. *Proc. Sixth Berkeley Symp. Math. Statist. Probab.* **2**, 583–602. Univ. California Press, Berkeley.
[48] Stein, C. (1986). *Approximate Computation of Expectations*. IMS, Hayward, California.
[49] Stein, C. (1992). A way of using auxiliary randomization. In *Probability Theory*, L. H. Y. Chen, K. P. Choi, K. Hu and J.-H. Lou, eds., Walter de Gruyter, Berlin - New York, 159–180.

DEPARTMENT OF MATHEMATICS, UNIVERSITY OF CALIFORNIA, LOS ANGELES CA 90095-1555
E-mail address: reinert@stat.ucla.edu

DIMACS Series in Discrete Mathematics
and Theoretical Computer Science
Volume 41, 1998

Annotated Bibliography of Perfectly Random Sampling with Markov Chains

David B. Wilson

1. Introduction and Scope

Random sampling has found numerous applications in physics, statistics, and computer science. Perhaps the most versatile method of generating random samples from a probability space is to run a Markov chain. But for how many steps?

In most cases one simply does not know how many Markov chain steps are needed to get a sufficiently random state. There is a large literature of heuristic algorithms for inferring when enough steps have been taken, but they are non-rigorous, and one never knows for sure that an adequate number of steps have been taken. These heuristic algorithms are beyond the scope of this bibliography, but the interested reader is referred to some lecture notes by Sokal and the MCMC Preprint Service.

In the past decade there has been much research on obtaining rigorous bounds of how many Markov chain steps are needed to generate a random sample. Sometimes these bounds are tight, sometimes they are unduly pessimistic. The interested reader is referred to a survey by Diaconis and Saloff–Coste and a survey by Jerrum and Sinclair; the size of this literature makes it beyond the scope of this bibliography.

In recent years there have been a large number of algorithms developed for sampling from the steady state distribution of suitably well-structured Markov chains, which require no *a priori* knowledge of how long the Markov chains take to get mixed. The algorithms determine on their own, during run time, how many steps to run the Markov chain. It is these algorithms that are the focus of this bibliography. Most of these algorithms return samples that are distributed *exactly* according to the stationary distribution of the Markov chain, but a few return samples that have some bias ε that the user can make as small as desired. Since the focus of this bibliography is on working computer algorithms, the symbol [SIM] is placed next to those articles that contain simulation results or give sample outputs. Each annotated entry (on the web version of this bibliography) contains links relevant to the paper, giving the article's abstract (click on the title) and authors' homepages when available, as well as links to online preprints. A few of the annotations were contributed by other people.

1991 *Mathematics Subject Classification.* Primary 60-04, secondary 60-00, 60J10.

Supported in part by an NSF Postdoctoral Fellowship, and in part by DIMACS. This survey was prepared in large part while the author was visiting DIMACS.

It has been pointed out that "stationary stopping times" may also be considered to be algorithms for sampling from a Markov chain's stationary distribution. However, typically these Markov chains have state spaces such as the symmetric group or the hypercube, for which one already knows how to effectively generate a random sample on a computer. In other cases the stopping rule may require an inordinate amount of computation to implement. Rather than to sample, the point of studying these stopping times is to understand interesting mathematical processes, such as shuffling a decks of cards. Since the literature on these stopping times is sizable, included are only those articles whose main theme is algorithmic. The reader interested in stopping times is referred to an article by Aldous and Diaconis and a survey by Lovász and Winkler.

Also relevant to the present bibliography is a literature on backwards compositions of random maps. The backwards composition of random maps can be used to define a "stochastically recursive sequence" of points from the state space. The articles from this literature study when this sequence of points converge almost surely, since convergence implies the existence of a stationary distribution of the Markov chain. (Existence can be nontrivial for infinite state spaces.) If one has the additional conditions that 1) the convergence occurs after a finite number steps, and 2) one can determine when this convergence has occurred, then one may use the technique of "coupling from the past", which is used in several of the articles listed below, to generate random samples from the state space. (There are numerous examples in which conditions 1 or 2 do not hold.) Diaconis is preparing a survey of the literature on stochastically recursive sequences, and appropriate links to it will be made when it is completed. In the meantime, some representative articles are listed below.

Section 2 contains the annotated bibliographic entries, and section 3 gives references to articles and surveys from related areas. Section 4 makes a number of comments about the articles in sections 2 and 3. Section 5 contains definitions of some the stochastic models refered to in the bibliography.

2. Annotated Bibliographic Entries

Andrei Broder, "Generating random spanning trees". *30th Annual Symposium on Foundations of Computer Science*, pp. 442–447, 1989.
David J. Aldous, "A random walk construction of uniform spanning trees and uniform labelled trees". *SIAM Journal on Discrete Mathematics*, 3(4):450–465, 1990.

These two articles give the same independently discovered random-walk based algorithm for generating random spanning trees of a graph. The algorithm uses a Markov chain on the set of spanning trees of an undirected graph to return a (perfectly) random spanning tree. Broder uses the algorithm to analyze the random walk on a ring. Aldous uses the algorithm to determine the properties of random trees and to compute some non-trivial probabilities pertaining to random walk in the plane. (There is another random tree algorithm based on computing determinants.)

Søren Asmussen, Peter W. Glynn, Hermann Thorisson, "Stationary detection in the initial transient problem". *ACM Transactions on Modeling and Computer Simulation*, 2(2):130–157, 1992.

This paper explores what is possible and what is not, and was the first paper to show that it is possible to obtain unbiased samples from the steady state distribution of a finite Markov chain by observing it, provided it is irreducible and one knows how many states it has. Equivalently, there is a universal randomized stationary stopping time that works for all (irreducible) Markov chains with a given (finite) number of states.

David J. Aldous, "On simulating a Markov chain stationary distribution when transition probabilities are unknown". *Discrete Probability and Algorithms*, volume 72 of *IMA Volumes in Mathematics and its Applications*, pp. 1–9. D.J. Aldous, P. Diaconis, J. Spencer and J.M. Steele, editors. Springer-Verlag, 1995.

This paper gives a generic Markov chain sampling algorithm that has some bias ε controllable by the user. While not exact, this algorithm was much more efficient than the previous one, and directly stimulated the development of two subsequent exact sampling algorithms. This paper gives the only nontrivial lower bound on the running time of an algorithm for exact sampling from generic Markov chains.

László Lovász, Peter Winkler, "Exact mixing in an unknown Markov chain". *Electronic Journal of Combinatorics*, 2, 1995. Paper No. R15.

This paper gives the first (universal) exact sampling algorithm that runs in time that is polynomial in certain parameters associated with the Markov chain. It gives a deterministic stationary stopping time that works when the Markov chain itself is not deterministic. The paper also contains a pretty lemma on random trees that is of independent interest.

Dana Randall, Alistair Sinclair, "Testable algorithms for self-avoiding walks". *Proceedings of the Fifth Annual ACM–SIAM Symposium on Discrete Algorithms*, pp. 593–602, 1994.

Gives a Markov chain algorithm for generating (approximately) random non-intersecting lattice paths. The algorithm determines on its own how long to run to generate a sample that is probably close to random, and assuming conjectures widely believed in physics, the running time is polynomial.

James G. Propp, David B. Wilson, "Exact sampling with coupled Markov chains and applications to statistical mechanics". *Random Structures and Algorithms*, 9(1&2):223–252, 1996. [SIM]

This paper gives an algorithm, monotone coupling from the past, for exact sampling with Markov chains on huge state spaces. Since monotone–CFTP relies on the Markov chain having special structure, it is not "universal" as several of the above algorithms are, but it is practical for a surprising number of previously studied Markov chains. Includes simulation results for the random cluster and dimer models.

Valen E. Johnson, "Studying convergence of Markov chain Monte Carlo algorithms using coupled sample paths". *Journal of the American Statistical Association*, 91(433):154–166, 1996. [SIM]

While technically not a paper on exact sampling, this paper investigates how the mixing time of a Markov chain may be inferred by running a large number of coupled simulations until they coalesce. The initial states of the Markov chains are chosen at random, and if the probability of rejection in rejection sampling is known, then rigorous estimates of the mixing time are given. Includes the (independently made) observation that for monotone Markov chains, only two coupled states need to be simulated. Includes simulation results. Additional articles that take a similar approach are available from Johnson's homepage.

Michal Luby, Dana Randall, Alistair Sinclair, "Markov chain algorithms for planar lattice structures (extended abstract)". *36th Annual Symposium on Foundations of Computer Science*, pp. 150–159, 1995.

This paper gives three new Markov chains for sampling certain dimer and ice systems. The focus of this paper is provable running time bounds. Monotone–CFTP may be applied to each of their Markov chains to get an exact algorithm; when this is done, their proofs may be interpreted as *a priori* bounds on the running time of CFTP, though in practice the exact algorithm runs much more quickly than the bounds suggest.

James G. Propp, David B. Wilson, "How to get a perfectly random sample from a generic Markov chain and generate a random spanning tree of a directed graph". *Journal of Algorithms*, to appear. This article combines two conference papers, appearing in *Proceedings of the Seventh Annual ACM–SIAM Symposium on Discrete Algorithms* and the *Proceedings of the Twenty-Eighth Annual ACM Symposium on Theory of Computing.* SIM

This article gives what are so far the best algorithms for exact sampling on generic Markov chains, as well as the first application of CFTP to a huge non-monotone state space. It gives a universal randomized stationary stopping time that is within a constant factor of optimal, and another algorithm that is faster when the Markov chain can be simulated starting from any state rather than just observed in action. Surprisingly, both algorithms are intimately related to the generation of random spanning trees of a weighted directed graph, and this paper gives tree algorithms that are both faster and more general than the Broder/Aldous algorithm. The algorithms uses various versions of CFTP (voter–CFTP, coalescing–CFTP, cover–CFTP, and tree–CFTP), and another technique called cycle popping. (An upcoming book by Lyons and Peres makes use of the cycle-popping algorithm when analyzing essential spanning forests of infinite graphs.) Includes a sample output from the tree algorithm.

David Eppstein, "Representing all minimum spanning trees with applications to counting and generation". Tech Report No. 95–50, U.C. Irvine, 1995.

Shows how to use the algorithms for random spanning tree generation to solve other random sampling problems. Per Eppstein's description: Shows how to find for any edge weighted graph G an equivalent graph EG such that the minimum spanning trees of G correspond one-for-one with the spanning trees of EG. The equivalent graph can be constructed in time $O(m+n \log n)$ given a single minimum spanning tree of G. As a consequence one can find fast algorithms for counting, listing, and randomly generating MSTs. Also discusses similar equivalent graph constructions for shortest paths, minimum cost flows, and bipartite matching.

Stefan Felsner, Lorenz Wernisch, "Markov chains for linear extensions, the two-dimensional case". *Proceedings of the Eighth Annual ACM–SIAM Symposium on Discrete Algorithms*, pp. 239–247, 1997. SIM

The authors show that monotone–CFTP may be used to sample random linear extensions of two-dimensional partially ordered sets, and give bounds on the running time. Includes structural results on the weak and strong Bruhat orders, and simulation results.

Wilfrid S. Kendall, "Perfect simulation for the area-interaction point process". *Proceedings of the Symposium on Probability Towards the Year 2000"*, 1996. To appear in *Probability Perspective*, C.C. Heyde and L. Accardi, editors, World Scientific Press. SIM

Shows how to do perfect simulations of the area-interaction point process, though the techniques extend to more general point processes. A unique feature of this application is that the Markov chain used is not uniformly ergodic, i.e. its mixing time is infinite. (The state space is an infinite partially ordered set with no top state, and one can find states from which the Markov chain takes arbitrarily long to equilibrate.) To deal with this problem, Kendall modified monotone–CFTP, replacing references to the (nonexistent) top state with references to a stochastically dominating process. Kendall also shows how to apply CFTP to repulsive point processes, which are anti-monotone. Includes simulation results.

Wilfrid S. Kendall, "On some weighted Boolean models". *Advances in Theory and applications of Random Sets*, D. Jeulin, editor, 1996. SIM

Following up on the previous article, shows how to apply CFTP to attractive birth-death processes and exclusion processes.

James A. Fill, "An interruptible algorithm for perfect sampling via Markov chains", 1997. An extended abstract appeared in the *Proceedings of the Twenty-Ninth Annual ACM Symposium on Theory of Computing*, and the full version is to appear in *The Annals of Applied Probability*.

Introduces a new algorithm for exact sampling with Markov chains, based not on CFTP, but rather on rejection sampling, with connections to strong stationary duality. Like CFTP, there are multiple ways to apply this algorithm, and this paper focuses on monotone Markov chains. This algorithm has the additional feature that it is immune to deadline-induced bias when the "deadlined resource" is the number of Markov chain simulation steps one is willing to run. Specifically, for Fill's algorithm the number of Markov chain simulation steps (a random variable) is independent of the output sample. [See also the remarks in the comments section.]

Peter W. Glynn, Philip Heidelberger, "Bias properties of budget constrained simulations". *Operations Research*, 38(5):801–814, 1990.

Among other things, this article shows that so long as the user insists on waiting long enough to get at least one random sample, a deadline will not introduce bias. This holds for any sampling procedure, whether the deadline is in terms of Markov chain steps, real time, or any other measure.

James A. Fill, "The move-to-front rule: A case study for two exact sampling algorithms". *Probability in the Engineering and Informational Sciences*, to appear in 1998 (issue 3).

Does a case study comparing the performance of CFTP and Fill's duality-based algorithm when sampling from the steady state distribution of the "move-to-front rule". Includes analysis of the mixing time when various sets of weights are used in the move-to-front rule.

O. Häggström, M.N.M. van Lieshout, J. Møller, "Characterization results and Markov chain Monte Carlo algorithms including exact simulation for some spatial point processes". Technical Report R-96-2040, Aalborg University, 1996. [SIM]

Shows how to apply monotone–CFTP to sample from the Widom–Rowlinson point process. Here there is no top or bottom state, but (in contrast to Kendall's chain) such states can be adjoined to the state space. (The authors instead use two other states, already in the state space, that are just as effective at testing coalescence.) The Widom–Rowlinson model can be marginalized to obtain the attractive area-interaction point process and the continuum random cluster model (with positive integral q). Sampling from the area-interaction process in this way turns out to be faster than (though not as generalizable as) Kendall's approach. The paper also describes a Swendsen–Wang type algorithm, and includes simulation results.

Jesper Møller, "Markov chain Monte Carlo and spatial points processes". To appear in *Proceedings Seminaire Européen de Statistiqe, "Stochastic geometry, likelihood, and computation"*, O. Barndorff–Nielsen, W.S. Kendall, and M.N.M. van Lieshout, editors, Chapman and Hall. [SIM]

Surveys the spatial point processes, including recent work on exact sampling, and includes simulation results.

James Propp, "Generating random elements of a finite distributive lattice". *Electronic Journal of Combinatorics*, 4(2), 1997. Paper No. R15.

An expository paper explaining the use of monotone–CFTP to sample from finite distributive lattices. Gives numerous examples of combinatorial interest.

Olle Häggström, Karin Nelander, "Exact sampling from anti-monotone systems". *Statistica Neerlandica*, to appear. [SIM]

Follows up on the trick for applying CFTP to repulsive point processes, showing how to apply CFTP to a large class of anti-monotone systems, including random independent sets, the Ising antiferromagnet, and the random cluster model with $q < 1$. Includes simulation results.

Michael Luby, Eric Vigoda, "Approximately counting up to four (extended abstract)". *Proceedings of the Twenty-Ninth Annual ACM Symposium on Theory of Computing*, pp. 682–687, 1997.

Gives a new Markov chain for generating random independent sets with activity λ. When the maximum degree $D \geq 4$ and $\lambda \leq 1/(D-3)$, they show that this Markov chain is rapidly mixing. CFTP may be efficiently applied to their Markov chain; when this is done, their proofs (but not the theorem itself) may be interpreted as *a priori* bounds on the running time of CFTP. Also contains some complexity-theoretic hardness results. [An article by Dyer and Greenhill extends and improves on this one, though itself is not connected to CFTP.]

D.J. Murdoch, P.J. Green, "Exact sampling from a continuous state space". *Scandinavian Journal of Statistics*, to appear. SIM

Describes a variety of scenarios for which CFTP may be applied to a (non-monotone) continuous state space, giving a sequence of algorithms that start out simple, but become increasingly sophisticated. The methods used are related to gamma-coupling and rejection sampling, and appear to be applicable to Bayesian parameter estimation. Includes a case study comparing the performance of each technique when used to sample from the posterior distribution of a set of pump reliability parameters.

W.S. Kendall, J. Møller,"Perfect Metropolis–Hastings simulation of locally stable spatial point processes", 1997. In preparation.

Haiyan Cai, "A note on an exact sampling algorithm and Metropolis Markov chains", 1997. Preprint.

Shows how to take a generic probability distribution on a space, together with a reference distribution from which it is already known how to sample, and construct a certain particular Metropolis-type monotone Markov chain. Coalescence occurs precisely when the rejection sampler with the same reference measure would accept.

S.G. Foss, R.L. Tweedie, "Perfect simulation and backward coupling". *Stochastic Models*, to appear.

Relates CFTP to more general stochastically recursive sequences, for which one has a sequence of random variables that with probability one coverage to a particular value with the right distribution, but for which one isn't necessarily able to determine when this convergence has happened. Studies the moments of the convergence time of stochastically recursive sequences, some of which may be finite with others being infinite.

Robert B. Lund, David B. Wilson, "Exact sampling algorithms for storage systems and networks". In preparation.

Show how to apply monotone–CFTP to sample from the steady-state distribution of certain storage systems. Rather than the linear data-flow used in traditional monotone–CFTP, the flow of data through the algorithm is two-dimensional.

Jesper Møller, "Perfect simulation of conditionally specified models". Technical Report R-97-2006, Department of Mathematics, Aalborg University, 1997. To appear in *Journal of the Royal Statistical Society, Ser. B.* SIM

Shows how to apply CFTP to sample from a class of Markov random fields, in which the individual variables (from a finite, countable, or continuous space) depend upon one another in a monotone or anti-monotone way. For the continuous distributions, the user specifies an ε, which controls not the bias (which is 0), but instead the numerical precision of the output. The application to the auto-gamma distribution, which includes the pump-reliability application above, appears to be faster than the approach of Murdoch and Green.

S.G. Foss, R.L. Tweedie, J.N. Corcoran, "Simulating the invariant measures of Markov chains using backward coupling at regeneration times". *Probability in the Engineering and Informational Sciences*, to appear. SIM

Gives a method for approximately sampling from the steady-state distribution of a Markov chain when the state of the chain takes on a certain particular value on a semi-regular basis. The Markov chain of interest is lifted to one that keeps track of 1) the original chain's state, and 2) the number of steps since the original chain's state took on that special value. The lifted chain is then truncated to force a return to the special state whenever the second coordinate gets too big. Using an independent set of coins for each value of time and each value of the second coordinate of the modified lifted chain, the authors show how to effectively test for coalescence. Includes discussion for how to choose the truncation parameter, though this not mechanized to the extent where the user could specify a desired maximum bias. Includes a case study on a two-server re-entrant queueing network.

Elke Thönnes, "Perfect Simulation of some Point Processes for the Impatient User", 1997. Preprint. SIM

Applies Fill's algorithm to sample from the Widom-Rowlinson model (and consequently from the attractive area-interaction point process as well). Includes simulation results, and compares performance (time and memory) with the CFTP algorithm for this model.

James Propp, David Wilson, "Coupling from the past: a user's guide". *DIMACS Series in Discrete Mathematics and Theoretical Computer Science, "Microsurveys in Discrete Probability"*, D. Aldous and J. Propp, editors, American Mathematical Society, 1998.

An expository paper describing a variety of very different situations for which coupling-from-the-past may be applied, and gives advice on recognizing additional applications. Includes some caveats for the practitioner.

Olle Häggström, Karin Nelander, "On exact simulation of Markov random fields using coupling from the past", 1997. Preprint. SIM

Describes a scheme for sampling random configurations of a Markov random field using CFTP. This scheme generalizes the CFTP algorithm for monotone and anti-monotone spin systems, and the authors provide a theorem bounding the rate of its convergence. Specific applications that are treated are random proper q-colorings (which had eluded other researchers) and the discrete Widom-Rowlinson model with multiple particle types. Includes simulation results.

J. van den Berg, J.E. Steif, "On the existence and non-existence of finitary codings for a class of random fields", 1998. Preprint.

Proves theorems about the infinite-volume Ising model and other Markov random fields. The existence results come from constructions that were inspired in part by monotone-CFTP.

J. N. Corcoran, R. L. Tweedie, "Perfect Sampling of Harris Recurrent Markov Chains", 1998. Preprint. SIM

Describes a method related to gamma-coupling for applying CFTP. One somehow finds a coupling so that 1) there's some chance of the whole state space eventually getting mapped to some small set, and 2) there's some chance of that small set getting mapped to a particular point. The main application considered is to monotone Markov chains on a finite interval of the reals. Includes simulation results.

"Annotated bibliography of perfectly random sampling with Markov chains".

In addition to appearing in this volume of the *DIMACS Series in Discrete Mathematics and Theoretical Computer Science*, a hypertext version of this bibliography, at `http://dimacs.rutgers.edu/~dbwilson/exact/`, will be updated as new articles appear.

3. Surveys and Representative Articles from Related Areas

Information on heuristic algorithms for assessing Markov chain convergence to the stationary distribution (among other things) can be found in these resources:

- **Alan D. Sokal**, "Monte Carlo methods in statistical mechanics: Foundations and new algorithms", 1989. Lecture notes from Cours de Troisième Cycle de la Physique en Suisse Romande. Updated in 1996 for the Cargèse Summer School on "Functional Integration: Basics and Applications".
- MCMC Preprint Service (`http://www.stats.bris.ac.uk/MCMC/`)

Surveys of rigorous bounds on the convergence rate of Markov chains

- **Persi Diaconis, Laurent Saloff-Coste**, "What do we know about the Metropolis algorithm?". *Proceedings of the Twenty-Seventh Annual ACM Symposium on Theory of Computing*, pp. 112–129, 1995.
- **Mark Jerrum, Alistair Sinclair**, "The Markov chain Monte Carlo method: an approach to approximate counting and integration". *Approximation Algorithms for NP-hard Problems*, D.S. Hochbaum ed., PWS Publishing, Boston 1996.

Articles/surveys on stopping times

- **David Aldous, Persi Diaconis**, "Shuffling Cards and Stopping Times", *American Mathematical Monthly*, 93(5):333-348, 1986.
- **László Lovász, Peter Winkler**, "Mixing Times", this volume.

Representative articles on stochastically recursive sequences

- **Gérard Letac**, "A contraction principle for certain Markov chains and its applications", *Contemporary Mathematics, "Random Matrices and Their Applications"*, 50:263–273, 1986.
- **Herman Thorisson**, "Backward limits", *The Annals of Probability* 16(2): 914–924, 1988.
- **A.A. Borovkov, S.G. Foss**, "Stochastically recursive sequences and their generalizations", *Siberian Advances in Mathematics*, 2(1):16–81, 1992.

4. Comments

It is tempting to assume (and several people have) that coupling-from-the-past is only efficient when the state space is monotone or anti-monotone. But the articles on random tree generation, the Murdoch–Green article, and the recent Häggström–Nelander preprint, all provide good examples of efficient application of CFTP when there is no monotone or anti-monotone structure.

One can show that CFTP, as it was originally formulated, can only work if the Markov chain is uniformly ergodic. Kendall's article showed how to circumvent this problem by extending CFTP to use an auxiliary Markov chain (rather than an i.i.d. process) to generate the "random coins" used to drive the Markov chain of interest.

If there is a fixed deadline at which time an experimentalist must abort any simulation in progress, it is possible for the deadline to induce bias in favor of samples that do not take long to generate. But as the Glynn–Heidelberger article (listed above) shows, deadline-induced bias is eliminated if the experimenter commits to obtaining at least one sample.

In some situations, such as studies of statistical mechanical models which have a certain "spatial mixing" property, the user may reasonably want only a few large samples. Then it would be desirable not to commit the user to obtaining at least one sample. To eliminate the deadline-induced bias in this case, one can design an algorithm so that the running time is independent of the sample produced. Achieving this independence can be trickier than simply decoupling the number of Markov chain steps from the output sample, since for these models it is not uncommon for there to be variations in the time to simulate individual steps of the Markov chain. For example, for the application of sampling from many point process models, there is typically not even an upper bound on the time to simulate one Markov chain step (though the expected time is finite). For point process models with an unbounded number of points, one can go so far as to argue that there is *no* exact sampling algorithm that is interruptible with respect to time, even though several of these models have algorithms that are interruptible with respect to the number of Markov chain steps.

5. Definitions of Selected Terms

Included here are the definitions of some terms used but not defined above. Depending upon the model, further background can be found in Sokal's lecture notes, Møller's survey, or Lyon's contribution to this volume.

area-interaction point process: Defined relative to the Poisson point process, the probability density of a configuration is given an extra weight that is exponentially small or large in the area of the union of circles (or other fixed shapes) centered about the points. When the extra weight is exponentially small (large) in the area, in a configuration with many points, the points will tend to be nearby (far away), giving the attractive (repulsive) area-interaction point process.

continuum random cluster model: Defined relative to a Poisson point process, clusters are formed by taking the union of circles (or other shapes) centered about the points. The probability density of a configuration gets an extra factor of q for each cluster that it contains. The discrete version of the continuum random cluster model is *not* the random cluster model, but rather is related to it in the same way that site percolation is related to bond percolation.

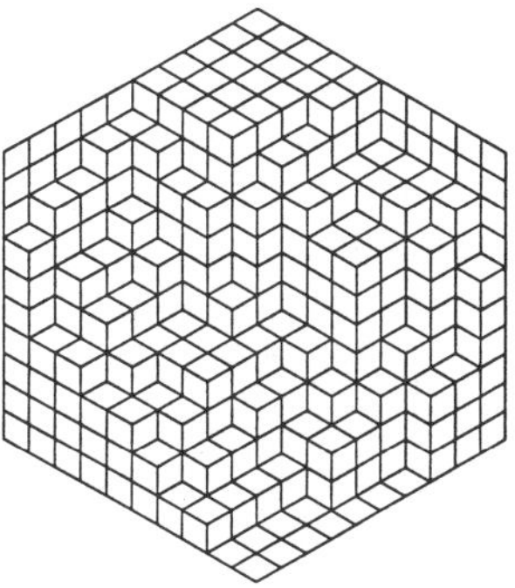

dimer model: A dimer configuration is a perfect matching of a graph, i.e. a pairing of the vertices of the graph so that each vertex is paired with exactly one other vertex, connected to it by an edge of the graph. Typically all configurations are equally likely, but sometimes each edge contributes a weight to those perfect matchings that contain it. (The term "dimer" comes from considering the vertices to be atoms. The pairing of vertices corresponds to the placement of bonds to make dimers.) Depicted here is a random dimer configuration of the hexagonal grid, generated by monotone-CFTP, and shown as a tiling. Each vertex/atom corresponds to a triangle, and each edge/dimer is shown as a rhombus.

hard core model: A hard-core configuration is an independent set of a graph, i.e. a set of vertices of the graph such that no two vertices are connected by an edge. The probability of an independent set gets an extra factor of λ, known as the "activity" or "fugacity", for each vertex that it contains. Shown here is a random independent set of the grid (with periodic boundary conditions), with $\lambda = 2$, generated by monotone-CFTP.

ice model: An ice configuration is an Eulerian orientation of a graph, i.e. an assignment of directions to the edges of a graph so that at each vertex the number of edges directed towards it equals the number of edges directed away from it. Some edges may be fixed at the boundaries of the graph, but given those constraints, all configurations are equally likely. (The term "ice" comes from an analogy with ice, where the vertices correspond to water molecules, the edges to hydrogen bonds, and the edge orientations determine which water molecule's hydrogren atom is involved in the hydrogen bond.)

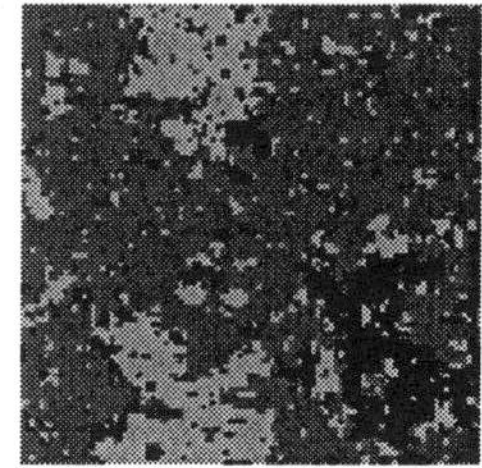

Potts model: Each vertex on a graph (typically a grid) is assigned one of q different colors. The probability of a configuration gets an extra weight for each edge whose two endpoints are the same color. (The weight is thought of as being related to an interaction energy and the temperature.) When the weights are at least one, the model is "ferromagnetic", if they are less than one, then the model is "antiferromagnetic". The case $q = 2$ is the Ising model. When the weight factor is zero, one obtains proper q-colorings. Shown here is a random $q = 3$ Potts configuration at the "critical temperature" on the 2D grid. It was generated by monotone-CFTP on the related random-cluster model.

random-cluster model: Also known as the Fortuin–Kasteleyn random cluster model, and bond-correlated percolation, it can be defined relative to ordinary bond percolation (each edge occuring independently with probability p), where the probability of a configuration gets an extra weight of q for each connected component it contains. When q is a positive integer, this model is closely related to the q-state ferromagnetic Potts model. When $q \to 0$ and then $p \to 0$, the result is a random spanning tree. When $q = 1$ one gets ordinary percolation.

spanning tree: A connected, acyclic subgraph of a given graph. If the graph is directed, then "spanning tree" refers to an "arborescence", in which a distinguished vertex is the "root" of the tree, and the edges of the tree are directed towards the root. A random spanning tree is such a tree (or arborescence) chosen uniformly at random, or if the graph is weighted, with probability proportional to the products of the weights of the edges contained in the tree. Shown here is a random spanning tree of the grid, generated by loop-erased random walk.

Widom-Rowlinson model: A model in which there are typically two (but possibly more) types of particles, where particles of different types can't be nearby. In the discrete case, there is a graph, each vertex may contain zero or one particles, and no edge has two different types of particles on its endpoints. The probability of a legal configuration is proportional to some constant (the "fugacity" or "activity") raised to the power of the number of particles. In the continuous case, particles lie in the plane (or other manifold), and particles of different type must be some minimum distance apart. The probability density of a configuration is what one would get by randomly assigning particle types to the points of a Poisson process, and conditioning on the configuration being legal.

Acknowledgements

The author thanks Andrei Broder, Wilfrid Kendall, Jesper Møller, and James Propp for their helpful comments.

Institute for Advanced Study, Princeton, New Jersey 08540
E-mail address: dbwilson@alum.mit.edu

Selected Titles in This Series

(Continued from the front of this publication)

12 **David S. Johnson and Catherine C. McGeoch, Editors,** Network Flows and Matching: First DIMACS Implementation Challenge

11 **Larry Finkelstein and William M. Kantor, Editors,** Groups and Computation

10 **Joel Friedman, Editor,** Expanding Graphs

9 **William T. Trotter, Editor,** Planar Graphs

8 **Simon Gindikin, Editor,** Mathematical Methods of Analysis of Biopolymer Sequences

7 **Lyle A. McGeoch and Daniel D. Sleator, Editors,** On-Line Algorithms

6 **Jacob E. Goodman, Richard Pollack, and William Steiger, Editors,** Discrete and Computational Geometry: Papers from the DIMACS Special Year

5 **Frank Hwang, Fred Roberts, and Clyde Monma, Editors,** Reliability of Computer and Communication Networks

4 **Peter Gritzmann and Bernd Sturmfels, Editors,** Applied Geometry and Discrete Mathematics, The Victor Klee Festschrift

3 **E. M. Clarke and R. P. Kurshan, Editors,** Computer-Aided Verification '90

2 **Joan Feigenbaum and Michael Merritt, Editors,** Distributed Computing and Cryptography

1 **William Cook and Paul D. Seymour, Editors,** Polyhedral Combinatorics